Compound Semiconductor
Surface Passivation and
Novel Device Processing

MATERIALS RESEARCH SOCIETY
SYMPOSIUM PROCEEDINGS VOLUME 573

Compound Semiconductor Surface Passivation and Novel Device Processing

Symposium held April 5–7, 1999, San Francisco, California, U.S.A.

EDITORS:

H. Hasegawa
Hokkaido University
Sapporo, Japan

M. Hong
Lucent Technologies, Bell Laboratories
Murray Hill, New Jersey, U.S.A.

Z.H. Lu
University of Toronto
Toronto, Ontario, Canada

S.J. Pearton
University of Florida
Gainesville, Florida, U.S.A.

Materials Research Society
Warrendale, Pennsylvania

CAMBRIDGE UNIVERSITY PRESS
Cambridge, New York, Melbourne, Madrid, Cape Town,
Singapore, São Paulo, Delhi, Mexico City

Cambridge University Press
32 Avenue of the Americas, New York NY 10013-2473, USA

Published in the United States of America by Cambridge University Press, New York

www.cambridge.org
Information on this title: www.cambridge.org/9781107414099

Materials Research Society
506 Keystone Drive, Warrendale, PA 15086
http://www.mrs.org

First published 1999
First paperback edition 2013

Single article reprints from this publication are available through
University Microfilms Inc., 300 North Zeeb Road, Ann Arbor, MI 48106

CODEN: MRSPDH

ISBN 978-1-107-41409-9 Paperback

CONTENTS

*Invited Paper

PART III: OXIDES—STRUCTURAL, TRANSPORT AND OPTICAL PROPERTIES

PART IV: COMPOUND SEMICONDUCTOR SURFACE PASSIVATION AND NOVEL DEVICE PROCESSING

*Invited Paper

PART V: ELECTRONIC DEVICES AND PROCESSING

*Invited Paper

PREFACE

This proceedings volume is the record from Symposium Z, "Compound Semiconductor Surface Passivation and Novel Device Processing," held April 5–7 at the 1999 MRS Spring Meeting held in San Francisco, California. It covers a wide range of activity in the field of passivation and control of compound semiconductor surfaces and associated novel processing techniques for electronic and photonic devices. The markets for GaAs and InP-based electronics cover wireless communication, mobile phones, defense applications where radiation-hardness is critical, automobile collision avoidance radar and satellites. On the photonics side, displays, communication systems, infrared and UV detectors and lighting applications are the prime markets. In all of these devices there is a critical need for control of the surface properties and for reliable long-term encapsulation/passivation of the surface. Compound semiconductors typically have relatively high surface recombination velocities compared to Si and are subject to disruption of the surface during device processing. Therefore, novel processing methods which minimize surface damage are very important. These two themes—surface passivation and novel device processing—formed the basis for the symposium. Summaries of these topics were provided by invited review papers, while contributed and poster papers described work in progress.

It was clear from the symposium that understanding and control of surface properties has progressed tremendously in the past few years. In this proceedings volume we divided the symposium into the following broad topic areas:

- fundamentals of surfaces and their passivation
- novel approaches for surface passivation and device processing
- oxides—structural, transport and optical properties
- compound semiconductor surface passivation and novel device processing
- electronic devices and processing

There was a lot of interest in the novel oxides developed for MOSFET gate dielectrics, which have enabled the first demonstration of GaAs, GaN and InGaAs MOSFETs.

The symposium was well attended, with a lively and informative poster session. This was the first symposium on this topic at MRS and it proved to be a major success.

H. Hasegawa
M. Hong
Z.H. Lu
S.J. Pearton

May 1999

ACKNOWLEDGMENTS

The outstanding success of the symposium was due to the efforts of the following people: the authors and speakers who presented their technical work at the Meeting and prepared the papers in this proceedings volume; the symposium organizers, who put together the program and saw that it ran smoothly; the session chairpersons (Z.H. Lu, H. Hasegawa, K.D. Choquette, M. Hong, D.Q. Deppe, and H. Cho); the staff of the Materials Research Society, who provided the organization for the Meeting; and most importantly, the sponsors listed below, whose financial backing enabled the organizers to cover the Meeting expenses. The editors of these proceedings extend their sincere appreciation to all who contributed to the success of the symposium.

Symposium Support

Plasma-Therm, Inc.
American Xtal Technology
Riber, Inc.
Johnsen Ultravac
Lucent Technologies, Bell Laboratories
University of Florida
University of Toronto
Hokkaido University

MATERIALS RESEARCH SOCIETY SYMPOSIUM PROCEEDINGS

MATERIALS RESEARCH SOCIETY SYMPOSIUM PROCEEDINGS

Prior Materials Research Society Symposium Proceedings available by contacting Materials Research Society

Part I

Fundamentals of Surfaces
and Their Passivation

THEORY OF THE SULPHUR-PASSIVATED InP(001) SURFACE

LAURENT J. LEWIS[a]
Département de physique et GCM, Université de Montréal, C.P. 6128, Succ. Centre-Ville,
Montréal, Québec, Canada H3C 3J7
CHANDRÉ DHARMA-WARDANA[b]
Institute for Microstructural Sciences, National Research Council of Canada
Ottawa, Ontario, Canada K1A 0R6

ABSTRACT

We present a detailed and comprehensive theoretical investigation of the sulphur-passivated (001) surface of InP. First, the ground-state structure is determined using density-functional methods, including full relaxation of the surface. The lowest-energy structure at 0 K is a striking (2×2) reconstruction with the S atoms displaced from the bridge sites to form short and long dimers, belonging to two distinct sublayers. This surface structure is used to calculate the backscattering Raman spectrum; the two peaks arising from surface-layer vibrations predicted by our calculations are observed. Next, our first-principles calculations are extended to the study of a number of other stable states of the surface that can arise upon annealing. For this purpose, we construct and relax several higher-energy states of the surface, and calculate the corresponding core-level photoemission spectra. A remarkable sequence of structures is found to unfold from the fully S-covered ground state as they become energetically accessible. The surface S atoms exchange with bulk P atoms, forming new (and strong) S–P bonds while dissociating pre-existing S–S dimers. The predicted core-level spectra are found to be entirely consistent with the experimental measurements; our calculations indicate that the annealed (at about 700 K) surface is a (2×2) structure containing two S and two P atoms per unit cell. Finally, we have used the predicted stable surface structures to calculate the photoemission and inverse photoemission spectra. They are found to agree well with experiment if the surface is assumed to consist of a mixture of the above ground-state and annealed structures.

INTRODUCTION

III–V compound semiconductors (GaAs, InP, etc.) are technologically important materials having many potential applications in devices such as solar cells, Schottky and laser diodes, and integrated optoelectronic circuits [1, 2]. These devices require chemically-stable surfaces or interfaces. This is assured in Si technology by surface oxidation; for III–V materials, however, the untreated surfaces are quite reactive due to the high density of surface states. Passivation methods employing sulphur overlayers have been developed to reduce the surface density of states of III–V compounds [3, 4], significantly improving the electronic properties of the surfaces.

An important step towards understanding passivation is the determination of the atomic structure of the surfaces. For InP(001)-S, it was suggested, on the basis of low-energy electron diffraction (LEED) and x-ray photoelectron spectroscopy (XPS) experiments [4], that the surface is terminated with a monolayer of sulphur, the S atoms occupying bridge sites and forming bonds to indium atoms only in a (1×1) pattern. A further study using x-ray absorption near-edge structure (XANES) [5] suggested that the In-S-In bond angle was close to $100°$ — smaller than the ideal tetrahedral In-P-In bond angle ($109.5°$). In the analysis used in these initial experiments [4, 5], no surface relaxation was however considered.

We have determined the ground-state structure of the InP(001)-S surface [6] using first-principles total-energy minimization methods [7]. As discussed below, the surface is found to

Mat. Res. Soc. Symp. Proc. Vol. 573 © 1999 Materials Research Society

have two S sublayers separated by 0.22 Å; the top S sublayer contains monomer pairs sitting close to bridge sites, while the lower sublayer contains strongly-dimerized S pairs. Thus the S atoms exist in *two types of chemical environments*. This structure is quite different from the currently-accepted description of the more extensively studied GaAs-S surface [8], which our calculations confirm.

Core-level (CL) spectroscopy can distinguish an atom in different bonding situations since the core-level positions of atomic lines shift as a function of the chemical environment [9]. The method is a powerful tool for probing surface structure; this however requires theoretical predictions of the CL excitation energies (CL-E), in particular when different structures "compete" because, e.g., of annealing, surface preparation, etc., which may lead to structures other than the ground state. We have calculated the CL-E from first-principles in order to follow the evolution of the InP(001)-S surface upon annealing. The theory predicts a number of stable structures besides the ground which become energetically accessible upon annealing. In particular, surface S atoms exchange with bulk P atoms, forming new strong S–P bonds while dissociating pre-existing S–S dimers. Thus, we are lead to a picture wherein the InP(001)-S surface is actually a system which could contain a mixture of S, S and P, or P-terminated InP(001) domains, depending on the kinetics imposed by the annealing conditions. Our results are confirmed by calculations of the photoemission (PE) and inverse photoemission (IPE) spectra. We give full details of these surface configurations below, but first present the theoretical framework underlying the present calculations.

COMPUTATIONAL FRAMEWORK

Total-energy minimization

The total-energy minimization calculations were performed within the framework of density-functional theory (DFT) in the local-density approximation (LDA), using plane waves (PW) to expand the electron wavefunctions, together with non-local, norm-conserving pseudopotentials (PP) [10]. The electron exchange-correlation energy is taken to be of the Ceperley-Alder form [11]. Only the Γ point was used to sample the reciprocal space and a 10 Ry energy cutoff was employed in the plane-wave expansions. The validity of our pseudopotentials for In, P and S was verified on various molecular configurations [12, 13, 14].

The semi-infinite crystal with with the InP(001) surface exposed to vacuum is modeled, for the total-energy minimization calculations, as a supercell slab containing six layers, each with four atoms. The first (topmost) layer contains four S (or P) atoms. The second contains four In atoms, as implied by the chemistry of the material and by recent experimental studies [4, 5]. Subsequent layers follow the zinc-blende structure, except the bottom one, which consists of four In atoms and two H atoms positioned in a (1×2) pattern such as to saturate the dangling bonds [13]. The atoms in the bottom three layers are held fixed in their bulk positions, while the other atoms are allowed to relax so as to minimize the total energy. Periodic boundary conditions are applied in the x, y, and z directions but a vacuum region of width 7.225 Å (equivalent to five bulk In-P interplaner distances) is used along z, i.e. normal to the surface. For the fixed layers we used the theoretical bulk lattice constant [13] of 5.78 Å which compares well with the experimental [15] value of 5.87 Å.

Core-level excitation energies

Using the relaxed structures provided by the PW-PP calculations, the S-$2p$ core-level spectra were determined using the all-electron full-potential–semi-relativistic (FP-SR) method [16]. In the core-level-excitation process, an electron in a core level (here S-$2p$) is knocked out by an incident photon and placed in an outgoing electron state. The "hole" left

4

behind has an effective positive charge which interacts with the remaining atomic and band electrons, as well as with the outgoing electron. Thus, the photoionization process involves electronic relaxation effects in the initial state as well as in the final state. In photoionization experiments, the time scale of the relaxation is usually short and electronic relaxation has occured by the time the kinetic energy of the outgoing electron is measured. Hence the measured core-level excitation energy involves initial-state as well as final-state relaxation effects; in our calculations, both are included.

We consider a photoemission experiment where the photon energy is $h\nu$, while the total energies of the initial and final states of the crystal are E_i and E_f. The measured kinetic energy E_{ke} of the photoelectron is such that $h\nu = E_f - E_i + E_{ke}$ and the measured CL-E, $\varepsilon_{2p} = E_f - E_i$, is a property of the initial and final states of the system. Evaluation of the electronic excitation effects via many-body methods is a numerically very demanding task. It is important to note that the DFT (Kohn-Sham) eigenvalues are not excitation energies; nevertheless, the total-energy difference (TED) between the ground state and the "excited" state (i.e., with a core hole) can be taken as a measure of the CL-E. Another method that can provide a reasonable estimate of the CL-E is the "Slater transition-state" method (STS) [17]. It is numerically simple and is closely related to DFT. In STS, the $2p$ core-hole energy includes both initial-state and final-state effects, and is shown (under certain assumptions) to be equal to the $2p$ eigenvalue of the "transition state", a state having a *half-occupied* core hole. We have used both TED and the STS method to calculate the CL energies. Full details of the calculations are given in Ref. [14]. The difference in the estimates from the two methods — about 0.5 eV — provides a theoretical "error bar".

Photoemission spectra

We have also calculated the photoemission (PE) and inverse photoemission (IPE) spectra of the energetically-probable surface structures obtained from our first-principles calculations, and compare them to the experimental data reported recently [18]. We have bench-marked our procedure using the (111) surface of silicon. As was the case for the CL spectra discussed above, the PE and IPE spectra were determined using the energy-minimized geometries obtained from first-principles. We thus constructed supercells containing typically 16 atomic layers which are then used to calculate the electronic band structure, here within the all-electron, full-potential-linear-muffin-tin-orbital (FP-LMTO) method [16].

The photoelectron current per unit solid angle in PE at photon energy $\hbar\omega$ is given by

$$J^{PE}(E_{kin}, \omega) \propto \sqrt{E_{kin}} \sum_i W_{if}. \tag{1}$$

where $E_{kin} = E_f - E_F - \phi$ (E_F is the Fermi energy and ϕ is the workfunction) and W_{if} is the transition probability between initial and final states; it is given by the Fermi golden rule:

$$W_{if} = \frac{2\pi}{\hbar} |\langle f \mid H_{int} \mid i \rangle|^2 \delta(E_f - E_i - \hbar\omega). \tag{2}$$

Likewise, for the IPE current,

$$J^{IPE}(E_{kin}, \omega) \propto [1/\sqrt{E_{kin}}] \sum_f W_{fi} \tag{3}$$

where, now, $E_{kin} = E_i - E_F - \phi$. The transition matrix elements of Eqs. 1 and 3 are evaluated directly from the Kohn-Sham eigenstates appropriate to the occupied/unoccupied levels. Our calculations indicate that the DOS profile provides a poor approximation to the measured spectra; this is of some importance since experimental PE/IPE results are often

5

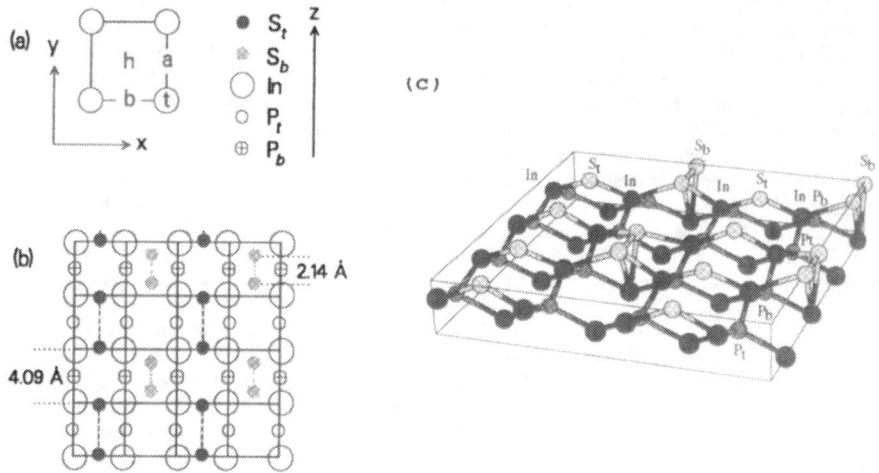

Figure 1: (a) Four possible sites for the S atom: bridge {b}; anti-bridge {a}; top {t} and hollow {h}. (b) Top view and (c) ball-and-stick model of the fully-relaxed S-passivated InP(001) surface.

presented as a mapping of the bandstructure of a material, which does not seem to be correct. Full details can be found in Ref. [19].

RESULTS

Ground-state structure

There are four possible arrangements for the S atoms which are consistent with the reported (1×1) pattern: "hollow", "bridge", "anti-bridge", and "top" sites [cf. Fig. 1(a)]. Of these, the bridge-site structure is the most favourable if the (1×1) pattern is imposed, just as in the case of GaAs-S [8]. The ground-state structure provided by our first-principles simulations, however, is the (2×2) reconstruction depicted in Figs. 1(b) and (c). The surface S layer splits into two sublayers, each containing half the total sulphur. The top sublayer (S_t) sublayer contains long dimers ("monomer pairs"), with the S atoms positioned close to bridge sites, but slightly displaced in the $[\bar{1}10]$ direction, yielding a S_t–S_t bond length of 3.82 Å. The bottom sublayer (S_b), in contrast, contains strongly dimerized (along the $[\bar{1}10]$ direction) S pairs, with a bond length of 2.14 Å. The sequence of atomic planes (SAP) in the z-direction, with S_t at the surface, may be presented as:

$$SAP = (In - P)_{bulk} - In - P_b - P_t - In - S_b - S_t \qquad (4)$$

The interplaner distances near the surface region can be compared with the bulk In-P distance d_0 of 1.445 Å. The S_t layer relaxes only 0.116 Å towards the bulk, while S_b is displaced 0.22 Å below the S_t layer. The In layer below S_b remains integral while relaxing inwards; the following P layer, echoing the splitting of the top S layer, separates into two sublayers. The interplaner distances, starting from the left-most In-atom are: In-P_b = 1.321 Å, while In-P_t = 1.532 Å, giving an average In-$\langle P_{b,t}\rangle$ distance of 1.427 Å. The other distances are

6

P_t-In = 1.299 Å, In-S_b = 1.169 Å, and In-S_t = 1.388 Å. We also recovered this ground state using a larger supercell with eight atoms per plane (x-dimension doubled). The surface thus loses its (1×1) periodicity in favour of a (2×2) pattern, with each short dimer surrounded by four long dimers, and *vice versa*. As we will see below, there are many metastable states on this surface.

The structure proposed on the basis of XANES [5] recovers the correct [$\bar{1}$10] orientation of the S-atoms, but not other details. According to our calculations, it does not in fact correspond to a local energy minimum, and lies some 0.5 eV above the ground state. However, a recent study [20] of the surface using scanning tunneling microscopy (STM) found that it has a locally ordered (2×2) structure, as predicted by our model. In contrast, LEED measurements, which depend on long range order, show a (1×1) pattern, in agreement with Tao et al. [4]. This is also consistent with our model: The In-layer just below the S-layer has a large electron-scattering cross section and retains a (1×1) aspect in our structure.

In order to confirm the existence of the lowest energy structure predicted by our calculations, we have determined the Raman spectrum of the surface, which can then be compared to the experimental Raman spectrum (longitudinal vibrational modes along z). In order to do this, we calculated the interplaner force constants in the "frozen-phonon" approach [21]. The lattice dynamics along [001] rigorously simplifies to a linear chain of masses, each node of the chain representing a rigidly-vibrating plane of atoms. The phonon modes were calculated using a slab of 1006 layers, with the top six layers having the sequence of atomic layers obtained above for the ground state. The calculated back-scattering Raman spectrum of the system [22] is shown in Fig. 2(a). The calculated *intensities* are, as usual, a qualitative measure of the relative strengths of the peaks; however, their *positions* are expected to be reliable to within a few wavenumbers. The "bulk" LO-phonon peak of InP is calculated to

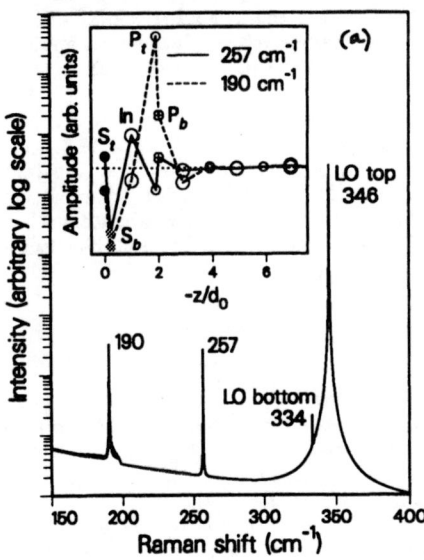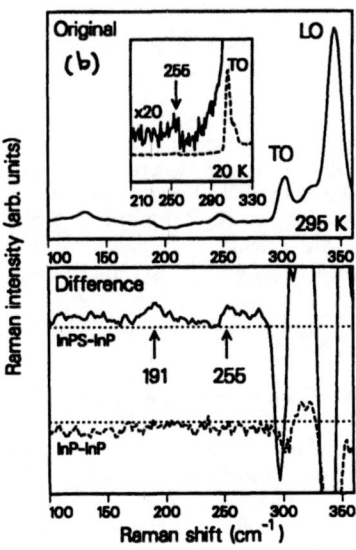

Figure 2: (a) Calculated Raman spectrum of the fully-relaxed InP(001)-S surface. The (longitudinal) vibrational amplitudes of the atomic planes are shown in the inset as a function of z/d_0 ($d_0 = 1.445$Å). (b) Experimental Raman spectra of InP(001)-S at 295 and 20 K (inset).

be at 346 ±4 cm^{-1} at 0 K, in excellent agreement with the experimental value at (20 K) of 349±1 cm^{-1}. The weak feature (note logarithmic intensity scale) at 334 cm^{-1} is the bottom of the LO-phonon branch and is not visible for periodic structures, but appears in our calculation since k-conservation fails in the presence of a surface.

Two modes which arise from the presence of the S-layer and the surface reconstruction were identified in the theoretical Raman spectrum, at 190±2 and 257±3 cm^{-1}. The phonon eigenvectors [displacements from equilibrium; cf. inset of Fig. 2(a)] indicate that for the mode at 190 cm^{-1}, the P_b and P_t sublayers displace in the *same* direction, with P_t having the largest amplitude, while the S_b and S_t sublayers also displace together but in the sense opposite to $P_{b,t}$. For the mode at 257 cm^{-1}, the S_b and S_t, as well as P_b and P_t, vibrate in *opposite* directions. Experimental observation of these two modes would evidently be strong evidence for the proposed structure, and in particular the distinctive splitting of the S and P layers produced by the (2 × 2) reconstruction.

Since the contribution of the surface sulphur layer to the total spectrum is of order $1/N$ (N = the number of layers traversed by the light probe), state-of-the-art light scattering techniques are called for. The Raman spectrum was measured in the quasi-backscattering configuration for a number of freshly prepared samples [20] at 295 K and 20 K. In order to isolate the very weak InP-S surface mode, the Raman spectrum of the bare InP(001) substrate recorded under identical conditions were subtracted from each InP(001)-S spectrum. Details of the experiment are given in Ref. [6].

The spectrum of an InP-S sample measured at 295 K is shown in the upper panel of Fig. 2(b); it is complicated by the presence of second-order Raman (SOR) lines below the transverse optic (TO) phonon. The inset shows the corresponding spectrum at 20 K, where the SOR are much reduced. The weak peak at 255 cm^{-1} is an order of magnitude weaker than the TO-phonon peak, which is itself weak in the back-scattering configuration. The lower panel of Fig. 2(b) shows "difference" spectra at 295K obtained by subtracting the untreated InP(001) spectrum from the InP(001)-S spectrum and averaging over 15 different runs for the same physical point on the surface. A peak-search routine, which applies a Poisson statistics analysis, detected peaks at 191±1 and 255±1 cm^{-1}, in remarkable accord with the theoretical (0 K) values of 190 and 257 cm^{-1}. The peaks are clearly seen in the top curve of the lower panel, whereas there is no such peak structure in the bottom curve, which is for the clean surface. Entirely equivalent results were obtained from other samples.

Our results are interesting in many respects; in particular, they show that the InP-S surface is quite different from that currently accepted for GaAs-S. Our structure, if used in calculations for GaAs-S, is in fact found to be a local minimum approximately 0.2 eV above the calculated ground state of GaAs-S. Clearly, results from GaAs-S studies cannot be simply carried over to the InP-S system, and *vice versa*.

Core-level shifts

We examine here the effect of annealing on the structure of the surface, and give a systematic explanation of the observed CLS, complementing the Raman measurements presented above. Annealing opens an energy window whereby structures other than the ground state become statistically or kinetically accessible. In addition, this sets in motion diffusion processes where surface atoms migrate towards the bulk and *vice versa*.

The sequence of surface structures we considered is given in Fig. 3, where (a) is the ground-state geometry discussed above. We give, for each, the energy with respect to the lowest-energy structure at that coverage, but do not attempt to compare the lowest-energy structures at different coverages since that would require addressing the issue of how the sulphur atoms which migrate to the bulk arrange themselves. The corresponding theoretical S-2p excitation energies (from TED) are reported in Table I; the S atom with the core hole

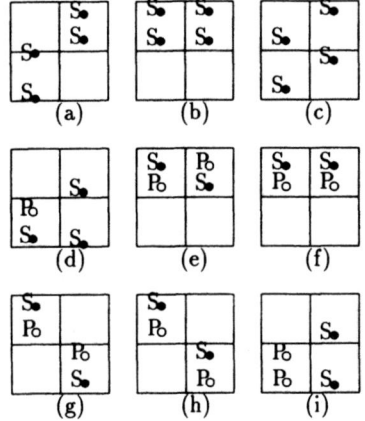

Configuration	Structure	ε_{2p} (ΔE_g)
Full S-coverage)		
(a) (ground-state)	2×2	(0.0)
S*-S	monomer	161.9
S*-S	dimer	162.9
(b) (unstable)	2×1	(0.86)
(c)	2×1	(0.30)
S*-	monomer	161.6
3/4-ml-S coverage		
(d)	2×2	(0.0)
S*-P	dimer	161.3
S*-S	monomer	160.6
1/2-ml-S coverage		
(e)	2×2	(0.0)
S*-P	dimer	162.9
(f)	2×1	(0.17)
S*-P	dimer	163.6
(g)	2×2	(0.135)
S*-P	dimer	163.0
(h)	2×2	(0.046)
S*-P	dimer	162.8
(i)	2×2	(1.0)
S*-S	monomer	161.6

Figure 3: Top view of the fully-relaxed InP(001)-(S,P) surface structures. (a)-(c) are full S-covered; (d) is 3/4-S-covered; (e)-(i) are 1/2-S-covered.

Table I: CL energies ε_{2p} (eV) for the surfaces of Fig. 3, and ground-state energies ΔE_g [per (2×2) cell, in eV] relative to the lowest-energy structure at each coverage. S* is the atom with the $2p$ core-hole.

is marked S*. Two different spectra are expected in structures with two inequivalent S atoms. Further, each S-$2p$ level is a relativistic doublet and hence each spectrum is expected to show a double-peak structure. The spin-orbit splitting was calculated using the method of Koelling *et al.* [23] in the all-electron FP-SR LDA scheme (see Ref. [14] for details). It was found to be equal to 1.29 eV ($\Delta \varepsilon = -0.86$ eV for $j=1/2$ and $+0.43$ eV for $j=3/2$). The two peaks of the doublet have intensities 1:2 since the level degeneracies are $2j + 1 = 2$ and 4. The lineshapes were assumed to have a Lorentz-Gauss (Voigt) profile; no attempt was made to provide a first-principles spectral lineshape. The relative intensities of the two peaks in the doublet, however, was forced by the degeneracies $2j + 1$ discussed above.

A (2×1) reconstruction has been reported on the basis of photoemission and LEED studies of annealed samples by Mitchell *et al.* [18] and was assigned to the structure of Fig. 3(b), containing S dimer rows. Our calculations show that this geometry is unstable, i.e., not even an energy minimum. However, local energy minima do exist, e.g., Fig. 3(c), which has just one type of S atoms [6], lying ∼0.07 eV per S atom above the ground state. LEED and STM may have difficulty in distinguishing P and S atoms on surfaces with partial (S,P) coverages as in, say, Fig. 3(d)-(i), while CL spectroscopy has no such difficulty since S-$2p$ and P-$2p$ excitations are well separated. A stringent test of the (2×2) structure is, therefore, to compare the *predicted* dimer and monomer CL spectra against experiment. This would distinguish the surface from the (1×1) structure analogous to the GaAs(001)-S [8], where the S atoms are at bridge sites, and from the (2×1) structure of Fig. 3(c), since they have only one type of sulphur.

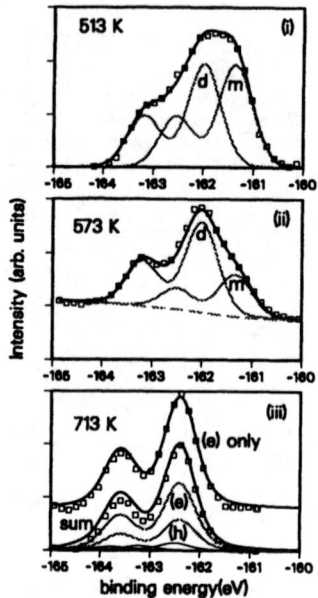

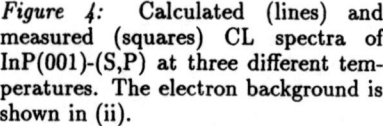

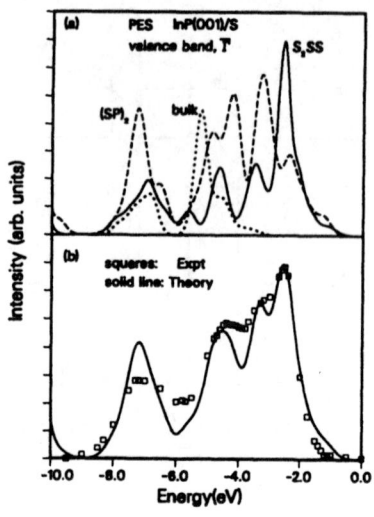

Figure 4: Calculated (lines) and measured (squares) CL spectra of InP(001)-(S,P) at three different temperatures. The electron background is shown in (ii).

Figure 5: Calculated (lines) and measured (squares) PE spectra of InP(001)-(S,P): (a) individual and (b) composite PE spectra, as discussed in the text.

Photoelectron-spectroscopy measurements (carried out at the Stanford Synchrotron Radiation Laboratory) were taken at three different temperatures — 513, 573 and 713 K — and are displayed in Fig. 4. Details of the experiment can be found in Ref. [14]. Since Fig. 3(a) is the most likely structure, we analyse the low-T spectrum as follows. We assume that long-dimer sulphur atoms ("monomer" – 'm') and short-dimer atoms ("dimer" – 'd') contribute equally to the spectrum. Hence we have two equal-intensity doublets with the same s-o splitting, lineshape and a 1:2 intensity ratio in each doublet, as discussed above. With these strong constraints, the experimental profile is fitted with the two doublets (four Voigt profiles) whose peak positions are adjusted from the theoretical values only to within the expected error of ±0.5 eV. The resulting fitted peak positions are designated the *experimental* peak values. The experimental s-o splitting is 1.18 eV; this will be used in displaying other data. On fitting the experimental data at 513 K, the $2p$ experimental binding energies of 'd' and 'm' S atoms were found to be −162.57 and −161.95 eV, comparing very well with the theoretical values of −162.9 and −161.9 ± 0.5 eV for the structure of Fig. 3(a). The experimental spectrum at 513 K is thus *clearly consistent* with the predicted (2×2) ground-state structure.

On heating, other configurations begin to compete with the ground state. The next stable, fully-S-covered structure is that of Fig. 3(c). This structure gives a single S-$2p$ CL-E calculated to be ∼161.6 eV, i.e., near the previous 'm' peak. If this is formed on heating, the 'd' intensity should decrease while the 'm' peak grows. Fig. 4(ii) shows that the monomer peak has, rather, decreased in intensity. Thus the structure of Fig. 3(c) does not form, because of other competing processes. Such a process is the surface-S↔bulk-P exchange;

there is strong evidence for such exchanges from X-ray photoelectron diffraction studies [24]. The analysis of the CL spectra, therefore, must include S coverages other than one.

The 3/4-S-covered structure of Fig. 3(d) is particularly interesting. When a monomer S atom in Fig. 3(a) is replaced by a P atom, the initially long S–P bond contracts to a tight S–P molecule, while the dimer S–S elongates to a monomer-pair. The structure of Fig. 3(d) however cannot account for the measured spectrum, probably because the fast S↔P exchanges leading, instead, to 1/2-S-covered surfaces. The stable 1/2-covered surfaces are shown in Figs. 3(e)-(i). In all these, the S–P dimers are tightly bound and *buckled* with the P atom projecting out. Unlike the ground state, any one of these structures, having only one kind of S atoms, will produce a *single* core-level-doublet spectrum. The structures are close to each other in energy and thus all contribute to the surface partition function, with a weight $Z^{-1}\exp(-\Delta E_g/k_BT)$. The weighted-sum theoretical spectrum for structures (e)-(i) at 713 K is displayed as 'sum' in Fig. 4(iii). However, the CL spectrum from Fig. 3(e) by itself, or (h) by itself, agrees better with experiment than the 'sum'. Thus, the annealed surface is *not* a simple weighted sum of the possible structures. One could imagine that the lowest energy structure nucleates first, thus determining further surface growth (a kinetic proces). In addition, the 1/2-S-covered structure of Fig. 3(e) is consistent with LEED (2×1) data from annealed samples [18], since S and P atoms could look alike under LEED.

The above analysis suggests that kinetic effects are relevant. Hence we have not tried a structural analysis of the spectrum of Fig. 4(ii): structures containing sulphur atoms on several subsurface P-planes may be relevant at these intermediate situations. To summarize, therefore, the above analysis of CL spectra confirm the geometry discussed above for the (2×2) ground state, while indicating that the higher temperature structure is likely to contain rows of S–P pairs.

Photoemission spectra

PE and IPE experiments have been carried out on annealed InP(001)-S surfaces characterized using LEED by Mitchell *et al.* [18]. Our calculations do not support the interpretation of the data as arising from a simple (1×1) pattern. As suggested by the CLS calculations, the annealed samples could well be an admixture of the ground-state structure [(a) in Fig. 3, hereafter referred to as S_2SS] and the lowest-energy 1/2-S-covered surface [(e) in Fig. 3; hereafter referred to as $(SP)_2$]. We verify this hypothesis in what follows.

Fig. 5(a) presents our calculation (as detailed above) of the PE spectra for both the S_2SS (full line) and the $(SP)_2$ (dashed line) structures. Also shown is the spectrum from bulk InP, projected onto the surface $\bar{\Gamma}$ point (dotted line). The experimental data of Mitchell *et al.* [18] are displayed as squares in Fig. 5(b). [The PE spectra of each phase (and the bulk) were separately calculated and aligned to have the same d-electron peak arising from In atoms (which are not significantly affected by surface reconstructions).] It is clear that neither the S_2SS nor the $(SP)_2$ spectrum alone can account for the experimental data, i.e., the measured spectrum likely contains features from both structures. Indeed a composite spectrum [full line in Fig. 5(b)] obtained by assuming that the region of the surface sampled by the light contains 30% S_2SS and 70% $(SP)_2$, yields a quite reasonable fit to the experimental data. Of course, the experimental spectrum depends on details of sample preparation and annealing history. However, an invariant property would be that a composite theoretical spectrum constructed from the spectra of three main annealing components — S_2SS, $(SP)_2$ and possibly P-terminated InP(001) — can always be found to match the essential features of a given experimental spectrum. We are therefore led to conclude that the samples examined by Mitchell *et al.* [18] using PE actually contain a mixture of S_2SS and $(SP)_2$ phases. This conclusion is reinforced by the observation that the width of the features in the experimental PE spectra of InP(001)-S is much broader than from cleaved GaAs or InP(110) surfaces [25].

11

CONCLUSION

We have presented a comprehensive study of the sulphur-passivated InP(001) surface, based on first-principles (density-functional theory) calculations of the energies of various possible states of the surface. The ground-state structure is found to be a striking new reconstruction containing S dimers and monomers. This phase has been confirmed by Raman scattering measurements. We have also examined the behaviour of the surface upon annealing in an attempt to provide a sound interpretation of finite-temperature core-level-shift and photoemission measurements. This analysis not only confirms that the ground-state structure we predicted is indeed correct, but also demonstrates that several structures may contribute to the observed spectra. In particular, exchange of surface S atoms with bulk P atoms is promoted at finite temperature (i.e., the surface is only partially S covered), resulting in a delicate process of chemical association and dimer dissociation at the surface, which plays a determining role in structural relaxation. Such aspects of the physics of surfaces must be considered in the interpretation of finite-temperature spectroscopic data. For the particular case of the InP(001)-S surface, it is felt that further theoretical and experimental study would benefit from better characterized surfaces. However, all the presently-available evidence — STM, Raman, CL, PE, and IPE spectroscopy, except perhaps LEED — are consistent with our picture of the as-prepared and moderately-annealed surfaces being a 2×2-reconstruction, essentially S_2SS for the former and mostly $(SP)_2$ for the latter.

ACKNOWLEDGEMENTS

We are grateful to very many people for contributions, discussions, comments, help, criticisms, etc., in particular G. Aers, R. Cao, J.M. Jin, F. Himpsel, D. Lockwood, Z.H. Lu, A. McLean, A. Pasquarello, and Z.J. Tian. This work was partially supported by grants to LJL from the Natural Sciences and Engineering Research Council of Canada and the "Fonds pour la formation de chercheurs et l'aide à la recherche" of the Province of Québec.

REFERENCES

[a] E-mail: lewis@physcn.umontreal.ca.

[b] E-mail: chandre.dharma-wardana@nrc.ca.

1. Science, "Is the future here for GaAs?", **262**, 1819 (1993).

2. T.J. Coutts and S. Naseem, Appl. Phys. Lett. **46**, 164 (1985).

3. W.M. Lau, S. Jin, X.W. Wu, and S. Ingrey, J. Vac. Sci, Technol. A **9**, 994 (1991); M. S. Carpenter *et al.*, Appl. Phys. Lett. **52**, 2157 (1988).

4. Y. Tao, A. Yelon, E. Sacher, Z.H. Lu, and M.J. Graham, Appl. Phys. Lett. **60**, 2669 (1992).

5. Z.H. Lu, M.J. Graham, X.H. Feng, and B.X. Yang, Appl. Phys. Lett. **60**, 2773 (1992); Z. H. Lu *et al.*, *ibid* 2932 (1993)

6. J.-M. Jin, M.W.C. Dharma-wardana, D.J. Lockwood, G.C. Aers, Z.H. Lu, and L.J. Lewis, Phys. Rev. Lett. **75**, 878 (1995).

7. M.P. Teter, M.C. Payne, and D.C. Allan, Phys. Rev. B **40**, 12255 (1989).

8. See e.g. H. Oigawa, J. Fan, Y. Nannichi, K. Ando, K. Saiki, and A. Koma, Jpn. J. Appl. Phys. **28**, L 340 (1989). T. Ohno, Surf. Sci. **255**, 229 (1991). M. Tanimoto, H. Yokoyama, M.Shinohara, and N. Inoue, Jpn. J. Appl. Phys. **33**, L 279 (1994). M. Sugiyama *et al.*, Phys. Rev. B **50**, 4905 (1994).

9. See *Synchrotron Radiation Research: Advances in Surface Science*, edited by Z. Bachrach (Plenum Press, New York, 1990), and J. Himpsel *et al..*, in *Photoemission and Absorption Spectroscopy of Solids and Surfaces with Synchrotron Radiation*, edited by M. Campagna and R. Rosei (North-Holland, Amsterdam,1990), p203; also, E. L. Bullock *et al..*, Phys. Rev. Lett, **74,**

10. L. Kleinman and D.M. Bylander, Phys. Rev. Lett. **48**, 1425 (1982).

11. D.M. Ceperley and B.J. Alder, Phys. Rev. Lett. **45**, 566 (1980).

12. J.-M. Jin and L.J. Lewis, Phys. Rev. B **49**, 2201 (1994); J.-M. Jin, L.J. Lewis, V. Milman, I. Stich, and M.C. Payne, *ibid* **48**, 11 465 (1993).

13. J.-M. Jin and L.J. Lewis, Surf. Sci. **325**, 251 (1995).

14. Z. Tian, M.W.C. Dharma-wardana, Z.H. Lu, A. Cao, and L.J. Lewis, Phys. Rev. B **55**, 5376 (1997).

15. J.C. Brice, in *Properties of Indium Phosphide* (INSPEC, London, 1991) p. 5, and references cited therein.

16. M. Methfessel, Phys. Rev. B **38**, 1537 (1988); M. Methfessel, D. Henig, and M. Scheffler, Phys. Rev. B **46**, 4816 (1992), and references therein.

17. J.C. Slater, *The self-consistent field for molecules and solids*, Vol IV, (McGraw-Hill, New York) 1974; R.M. Dreizler and E.K.U. Gross, *Density Functional Theory*, section 4.4, (Springer-Verlag, New York) 1990.

18. C.E.J. Mitchell, I.G. Hill, A.B. McLean and Z.H. Lu, Progress in Surface Science, **50**, 325 (1995).

19. M.W.C. Dharma-wardana, Z. Tian, Z.H. Lu, and L.J. Lewis, Phys. Rev. B **56**, 10526 (1997).

20. X.R. Qin, Z.H. Lu, J.G. Shapter, L.L. Coatsworth, K. Griffiths, and P.R. Norton, J. Vac. Sci. Tech. A **16**, 163 (1998).

21. K. Kunc, in *Electronic Structure, Dynamics, and Quantum Structural Properties of Condensed Matter*, ed. by J.T. Devreese and P. Van Camp (Plenum, New York, 1985) p. 227. K. Kunc and R.M. Martin, Phys. Rev. Lett. **48**, 406 (1982).

22. M.W.C. Dharma-wardana, G.C. Aers, D.J. Lockwood, and J.-M. Baribeau, Phys. Rev. B **41**, 5319 (1990); *Light Scattering in Semiconductor Structures and Superlattices*, Ed. D.J. Lockwood and J.F. Young (Plenum, New York, 1991), p. 81.

23. D.D. Koelling and B.N. Harmon, J. Phys. C (UK) **10**, 3107 (1977)

24. Z.H. Lu and R. Cao (unpublished).

25. A. B. McLean (private communication)

IN SITU SURFACE PASSIVATION OF GAAS BY THERMAL NITRIDATION USING METALORGANIC VAPOR PHASE EPITAXY

Jingxi Sun, F. J. Himpsel*, A. B. Ellis**, T. F. Kuech
Department of Chemical Engineering,
*Department of Physics
**Department of Chemistry
University of Wisconsin-Madison
Madison, WI 53706

ABSTRACT

An ammonia-based, in situ passivation of GaAs surfaces conducted within a metalorganic vapor phase epitaxy reactor is present. The shift of the GaAs surface Fermi level, and hence the surface charge density, resulting from this in situ passivation, has been studied using photoreflectance (PR) spectroscopy. Samples consisting of an undoped GaAs layer on highly doped n-GaAs (UN^+) and p-GaAs (UP^+) structures allow for the exact determination of the surface Fermi level position using PR. These structures were grown by MOVPE and in situ thermal nitridation was performed after growth within the MOVPE system without exposure to the air. After nitridation, the surface Fermi level can be shifted by ~ 0.23 eV towards the conduction band edge for UN^+ structures and by ~ 0.11 eV towards the valence band edge for UP^+ structures from the normally mid-gap 'pinned' positions.

INTRODUCTION

Compound semiconductors, such as GaAs and InP, are widely used for high-speed semiconductor devices as well as other microelectronic and optoelectronic devices. Most III-V compound semiconductors, particularly GaAs, possess a high density of midgap surface states that have complicated the development of many potential applications. Surface passivation is therefore an important process in the fabrication of GaAs-based electronic devices. The nitridation of GaAs surfaces can result in the formation of a thin nitride-based surface layer. This thin nitride-based layer can act as an effective surface passivation layer which can reduce the surface state density and hence improve the device performance [1, 2]. The Fermi level position at GaAs surfaces is a crucial property, however, the shift of surface Fermi level due to thermal nitridation has not yet been systematically studied.

In this study, we report on an in situ surface passivation of GaAs by thermal nitridation within a metalorganic vapor phase epitaxy (MOVPE) system. In situ nitridation of the GaAs surface after growth allows for a controlled chemical process on a surface free of oxide contamination. The in situ nitridation also provides the possibility of depositing subsequent layers, such as silicon nitride or silicon dioxide, immediately after nitridation of the GaAs surface. We have studied the shift of GaAs surface Fermi level, due to in situ thermal nitridation, using photoreflectance spectroscopy (PR).

PHOTOREFLECTANCE SPECTROSCOPY

PR has been shown to be a very useful method for determining the surface electric field and hence the surface Fermi level [3]. The special structures, proposed initially by Van Hoof et al. [4], are used to simplify the data analysis and ascertain the GaAs surface Fermi level position in

15

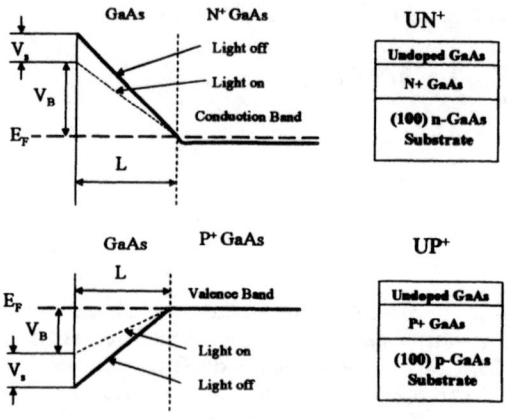

Figure 1. Schematic diagrams of UN+ and UP+ structures

this work. These structures consist of a thin layer of undoped GaAs on highly doped n- or p-type GaAs (UN+ and UP+ structures), resulting in a nearly constant surface electric field in the undoped layer. The diagrams of these structures and related surface band structures are shown in Figure 1. The surface electric field from the Franz-Keldysh Oscillations (FKOs) observed in PR spectra, and hence the surface barrier height, V_B, can be accurately determined [5,6] using these structures providing the thickness of the undoped GaAs layer (L). The PR spectra were analyzed to get the surface electric field using the following equations:

$$n\pi = \varphi + (4/3)[(E_n - E_0)/\hbar\theta]^{3/2} \quad (1)$$

$$F = (\hbar\theta)^{3/2}/(e\hbar/\sqrt{2\mu}) \quad (2)$$

where E_n is the photon energy of nth extrema, E_0 is the energy gap of GaAs, $\hbar\theta$ is electro-optic

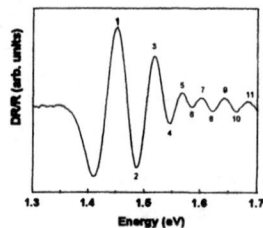
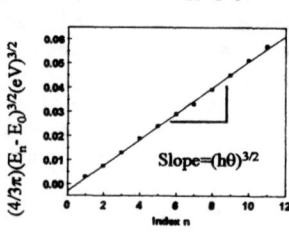

energy, F is the surface electric field and μ is the reduced effective mass. Figure 2 show the PR spectra from the UN+ structure and the data analysis [3]. Many FKOs can be seen from this data due to the strong, well-defined constant electric field within the high quality undoped layer in this structure. The surface electric field is determined by plotting $(4/3\pi)(E_n-E_0)^{3/2}$ as a function of index n. The solid line is a least square fit to a linear function that allows accurate evaluation of the surface electric field. During photoreflectance spectroscopy, the surface photovoltaic effect reduces the surface electric field from its non-illuminated value [3], as shown in Figure 1. The surface photovoltage (V_s) can become negligible during PR measurements at sufficiently high temperatures (> 400K) [6].

Figure 2. Photoreflectance spectrum of the UN+ structure and a plot used to determine the surface electric field.

EXPERIMENTAL PROCEDURES:

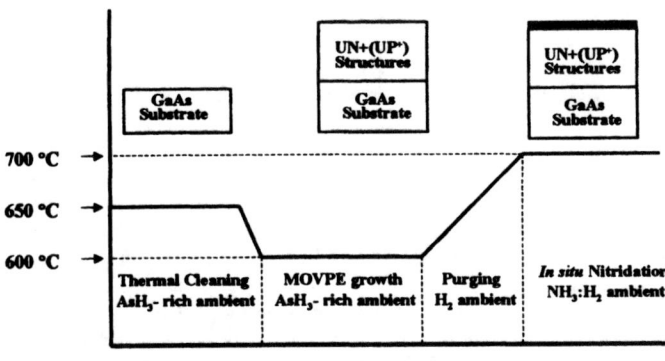

Figure 3. Schematic diagram of the in-situ thermal nitridation process

All GaAs samples were grown in a conventional horizontal low-pressure (78 Torr) MOVPE reactor [7] using trimethyl gallium (TMGa) and arsine (AsH₃). Disilane (Si₂H₆) and carbon tetrachloride (CCl₄) were employed for n-type and p-type doping, respectively. The UN⁺ (UP⁺) structures consist of an undoped layer and a 0.5 μm highly doped (1.0 x 10¹⁸ cm⁻³) n-type (UN⁺) or p-type (UP⁺) buffer grown on a n⁺ or p⁺ singular GaAs (100) substrate. The *in situ* thermal nitridation was performed immediately after the layer growth. While the reactor temperature was changed to the nitridation temperature, the reactor was purged with H₂. Once at the nitridation temperature, the surface was exposed to a feed ratio of electronic grade NH₃-to-H₂ of 1-to-21. The nitridation time was fixed at 5 minutes at a constant temperature of 700 °C. The *in situ* nitridation process is shown in Figure 3. The PR apparatus followed the design of Shen and co-workers [8]. The pump light was a 1 mW HeNe laser, mechanically chopped at 400 Hz. The probe light source was provided by a tungsten-halogen bulb and monochromator combination. The reflected probe light was focused on a large-area Si photodiode through a cutoff filter and captured by a lock-in amplifier. The samples were held in a vacuum chamber at ~ 10⁻⁶ Torr during the PR measurements in order to eliminate any effects caused by the room ambient conditions.

RESULTS & DISCUSSION

The room-temperature PR spectra from UN⁺ and UP⁺ samples, with and without nitridation, are shown in Figure 4. The period of the FKOs in the PR spectra is reduced as a result of the nitridation treatment for both the UN⁺ and UP⁺ samples. This reduction in the period of FKOs can be attributed to the shift of the surface Fermi level from its normal 'pinned' positions on the untreated GaAs surface towards the conduction band edge (UN⁺) or the valence band edge (UP⁺), resulting in the reduction of the surface electric field. In order to obtain the surface Fermi level position, the PR spectra were obtained at a sample temperature of 420 K, with a greatly reduced or negligible photovoltaic effect. The surface barrier heights, determined from the PR spectra measured at 420 K, are reported in Table 1. We find that the surface Fermi level can be shifted by ~ 0.23 eV towards the conduction band edge for UN⁺ structures and ~ 0.11 eV towards the valence band edge for UP⁺ structures from the normal 'pinned' positions due to nitridation.

The chemical and physical changes on the GaAs surface due to exposure to ammonia at high temperatures will be a function of the reaction conditions, such as gas phase composition and

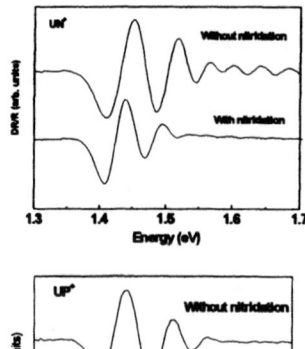

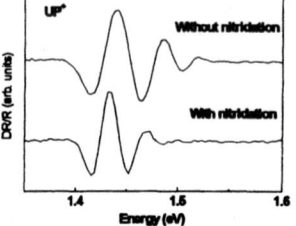

Figure 4. Room-temperature PR spectra for UN⁺ and UP⁺ structures indicating a shift in Fermi level on the surface

temperature, as well as the surface state prior to nitridation reaction. The MOVPE GaAs surface, in the absence of ammonia, has been studied in some detail. Two types of surface structures are typically reported depending on the surface stoichiometry. These surface structures are the Ga-rich reconstructions, such as c(8x2) or (2x6) reconstruction, and the c(4x4) As-rich reconstruction [9, 10, 11, 12]. These surface structures consist of a characteristic arrangement of arsenic or gallium dimers on the surface. The c (8x2) GaAs surface is terminated with Ga-Ga dimers [9]. The c(4x4) As-rich reconstruction consists of a double layer of As atoms on the GaAs surface [9,10,13]. This latter structure is thought to be present during typical MOVPE GaAs growth conditions due to the high As and AsH_3 partial pressures near the surface. During the dynamic conditions of growth, no long-range order is typically present on the GaAs surface [11], yet the local bonding may be similar to that seen in these reconstructions. For the purposes of this paper, we will assume that the local structure, i.e. Ga-Ga or As-As dimers, under these Ga-rich and As-rich conditions are also present prior to reaction with ammonia. The process conditions used in this study should naturally lead to either of these surface structures.

Table 1. Surface barrier heights of samples nitrided at 700 °C and without nitridation

Surface Barrier Height (volt)			
UN⁺ Structures		UP⁺ Structures	
Without nitridation	With nitridation	Without nitridation	With nitridation
0.76±0.01	0.53±0.03	0.51±0.01	0.40±0.01

The nitridation was carried out on surfaces that were held after growth in an H_2-only ambient. Because of the relatively high arsenic activity required to stabilize the GaAs surface, these annealing conditions prior to exposure to NH_3 lead to the preferential desorption of arsenic from the surface and the development of a Ga-rich surface stoichiometry [12]. The initial nitridation of GaAs might be attributed to the direct reaction of ammonia on a Ga-rich surface, formed by the desorption of arsenic. The NH_3 can adsorb on surface Ga species and reactively dissociate on this surface. The dissociation of adsorbed NH_3 and subsequent nitridation of GaAs at the adsorbate-GaAs interface has been previously proposed on the Ga-rich surface [14]. The

18

proceeding nitridation reaction, after this initial nitridation, would occur through the exchange and desorption of arsenic atoms from below the nitrided surface. Figure 5 show the proposed *in situ* nitridation processes.

The high density of mid-gap surface states on the GaAs (100) surface leads to effective pinning of surface Fermi level near the mid-gap of GaAs. The position of the 'pinned' surface Fermi level depends on the surface state energy distribution within the band gap and the free charge carrier type of the bulk material. The results presented here indicate that the *in situ* thermal nitridation at 700 °C in a NH_3:H_2 atmosphere can lead to a shift in the surface Fermi level by ~ 0.23 eV towards the conduction band edge in the UN^+ structures and by ~ 0.11 eV towards the valence band edge in the UP^+ structures from the normal 'pinned' positions. A shift in the surface Fermi level, resulting from the chemical changes on the surface, indicates that the surface state density within the band gap has been reduced or its distribution has been modified. This change in surface state distribution will have several implications for a device depending on the specific device application. In general, the motion of the Fermi level from the mid-gap position reduces the surface recombination velocity [15]. This nitridation process can therefore be used effectively in applications that require a reduced surface recombination velocity. The application of this passivation technique to the more demanding formation of a gate dielectric in a field effect transistor is more uncertain. There are two important issues in this application: surface roughness and the specific surface state density distribution. Our studies have shown that the surface state density, as indicated by the shift in surface Fermi level position, is modified. Further studies would be required in order to ascertain the detailed nature of the surface state density in order to evaluate its efficacy as a gate dielectric.

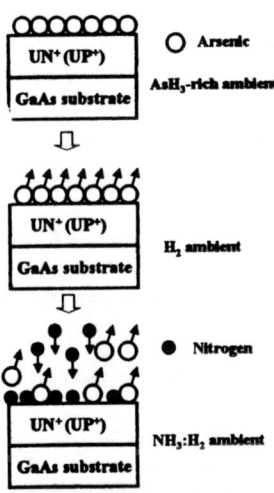

Figure 5. Schematic diagram of the model for in situ thermal nitridation

In conclusion, *in situ* thermal nitridation of GaAs using MOVPE was studied on n- and p-type materials. This nitridation reaction results in the shift of the surface Fermi level from normal 'pinned' positions towards the band edges. High process temperatures as well as the formation of a GaN-based surface layer on the GaAs surface can lead to a electronically passivated surface.

ACKNOWLEDGMENTS

This work is supported by National Science Foundation. The facility support of the UW-MRSEC is also gratefully acknowledged.

REFERENCES

1. Lisa A. DeLouise, J. Vac. Sci. Technol. A10 (1992) 1637.
2. Atsushi Masuda, Yasuto Yonezawa, Akiharu Morimoto and Tatsuo Shimzu, Jpn. J. Appl. Phys. 34 (1995) 1075.

3. F. H. Pollak and H. Shen, Material Science & Engineering R 10 (7-8) (1993).
4. C. Van Hoof, K. Deneffe, J. De Boeck, D. J. Arent, and G. Borghs, Appl. Phys. Lett. 54 (1989) 608.
5. D. E. Aspnes, Phys. Rev. B10 (1974) 4228.
6. X. Yin, H-M. Chen, F. H. Pollak, Y. Chan, P. A. Montano, P. D. Kirchner, G. D. Pettit, and J. M. Woodall, J. Vac. Sci. Technol. A10 (1992) 131.
7. T. F. Kuech, E. Veuhoff, and B. S. Meyerson, J. Crystal Growth 68 (1984) 48.
8. H. Shen, P. Parayanthal, Y. F. Liu, and F. H. Pollak. Rev. Sci. Instrum. 58 (1987) 1429.
9. D.K. Biegelson, R.D.Bringans, J.E. Northrup, and L.-E. Swartz, Phys. Rev. B41 (1990) 5701.
10. F. Reinhardt, W. Richter, A.B. Muller, D. Gutsche, P. Kurpas, K. Ploska, K.C. Rose and M. Zorn, J. Vac. Sci. Technol. B 11 (1993) 1427.
11. D. W. Kisker, G. B. Stephenson, P. H. Fuoss, F. J. Lamelas, S. Brennan and P. Imperatori, J. Crystal Growth, 124 (1992) 1.
12. F. J. Lamelas, P. H. Fuoss, P. Imperatori, D. W. Kisker, G. B. Stephenson and S. Brennan, Appl. Phys. Lett. 60 (1992) 2610.
13. A.P. Payne, P.H. Fuoss, D.W. Kisker, G.B. Stephenson and S. Brennan, Phys. Rev. B49 (1994) 14427.
14. X. Y. Zhu, M. Wolf, T. Huett, J. M. White, J. Chem. Phys. 97 (1992) 5856.
15. Peter T. Landsberg, Recombination in Semiconductors, Cambridge University Press, Cambridge, 1991, 210.

STRUCTURE OF SINGLE-CRYSTAL Gd$_2$O$_3$ FILMS ON GaAs(100)

A.R. KORTAN, M. HONG, J. KWO, J.P. MANNAERTS, N. KOPYLOV
Bell Laboratories, Lucent Technologies, Murray Hill, NJ 07974-0636,

ABSTRACT

We have studied the single-crystal Gd$_2$O$_3$ films grown epitaxially on GaAs(100) substrate with single-crystal x-ray diffraction. The sesquioxide Gd$_2$O$_3$ forms two hexagonal phases, one monoclinic and one cubic phase in bulk form. In our studies of different thickness films, we have found that the Gd$_2$O$_3$ grows only in the cubic phase with a unique epitaxial orientation. The two-fold (110) planes of the Gd$_2$O$_3$ are oriented parallel to the four-fold GaAs(100) surface, while alligning its [001] and [$\bar{1}$10] axes with the [011] and [01$\bar{1}$] axes of GaAs within the plane, respectively. The film chooses only one of the two such possible orientations, which can be explained by the local bonding configuration at the interface. We find evidence for an elastic strain in the films less than 50 Å thick.

INTRODUCTION

One of the most challenging issues of the present day semiconductor technology is the identification of reliable passivation layers. This is becoming increasingly more critical with device sizes shrinking to atomic dimensions. In the case of GaAs, the lack of a reliable oxide has imposed very restrictive rules in device designs. Earlier, we discovered an amorphous oxide mixture of Ga$_2$O$_3$(Gd$_2$O$_3$)[1] as an excellent candidate for GaAs(100) passivation. More recently, we discovered[2,3] that the elimination of Ga$_2$O$_3$ in the MBE film growth have resulted in an epitaxial single crystal Gd$_2$O$_3$ film growth which had excellent surface passivation properties[4]. This is particularly important because, when the gate dielectrics become only several atoms thick, a thermodynamically stable and defect free single-crystal oxide film intuitively should have much better defined characteristics and optimized properties compared to an amorphous oxide layer. Here, we summarize our work on identifying the structure of these Gd$_2$O$_3$ films.

STRUCTURE OF RARE-EARTH SESQUIOXIDES

In rare-earth (RE)$_2$O$_3$ sesquioxide structures, cations are typically six and seven-fold coordinated with oxygen atoms. In a high temperature hexagonal phase A ; all cations are sevenfold coordinated, with four oxygen atoms closer than the other three[5]. At a lower temperature the hexagonal A phase transforms into a monoclinic B phase when cations shift to a new position with mixed seven and six-fold coordinations. Upon further cooling B phase transforms into a cubic C phase. In this cubic phase all cations become only six-fold coordinated. These phases exhibit a dependence on the cation size. While for the first two elements of the series La and Ce only the A phase exists, the B phase exists for Sm and heavier rare-earths. There are at least two more phases that exist at temperatures above 2000 C, whose structures are not known. The richness of this phases indicate that, from energetics point of view, the cations can reposition themselves relatively easily.

The Gd metal is a trivalent atom with an unfilled shell of 4f^{7}5d^{1}6s^2 and has the common oxide form of Gd$_2$O$_3$ sesquioxide. Being in the middle of the lathanide series, it is reported to have all three phases. The hexagonal A phase has a narrow range of stability in temperature, but B and cubic-C phases exist over a broader range of temperatures.

The room temperature stable phase of Gd$_2$O$_3$ is the cubic-C phase. This is isomorphous to the α-Mn$_2$O$_3$, which has been the subject of a long extensive research. The structure of cubic-C phase is also found in mineral Bixbyite and is often referred by that name. This structure consist of some specific packing of two-different kinds of octahedral coordination polyhedra of the six-fold coordinated cations, as shown in Figure 1. These are, a regular

21

octahedron and an irregular octahedron that is used in joining regular octahedrons. Here the bond lengths assume values in the range 2.18-2.58Å. Even though the early work of Zachariasen[6] and Pauling[7] on α-Mn$_2$O$_3$ have identified a cubic structure, later refinements[8] have found evidence for deviations in the unit cell from a cubic structure towards an orthorhombic unit cell in the space group Pcab. Here, even the presence of impurities was found to play a role in stabilizing a particular space group.

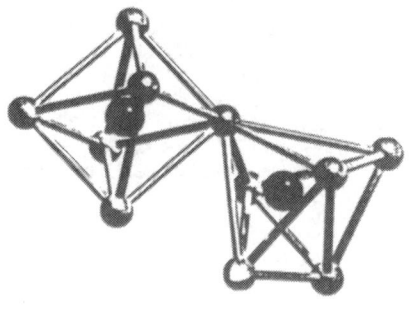

Figure 1. Two basic octahedral building blocks of the cubic-phase Filled (empty) circles represent Gd (oxygen) oxygen atoms.

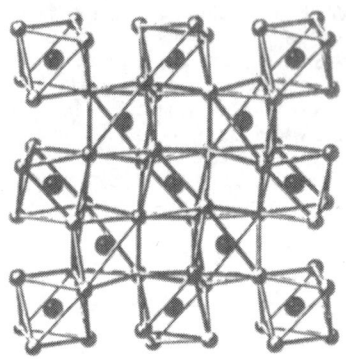

Figure 2. The (100) plane of the Gd$_2$O$_3$ cubic-phase.

Studies carried out on other rare-earth sesquioxides revealed slight deviations from the ideal cation positions. Similar studies have not been carried out for the bulk Gd$_2$O$_3$, but similar modifications in the idealized Gd positions of the cubic lattice can be expected. For the purpose of our thin film studies here, we ignore such higher order corrections, because such modifications if they exist, would be masked by finite size and strain caused effects.

Figure 2 shows a planar section from the cubic Gd$_2$O$_3$ unit cell of a=10.81Å. In this figure there are nine regular octahedrons, and four interstitial irregular octahedrons that join the regular ones. Here every Gd cation has 6 neighboring oxygens, and every oxygen has four nearest neighbour Gd. Every Gd donates 3x (1/6) electrons to a Gd-O bond, and every O donates 6x (1/4) electrons to the same bond, which adds upto 2 electrons per bond, and accounts for the charge balance.

X-RAY DIFFRACTION MEASUREMENTS

The measurements we report here were carried out on a 12kW rotating anode Cu Kα source equipped with a triple-axes, four-circle ganiometer. The resolution was defined by a pair of flat graphite crystals used as monochromatizer and analyser, yielding a longitutinal resolution of 0.01 Å^{-1}, and a transverse resolution of 0.005 Å^{-1}. The large Z value of Gd was particularly useful for allowing us to measure films as thin as several Angstroms. The triple axes geometry separated the θ and 2θ components for a clear identification of mosaic scan features, like dislocation caused broadening, layer tilts, and layer spacings distribution. The samples were mounted and held on a far edge in order to minimize the strains, and measurements were carried out in open air. As reported earlier[3], these films are not sensitive to ambient air.

The RHEED observations made during the growth were suggesting a two-fold single crystal growth for the Gd$_2$O$_3$ films. In order to determine the structure of these films with x-rays, we decided to start with a reasonably thick sample; 185Å, so that reasonable count rates were expected in our reciprocal space mapping. Figure 3 shows the single crystal scan made along the surface normal of the substrate. Here, we observed a diffraction peak near $2\theta =$ 47.5°. In order to associate this peak with a particular structure, we examined all known Gd_2O_3 structures. A good agreement is found with the (440) peak of the α-Gd_2O_3 cubic structure[7,8]. The measured lattice constant of 10.83Å agreed well with the bulk form of this cubic phase which has a 10.813 Å lattice constant, belongs to the space group Ia3, and is an isomorph of the Mn_2O_3 structure. This assignment fixed the [1 $\bar{1}$ 0] direction of the film parallel to the surface normal, or [001] direction of the GaAs substrate. In order to find the relative in-plane orientation, we searched for {222} strong reflections of the film on a cone centered around the [110] Gd_2O_3 axis or [001] GaAs axis, i.e. 360° ϕ-scan with a 35.264° χ-tilt (the angle between film [111] and [110]). As shown in figure 4, this scan revealed two {222} peaks at 0° and 180° phi angle at the right lattice spacing. This established an in-plane epitaxial relationship as $[001]_{Gd2O3}$ // $[011]_{GaAs}$ and $[1\bar{1}0]_{Gd2O3}$ // $[0\bar{1}1]_{GaAs}$. With the two-fold [110] film axis alligned along the substrate four-fold axis, epitaxial films could satisfy the epitaxial conditions in two degenerate orientations. Interestingly, the Gd_2O_3 films, were only growing in one of these orientations with no trace of the other orientation.

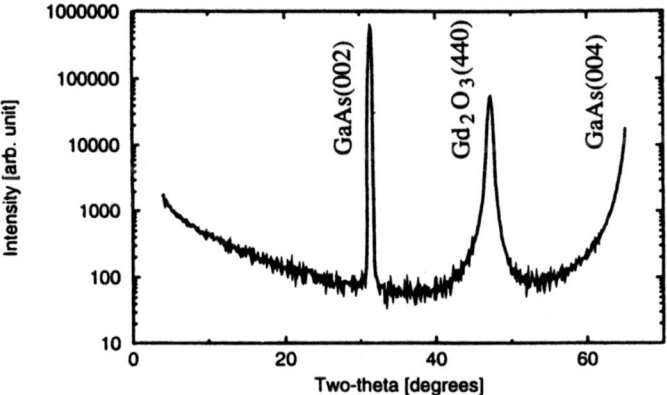

Figure 3. Single crystal scan along the surface normal reveals the (440) peak of the cubic phase of Gd_2O_3.

We also searched for evidence for a secondary phase formation. Scans along all the major zone axes, and specific lattice spacing that belonged to the monoclinic and the hexagonal phases were carried out. There were no evidence for these phases, in agreement with our in-situ reflection high-energy electron diffraction (RHEED) observations[2,3]. We next searched for peaks that belong to the cubic α-phase. Remarkably, all the diffraction peaks were observed at their predicted positions. Our measured intensities were in good agreement with the bulk values listed in powder diffraction files[9]. These values are shown in Table 1. The observed small deviations in some of the intensities have remained within our measurement statistics and analysis errors. Here, no geometrical corrections were made to our data due to the sample geometry. This approximation is valid in a range of diffraction angles where the sample volume does not change significantly.

With the structure identified, we next studied the effect of film thickness on this structure. In the single crystal geometry, we have studied 185, 45, 25 and 18Å thick Gd_2O_3 films.

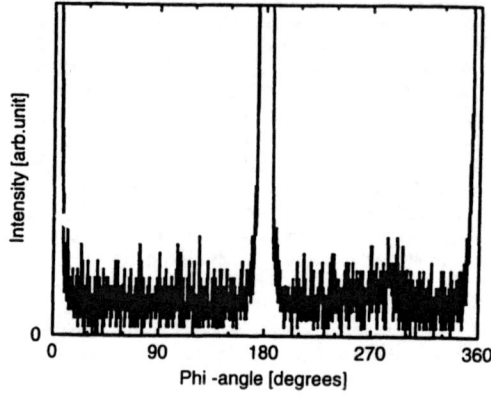

Figure 4. The 360° phi-scan about the surface normal finds two (222) peaks for Gd_2O_3 confirming the single crystal epitaxial growth.

Table 1

hkl	11-604	12-797	43-1014	Measured
211	10	12	5	7
222	100	100	100	100
321		2	1	2
400	35	35	32	29
411		6	3	5
420		2	1	2
332	5	6	2	4
422		2	<1	0.6
431	10	8	4	9
521		2	1	2
440	40	40	35	34

Table 1. Comparing the intensity values to the observed bulk values of the powder diffraction files finds good agreement.

Figure 5 shows how the Gd_2O_3 structure evolves with the film thickness. The intensity of this diffraction peak increased while the peak widths decreased with the film thickness. In addition, the peak position shifted towards larger diffraction angles for thicker films, typical to a strain-relaxation in the film. These features were typical characteristics of a strained layer epitaxial film growth. The peak widths in these 2θ-θ scans are a direct measure of the inverse correlation lengths perpendicular to the surface. In a defect free unstrained epitaxial film, this correlation length corresponds to the film thickness, and the peaks broadened only by the finite size effect. In a strained and/or defective film, however, a distribution of layer spacings may introduce an additional peak broadening. One should therefore be careful in the analysis of the intensity distribution around the diffraction peaks in reciprocal space.

The layer spacing, as determined from the position of the peak near 2θ = 47.5, varies from 2.015Å to 1.914Å, depending on the film thickness (Fig 6a). This change in layer spacing indicates a 5.5% strain relaxation in the film as shown in Fig. 6b. Here the strain assumes a zero value at the bulk layer spacing of 1.911Å or (440) of α - Gd_2O_3, as we will explain later. The widths of the peaks in Fig. 5 are plotted in Fig. 6c. The 18Å film exhibit a defect free region of about 9Å. The 25Å film gives a 26Å correlation length, meaning an essentially defect free film, while 45Å and 185Å films exhibit 34Å and 122Å correlation lengths, respectively.

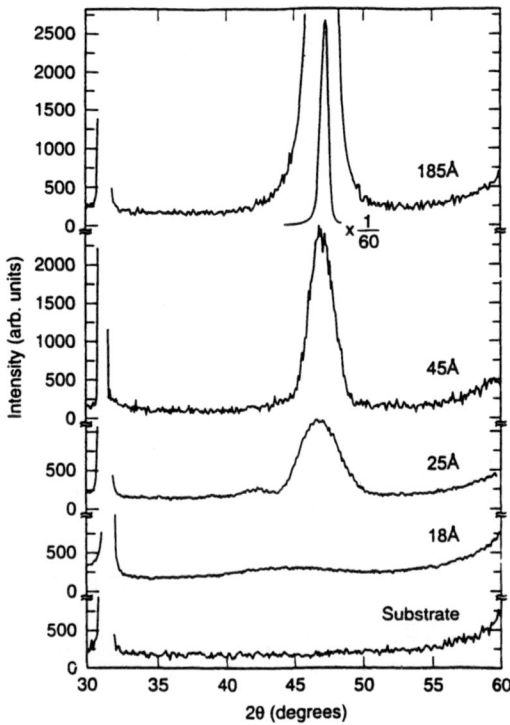

Figure 5. Two-theta scans along the surface normal show that the $Gd_2O_3(440)$ peak gets sharper and shifts to larger angles with increasing film thickness, suggesting strain relaxation.

The mosaic scans of the Gd_2O_3 (440) peak revealed valuable information about the nature of defects in the film. For a film with defect free atomic layers, perfectly alligned with the substrate layers one would expect to obtain a peak width identical to that of the substrate peaks, that is limited by the instrument resolution. Presence of defects, especially misfit dislocations,introduces a mosaic broadening. Figure 7 shows the mosaic scans of these four different films. The 18Å film shows a single sharp peak with a width of FWHM=0.39°, which is slightly above the instrumental resolution of 0.25°, while the 25Å film displays a sharp peak (FWHM=0.36°) superimposed on a broader peak. For thicker films (45Å and 185Å films), the sharp component disappears, but the broad component persists. The broad component has a width of 3.8° when it first appears in the 25Å film, but gets narrower with increasing thickness; 3.25° and 1.82° for 45Å and 185Å films, respectively.

This we interpret as a fully strained epitaxial growth for the films with thicknesses not exceeding 18-25Å. The broad peak is most likely associated with misfit dislocations[10], with its width resulting from the strain fields that exist around individual dislocations. For thicker films, misfit dislocations and their strain fields would be buried deep at the interface, and the peak widths would gradually decrease, in agreement with our observations[11].

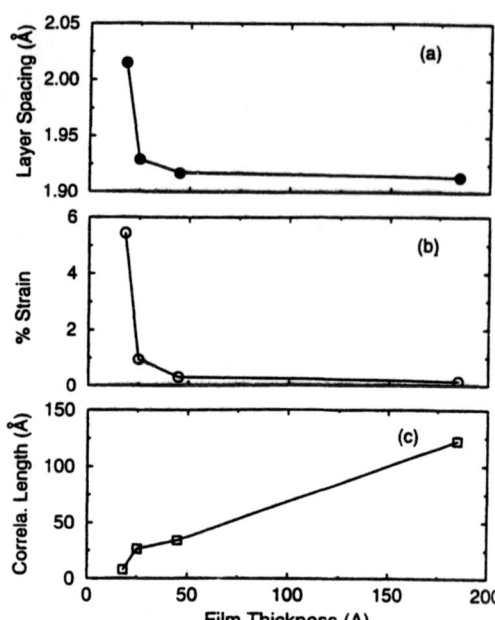

Figure 6. The thickness dependence of the $Gd_2O_3(440)$ peak shown in figure 5.

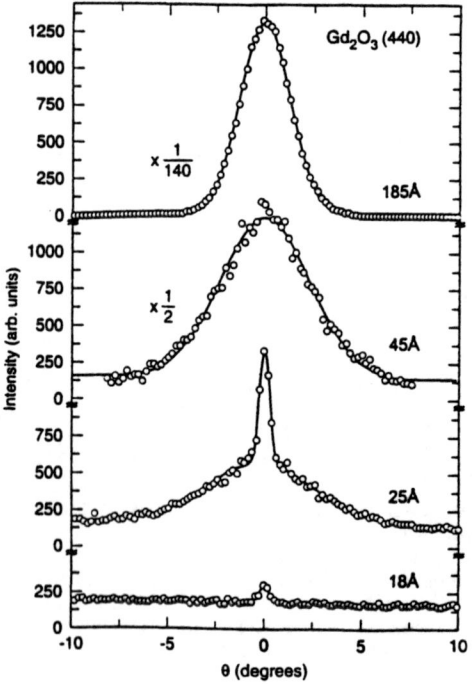

Figure 7. The sharp peaks in the mosaic scans indicate that films are elastically strained for 18, 25, and 45angstroms thicknesses.

26

DISCUSSION

Remarkably, the cubic α-phase of Gd$_2$O$_3$ has a very large unit cell a=10.79 Å when compared to that of the GaAs a=5.653 Å. Figure 8 shows the two lattices over a distance of lattice matching. In one direction; [011] of GaAs it takes about 32.58Å (with a +1.9% lattice mismatch), and in another; [01$\bar{1}$] GaAs direction it takes about 15.35Å (with a -3.9% lattice

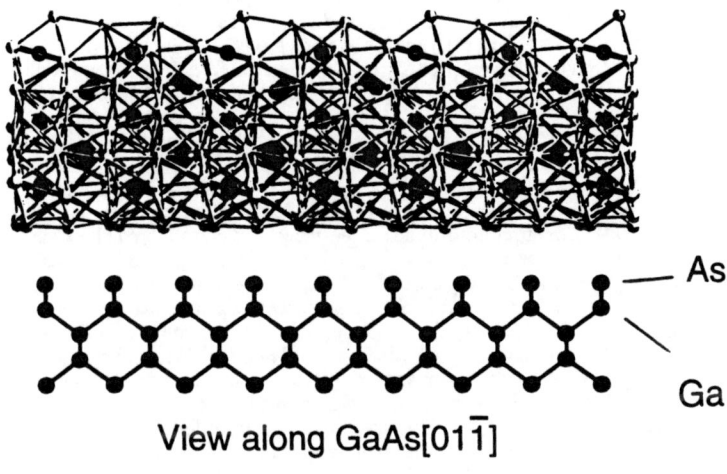

— As

Ga

View along GaAs[01$\bar{1}$]

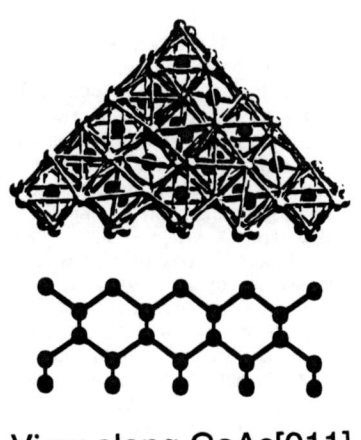

View along GaAs[011]

Figure 8. The Gd$_2$O$_3$ epitaxial system over the lattice matching lengths for two orthogonal directions.

27

mismatch) to match the unit cells. It is unlikely that interatomic forces acting over such long distances can be responsible for this super-cell epitaxy. This is especially true during the initial stages of the growth, when there are too few Gd and O atoms on the surface even to complete a single unit cell.

Since the epitaxial growth starts with nucleation at many different sites across the substrate surface, the film growth will not be coherent but will incorporate so-called antiphase boundaries. For a system with a long unit-cell matching distance, the number of possible nucleation sites or density of antiphase boundaries is expected to be very large. In addition to this large degree of degeneracy for possible nucleation sites, in Gd_2O_3 /GaAs(100) case there is a rotational degeneracy; the (110) two-fold planes of the film, in principal, can satisfy the same epitaxial relations in two different orientations on the four-fold symmetric (100) planes of the substrate, but the system chooses only one orientation, and grows with a highly structural perfection.

The II-VI compound growth on GaAs has successfully been studied by the consideration of valence electrons. The excess of valence electrons at the interface can change the tetrahedral configuration of the bonding at the interface, and the number of bonds can be reduced from four to three or two. In zinc blende structure, GaAs has four valence electrons per atom. In the Ga_2O_3 sesquioxide, Ga atom contributes three valence electrons and the O atom contributes six electrons to a Ga-O bond. This results in an extra electron, which makes the zinc blende tetrahedral configuration in Ga_2O_3 unstable. A tetrahedral lattice, however, can still be formed if one-third of Ga sites are left vacant. When we include these vacancies, the valence electron counting finds $(3x2+6x3)/6 = 4$ electrons per site on the average[12]. Another trivalent sesquioxide, Gd_2O_3 should behave identical to the Ga_2O_3 growth. The initial few layers may have no difficulty in continuing the growth of the substrate tetrahedral lattice, by leaving some Gd sites vacant. But eventually the growth will transform to the equilibrium α-phase which has a non-tetrahedral coordination. This transition may introduce some partial disorder. This may become important for very thin layers, in agreement with our observations, The integrated intensities of Gd_2O_3 (440) for the 18Å film fall significantly short of the 25Å film as can be seen in Fig. 5. This can be attributed to a film with a large volume fraction in a disordered state. Presence of a disordered transition region may also partially relieve some of the strain caused by the mismatch.

The prefered allignment along only one of the two possible orientation during the single crystal epitaxial growth can be explained by a strong local ordering that imposes a deterministic growth by favoring certain binding sites. A look at the interfacial structure and the atomic decoration of the unit cells in Fig. 8 reveals that, indeed there may be such a short range order. Along the [011] short super-cell direction, the substrate and Gd_2O_3 film consist of rows of atoms with very similar spacings. X-ray measurements do not reveal the interfacial structure. In order to be able to construct the interface structure and understand the importance of these rows, let's examine some bond strengths. Among the Gibbs free energy of formations for all possible pairs in Ga, As, Gd, and O, Gd_2O_3 is the strongest with -1739 kJ/mol, followed by Ga_2O_3 with -998 kJ/mol, and the rest of all possible binaries are negligible in comparison. Based on this information, the epitaxial growth should preferentially start first with a layer of oxygen atoms bonding to a Ga terminated surface. The oxygen atoms in this layer occupy the same positions as the As atoms would, as governed by the directional or the covalent nature of the bonding in GaAs substrate. This immediately explains the similarity of the rows of atoms between the two lattices. Indeed oxygen atoms in As sites at the interface are shared by the two lattices, would enforce this local order and facilitate the single crystal growth. If the GaAs surface was initially As terminated, oxygen would directly replace and substitute As, liberating it from the interface, and this extra layer of As would either dissolve in the growing film, segregate to the surface or would sublimate, depending on the growth conditions. High mobility and volatility of As is a well known fact, as has been observed in GaInP growth on GaAs[13].

With the O atoms at the interface assigned to the As positions, we can now explain the single crystal growth. This fixes the positions of the oxygen atoms in the Gd_2O_3 lattice growing on this template. The (110) plane of the bulk phase of the α-Gd_2O_3 has a similar configuration for the oxygen atoms. Only in every ninth row, there is an extra row of oxygens when compared to the GaAs(100) surface As sites. So far, we have explained the interfacial structure and the epitaxial growth but have not explained why the system chooses only one orientation out of two possible ones for the epitaxial growth. For this, we have to look at the second layer of atoms and beyond. Here, Gd atoms form the same rows as the O atoms, but now there are one-quarter less Gd atoms compared to the O atoms. As a result, oxygen atoms are rearranged along the rows to accommodate these missing Gd atoms. This significantly distorts the bonds and displaces the O atoms from their tetrahedral sites toward their equilibrium positions in the bulk Gd_2O_3 structure. Since, every O atom is bonded to two Ga atoms at the interface, its displacement would involve bond-stretching and bond-bending. The two degenerate growth orientations, therefore, distinguish themselves as being either predominantly bond-bending or bond-stretching. Since displacement of oxygen atoms along the rows shown, would involve primarily a bond-bending, this should be energetically favored over the alternative degenerate orientation which primarily favors stretched Ga-O bonds, as shown in figure 8.

CONCLUSION

Ordinarily, when we search for a material that is likely to grow an epitaxial film on a substrate, we look for a material that has a matching unit cell to that of the substrate. But the list of such lattice matching materials is indeed very short, and this has always been a problem in heteroepitaxial film growth for device fabrication.

The epitaxial system of the Gd_2O_3 on GaAs(100) is very interesting because, there is no simple match between the unit cells of the substrate and the film, as we are used to seeing in other systems. In other words, we would not have been able to guess this heterostructure a priori. Here matching conditions are met at second and third order unit cell dimensions. The epitaxy is enforced by the local order of the anions that are very similar across the interface, not by the simple unit cell dimensions. This opens up a whole new possibilities for new heteroepitaxial systems.

These films are also very interesting for the studies of dislocations. These Gd_2O_3 films experience anisotropic forces along the two inplane directions. Along the [011] GaAs direction it is under a large tensile strain (-3.9%), while in the orthogonal direction it experiences a compressive strain (1.9%). For the films exceeding few tenths of angstroms thickness, the critical thickness is exceeded and the misfit dislocations are created, which are most likely be uniaxial. All these important and interesting issues may find answers in further studies of this system.

ACKNOWLEDGMENTS
We thank A.Y. Cho, D. Murphy and R.M. Fleming for valuable discussions.

REFERENCES

[1]   M. Hong, M. Passlack, J.P. Mannaerts, J. Kwo, S.N.G. Chu, N. Moriya, S.Y. Hou, and V.J. Fratello, *J. Vac. Sci. Technol.* **B 14** (30), May/June, 2297, 1996.

[2]   M. Hong, J. Kwo, A.R. Kortan, J.P. Mannaerts, A. M. Sergent, *Procedings of MRS Fall Meeting,* 1998, Symposium D.

[3]   M. Hong, J. Kwo, A.R. Kortan, J.P. Mannaerts, A. M. Sergent, *Science* xx, xx, 1999.

[4]   T. Hashizume, K. Ikeya, M. Mutoh, H. Hasegawa, *Applied Surface Science,* **123** 599, (1998).

[5]   "Progress in the Science and Technology of the Rare Earths" Edited by LeRoy Eyring Volumes, **2, 3, 4, 5,** *Pergamon Press and North Holland Publishing,* (1982).

[6]   W. Zachariasen, *Z. Kristallogr.,* **67,** 455 (1928).

[7]   L. Pauling, M.D. Shappell, *Z. Kristallogr.,* **75,** 128 (1930).

[8]   S. Geller, *Acta Cryst.* **B27** ,821, 1971.

[9]   D. Grier, G. McCarthy, North Dakota State University, Fargo, North Dakota, *ICDD Grant-in-aid,* Volume [CD]: 1264.27, (1991).

[10]   R. Hall and J.C. Bean, *in "Strained-Layer Superlattices: Materials Science and Technology",* edited by T.P. Pearsall, page 1, Academic Press, (1991).

[11]   I.J. Fritz, S.T. Picraux, L.R. Dawson, T.J. Drummond, W.D. Laidig, N.G. Anderson, *Appl. Phys. Lett.,* **46** ,967, (1985).

[12]   Yu.G. Sidorov, S.A. Dvoretsky, M.V. Yakushev, N.N. Mikhailov, V.S. Varavin, V.I. Liberman, *Thin Film Solids,* **306** , 253, (1997).

[13]   O. Dehaese, X. Wallart, O. Schuler, F. Mollot, *Japan. J. Appl. Phys.,* **36** , 6620, (1997).

STRUCTURE OF CHEMICALLY PASSIVATED SEMICONDUCTOR SURFACES DETERMINED USING X-RAY ABSORPTION SPECTROSCOPY

A.P. HITCHCOCK*, T. TYLISZCZAK*, Z.H. LU**, P. BRODERSEN*, M.W.C. DHARMA-WARDANA***
*Brockhouse Institute for Materials Research, McMaster University, Hamilton, ON L8S 4M1
**Dept. of Metallurgy and Materials Science, University of Toronto, Toronto, ON M5S 3E4
Institute for Microstructural Sciences, National Research Council, Ottawa, ON K1A 0R6

ABSTRACT

The structure of monolayer-passivated single crystal semiconductor surfaces has been studied using synchrotron radiation X-ray absorption fine structure spectroscopy (XAFS). The near edge and extended fine structure signals, supported in some cases by first-principles calculations, have been used to investigate Ge(111)-Cl; GaAs(111)-Cl; GaAs(111)A-S, GaAs(111)B-S and GaAs(001)-S. The use of a solid state Ge X-ray fluorescence array detector has led to significant improvements in data quality and thus structural accuracy. The relationship between the derived surface structures and the development of improved passivated surfaces is discussed.

INTRODUCTION

Surface cleaning and passivation [1-5] are especially critical for material growth and device manufacturing of III-V compound semiconductors, where the native oxides are highly defective. Preparation of oxide-free surfaces which are air-stable for at least a few hours is essential to many processes involved in optoelectronic and microelectronic device manufacturing, such as epitaxial growth and overgrowth [3-11]. Furthermore, well-ordered and air-stable semiconductor surfaces provide a unique opportunity to integrate semiconductor and organic materials through molecular self-assembly [12,13]. In many cases chemical reactions in solution can be used to achieve oxide-free surfaces that are well-ordered, and sufficiently passive to allow air-transfer without degradation. For example an air-stable Ge(111) surface with Ge monochloride surface termination can be obtained by reaction of an oxide-stripped Ge(111) surface with dilute hydrochloric solution [14]. This surface has been successfully used to grow alkyl films through molecular self-assembly [12]. Many other examples of the application of surface passivation in electronic and optoelectronic devices can be found in these proceedings.

In order to develop improved recipes for forming passive surfaces it is helpful to have tools which can determine the structure of chemically passivated semiconductor surfaces. X-ray absorption spectroscopy [15-17] is particularly useful. Some of the advantages of XAFS include: local structure sensitivity with unambiguous identification of the central atom, directional sensitivity via polarization dependence, quantitative first shell bond lengths, availability of accurate and convenient computational codes, capability to study all environments relevant to semiconductor technology (surface adsorbates, thin films, dopants, buried interfaces etc). Limitations of XAFS include limited accuracy in bond lengths and co-ordination numbers, insensitivity to differentiate long range from mid-range order, limited sensitivity to bond angles. Many of these limitations can be overcome by combining XAFS with other structural techniques such as X-ray standing wave [18] and diffraction.

Application of X-ray absorption to semiconductor surface adsorbate studies generated in a UHV environment was first reported by Citrin et al [19] in the context of the development of the Auger-yield surface-XAFS detection method. Fluorescence detection has been applied to XAFS studies of semiconductor materials by a number of groups [17,20-22]. It is particularly powerful in the study of dilute dopant systems. Partial or total electron yield (TEY) detection techniques, which intrinsically have a very high surface sensitivity (a few 100 nm in TEY versus several microns for FY at the 2-4 keV energy range used here), might be expected to be well suited to studies of monolayer adsorbate systems, such as those important in semiconductor passivation. The intrinsic surface sensitivity is why we devoted a lot of our effort in the general area of semiconductor XAFS to various types of TEY detection [23,24]. However, especially at higher energy edges, the background from the substrate poses significant problems to electron detection, particularly TEY. Also, diffraction artifacts in TEY of single crystal semiconductors are often as challenging as those in FY detection [25]. For these reasons energy dispersive FY detection offers numerous advantages to TEY, especially when an array detector with high angular acceptance, high efficiency and high energy resolution is used.

Figure 1 compares the fluorescence yield (FY) and total electron yield (TEY) signal from Cl/GaAs(111) recorded simultaneously. Relative to the TEY spectrum, FY detection gives an enormous improvement in signal to background. This makes FY much less sensitive than TEY to systematic noise, even though the TEY signal is much larger, in part because Auger yields are ~10 times greater than FY in this regime, and in part because of signal amplification by secondary electron generation. Clearly, the ability of energy filtered X-ray fluorescence detection to collect almost all of the adsorbate signal with very high discrimination against substrate signal makes it superior for most adsorbate and thin film studies. FY has significant disadvantages with concentrated thick film or bulk samples where absorption saturation [26] distorts the signal, or where there is overlap of substrate and analyte fluorescence, as in InP-S.

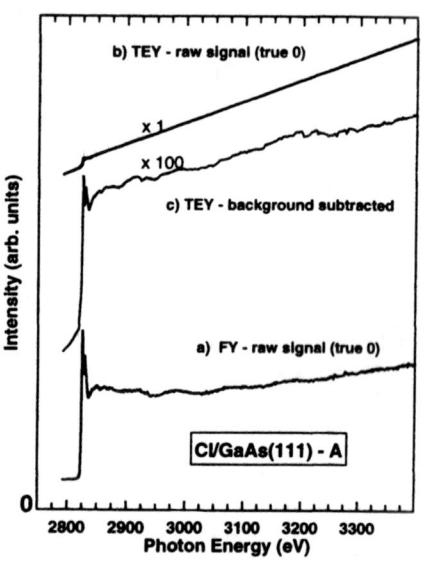

Fig.1 Comparison of simultaneously recorded fluorescence yield (FY) and total electron yield (TEY) Cl 1s X-ray absorption spectra of GaAs(111)A-Cl. Curves (a) and (b) are raw data with a true zero. Curve (c) is the TEY after background subtraction and 100-fold amplification.

Here we report studies of Ge(111)-Cl; GaAs(111)-Cl; GaAs(001)-S, GaAs(111)A-S, and GaAs(111)B-S passive layers prepared by solution chemistry. Although not directly relevant to passivation of compound semiconductor surfaces, the Cl/Ge(111) and Cl/GaAs(111) results [27] illustrate our procedures, including the use of first-principles structure calculation. The results on GaAs-S extends earlier studies of this surface by many different techniques including theory [28], RHEED [29,30], XPS [31,32], XSW [30] and XAFS [31,33,34]. Surfaces prepared using brief exposure at 25°C to a $(NH_4)_2S_x$ solution followed by a water rinse, without annealing, are found to produce a surface with XAFS characteristics (thus local structure), similar to that prepared using more elaborate *in situ* S MBE and annealing procedures [34]. In the case of *ex-situ* wet chemical cleaning/passivation, the resulting surface is quite often covered by physisorbed hydrocarbons. Thus many conventional surface analytical tools such as LEED and STM cannot determine the structure of these complex surfaces. Although in-situ annealing can be used to desorb most physisorbates, annealing can also significantly distort the as-prepared surface through diffusion and phase transformation. The temperature for diffusion and phase transformation on a III-V surface is quite low, ~ 200°C for S-InP(100) for example.

EXPERIMENTAL

Synchrotron Measurements

The experiments were carried out on the double-crystal (InSb) monochromator beam line [35] at the Synchrotron Radiation Center (SRC) of the University of Wisconsin-Madison. The measurements were carried out in a vacuum chamber with a pressure in the low 10^{-7} torr range. The photon beam size was ~1x2 mm^2. The sample size was ~15x30 mm^2. The photon incidence angle relative to the surface plane was 20°, fixed for all measurements. The sample holder can be rotated in one axis parallel to the photon incidence. For this work we measured the polar angle polarization dependence by recording spectra in two orientations: $E_\perp$, with the photon beam incident 20° off the surface plane but with the sample rotated so that the photoelectric vector (E) is approximately "perpendicular" (i.e., 70°) to the sample surface plane, and $E_\parallel$, with the photon beam incident 20° off the surface plane but with the sample rotated so that E is exactly parallel to the sample surface. Azimuthal re-orientation of the sample - (011) versus (0-11) for GaAs(001)-S - was achieved by removing and remounting the sample. The X-ray fluorescence signal was measured using a 9-element array of liquid nitrogen cooled, solid state Ge detectors. The energy window for each detector was individually optimized to minimize interference from Rayleigh scattering or signal from substrate fluorescence. TEY was measured from the sample current, with a counter electrode being an upstream ring biased with +90 V.

Sample Preparation

Ge(111)-Cl and GaAs(111)A-Cl: The Ge(111) wafer was undoped and the surface was protected by a thin oxide film in epi-ready condition. The epi-ready sample is immersed in an aqueous hydrogen chloride solution (HCl (38 wt%):H_2O=1:1) for 5 minutes followed by removal of the solution with a jet of clean nitrogen without a water rinse. After this treatment, the surface is found to be free of oxide, with a (1x1) bulk-like structure, and with the surface terminated by Ge-Cl bonds, as established by measurements using X-ray photoelectron spectroscopy, low-energy electron diffraction, and Cl K near-edge structure [12]. The freshly prepared sample was transferred to the analysis chamber within ~5 minutes.
S on GaAs: Fresh epi-ready surfaces of undoped wafers were passivated at 25°C in a concentrated aqueous $(NH_4)_2S_x$ solution for 5 min, rinsed with deionized water (~5 min), then blown dry with N_2.

RESULTS AND DISCUSSION

Passivation with Hydrogen Chloride

Figure 2 compares the Cl 1s X-ray absorption spectra of Ge(111)-Cl and GaAs(111)A-Cl recorded with $E_\perp$ and $E_\parallel$ polarization (NB X-rays are incident along the surface normal in $E_\parallel$). The spectra plotted are as-recorded data with a pre-edge linear background subtraction and edge-jump unit normalization. In each case the lowest energy feature, attributed to a Cl 1s → σ*(Cl-E) (E = Ge or Ga) resonance, is highly polarized, indicating that the chloride species is attached to the surface with near normal orientation. Analysis of the angular distribution of the resonance intensity [2] indicates the angle of the bond relative to the surface normal is 0±5°. In contrast, the second feature at 2829 eV is strongest in $E_{\parallel}$ polarization, indicating it arises from an in-plane electronic transition, or possibly backscattering from atoms located close to orthogonal to the Cl-E bond. The higher energy features are interpreted as the first extended fine structure signal, better associated with backscattering from neighbors rather than a specific electronic transition.

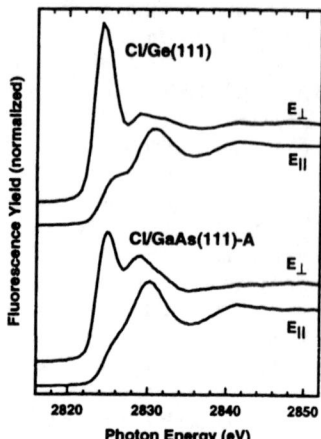

Cl/Ge(111): The Cl-Ge bond-length was derived from analysis of the extended fine structure (EXAFS) signal. **Figure 3** plots the long range XAFS signal and the magnitude of the Fourier transform of the Cl 1s EXAFS obtained at the two different polarizations. The $E_{\perp}$ Fourier transform shows a strong first-shell signal at about 2 Å corresponding to the Cl-Ge bond. In contrast, the transform of the $E_{\parallel}$ data is essentially featureless. The dramatic difference in the EXAFS intensity in the two different polarization directions clearly indicates a strongly anisotropic local geometry with the first-shell neighboring atom to Cl located in the atop site, aligned along the surface normal. This extends the information about orientation provided by the near-edge data. Quantitative structural parameters (Cl-Ge bond-length, coordination

Fig. 2 Cl 1s X-ray absorption near edge spectra (XANES) in $E_{\perp}$ and $E_{\parallel}$ polarization of Ge(111)-Cl and GaAs(111)A-Cl surfaces prepared by chemical reaction. Offsets have been used for clarity.

number, Debye-Waller factors) were derived using three approaches. Single-shell analysis of the back-Fourier-filtered first-shell signal (1.1 – 2.5 Å) from the $E_{\perp}$ EXAFS using tabulated Cl-Ge backscattering amplitudes and phase shifts based on Feff 3.0 calculations ([36]) yields R_{Cl-Ge} = 2.20 Å and a polarization-corrected coordination number of 1.05. The Fourier transform of the $E_{\parallel}$ data shows no back-scattering atoms within the first-shell. The second approach matched the unfiltered k-space EXAFS to a more sophisticated Feff 6.01 [36] simulation based on a surface geometry optimized using a high level quantum calculation. In the third approach the experimental R space data was fit to the Feff 6.01 calculation of the first shell scattering. This gives R_{Cl-Ge} = 2.180(5) Å when the k-space origin is optimized (E_o = 3.2 eV), or R_{Cl-Ge} = 2.161(7) Å if E_o = 0, a Debye temperature of 365(33) K and unit co-ordination number. The average, 2.17(1) Å, is taken to be the Cl-Ge distance in Ge(111)-Cl.

How does this value compare to that predicted by other methods? The sum of the covalent radii of Ge (1.22 Å) and Cl (0.99 Å) [37] is 2.21 Å, in reasonable agreement. However, the early EXAFS study by Citrin et al. [19] of a Ge(111)(2x8)-Cl surface prepared in UHV reported a Ge-Cl bond-length of 2.07 (3) Å. The disagreement is outside mutual error bars. The structure was also determined by a first-principles total energy minimization based on density functional methods [7,38], with local-density approximation (LDA) and nonlocal, norm-conserving pseudopotentials [39]. The cluster, which consisted of 17 layers (including 5 vacuum), was periodic in the <111>-direction and the surface unit cell was periodic in the x-y plane. The fully optimized structure contained six relaxed Ge layers and the Cl-overlayer. Relaxation pulled the surface Ge layers towards the bulk. The final optimized LDA structure predicts the Cl in the atop site with a Ge-Cl bond-length of 2.13(1) Å (uncertainty is the spread in the calculated Cl-Ge bond-lengths). This is rather shorter than our experimental value, but larger than that of 2.09 Å found using a similar theoretical method [40]. It is possible that the Ge(111)(2x8)-Cl structure differs from that which is produced in an aqueous-phase chemical reaction which replaces Ge-O bonds with Ge-Cl bonds (although a (1x1) LEED pattern was observed after Cl adsorption, as found

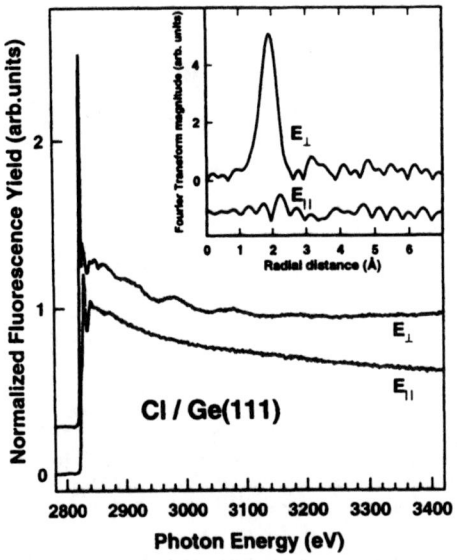

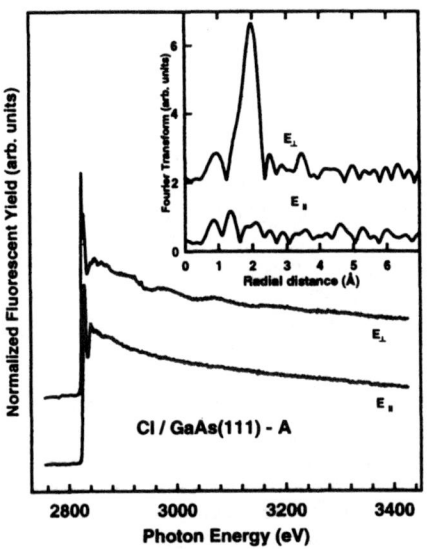

Fig. 3 Cl 1s extended X-ray absorption (XAFS) spectrum of Ge(111)-Cl . The insert compares the pseudo-radial distributions derived from Fourier transform analysis of the isolated XAFS with the E-vector along the surface normal ($E_\perp$) and in the surface ($E_\parallel$).

Fig. 4 Cl 1s XAFS of GaAs(111)A-Cl. The insert compares the pseudo-radial distributions derived from Fourier transform analysis of the isolated XAFS with the E-vector along the surface normal ($E_\perp$) and in the surface ($E_\parallel$).

for the present chemically passivated surface [12]). A repeat of the XAFS study of UHV prepared Cl/Ge(111) could help to determine if there is a real difference in the structure for chemically passivated Ge(111)-Cl and UHV-deposited Ge(111)-Cl.

<u>GaAs(111)A-Cl</u>: So far S-passivation has been the most successful method for protecting GaAs surfaces (see below). However, S is a known n-type donor and it will diffuse into the substrate at high temperature [41], which may not be wanted in certain applications. Recently a new passivation method using Cl-termination has been demonstrated [42]. Here we summarize our study of GaAs(111)A-Cl [27], which shows an XAFS signature remarkably similar to that for Ge(111)-Cl. **Figure 4** shows the Cl 1s XAFS and Fourier transforms for $E_\perp$ and $E_\parallel$ polarization. The highly anisotropic nature of this Cl environment is as evident from the extended signal as it is from the XANES - indeed the radial distribution is very close to that for Ge(111)-Cl. A single-shell analysis of the back-Fourier-filtered first-shell (1.1 – 2.5 Å) from the $E_\perp$ XAFS using tabulated Cl-Ga backscattering amplitudes and phase shifts [36] yields a Cl-Ga bond-length of 2.25 Å and a polarization-corrected coordination number of 0.8. The $E_\parallel$ Fourier transform shows no back-scattering atoms within the first-shell. Based on these experimental results, we conclude that Cl forms a mono-chloride bond with Ga along the (111) surface normal. In addition to being the most logical site from the viewpoint of chemical bonding, the Cl XAFS rules out structural models with the Cl atom bonded to As since this would give rise to significant first shell Cl-Ga backscattering signal in the $E_\parallel$ transform.

Figure 5 compares the experimental extended fine structure signal with an Feff 6.01 [36] calculation for atop-bonding of Cl to Ga. The multiple scattering XAFS calculation of an 85 atom 6-layer cluster with contributions of multiple scattering paths up to 7 Å was performed for a range of Cl-Ga bond-lengths and the results compared to experiment in order to find the optimum distance. Excellent agreement between calculated and experimental XAFS was obtained with a Cl-Ga bond-length of 2.175 Å, with similar quality fits over a 0.02 Å range. The calculated curve fits the experimental data better at high wavenumbers (≥ 4 Å^{-1}) than at low wavenumbers (≤ 4 Å^{-1}). The relatively poor fit below 4 Å^{-1} is likely associated with the inaccuracy of the current Feff multiple-scattering calculation

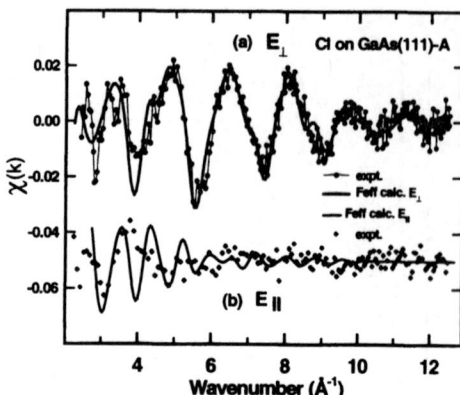

Fig. 5 Comparison of experimental XAFS (solid dots and crosses) and FEFF 6.01 calculated XAFS (solid lines) of GaAs(111)-Cl at $E_\perp$ (a) and $E_\parallel$ (b) polarizations. An orientation dependent Debye temperature was used [27].

at predicting near edge (XANES) structure in open framework systems such as surfaces of covalent semiconductors. The Cl-Ga bond is highly directional along the surface normal and there is a large difference between bond-stretching and bonding-bending constants [43]. This implies a large difference in the Debye temperatures at these two directions - the optimization suggests 350 K for $E_\perp$ and 250 K for $E_\parallel$ polarization.

The derived Cl-Ga bond length compares favorably with that of 2.25 Å, estimated from the sum of the Pauling covalent radii of Ga (1.26 Å) and Cl (0.99 Å) [44]. It is shorter than the Cl-Ga bond-length of 2.32 Å predicted from a cluster calculation of GaAs(111)A-Cl [45]. We believe that our experimental result is more accurate than that predicted by their calculation. It has been demonstrated that cluster model calculations have difficulties in accurately predicting surface physical parameters such as bond-lengths because the ratio of surface to bulk atoms is too large in the cluster [46]. A more accurate theoretical method is the SLAB method which considers translational symmetry of the surface plane. For example the SLAB calculated bond-length of 2.17 Å for Cl-terminated Ge(111) agrees well with experiment whereas a cluster model [45] calculated a much larger Cl-Ge bond-length of 2.30 Å.

Passivation of GaAs with Ammonium Sulfide

The utility of S adsorption as a practical tool for passivation and improving growth in GaAs processing was first identified in the mid 1980's [18,47]. Recipes used for S passivation include chemical reaction in solutions of aqueous NaS [48] and $(NH_4)_2S_x$ [42], deposition from electrochemically generated S [49], and atomic S vapor generated by thermal evaporation in a UHV environment [30,50]. For the GaAs(001) surface, early energy minimization calculations [28] predicted that the Ga-S-Ga bridge site (oriented along [011]) was most stable, although formation of S-dimers along [011] was also a possible candidate. Studies in recent years by a number of techniques have produced rather conflicting results, with some supporting the Ga-S-Ga bridge site [30,31,33], others implicating multiple sites, including substitutional sub-surface sites [10,32], and others proposing S-dimers. Some of the confusion may arise from the fact that the surface structure is a sensitive function of the preparation technique and in most cases,

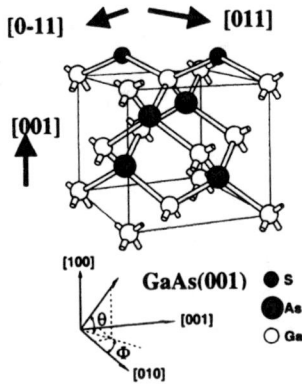

[0-11] ← → [011]

[001]

[100]

GaAs(001) ● S
[001] ● As
○ Ga

[010]

θ

Φ

Fig. 6 (upper) Structure of GaAs oriented along the (001) direction, with the proposed Ga-S-Ga bridge site indicated. (lower) Structure of GaAs(111)B surface (B has As outermost).

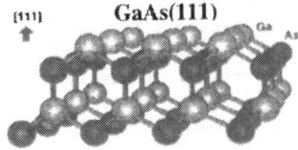

[111] GaAs(111)   Ga   As

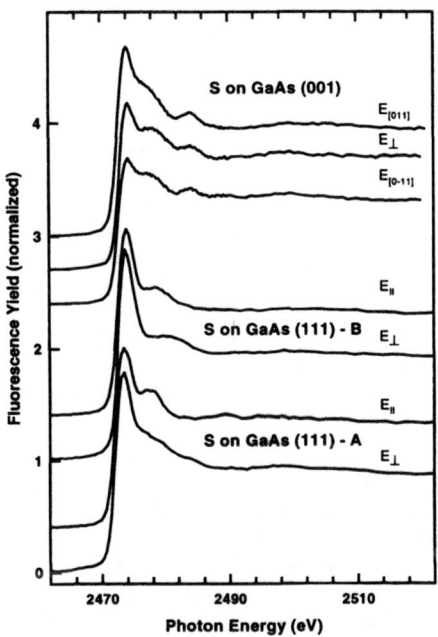

S on GaAs (001)

$E_{[011]}$

$E_\perp$

$E_{[0-11]}$

$E_\parallel$

S on GaAs (111) - B $E_\perp$

$E_\parallel$

S on GaAs (111) - A $E_\perp$

Fluorescence Yield (normalized)

Photon Energy (eV)

2470   2490   2510

Fig. 7 Near edge region of S 1s XAFS in $E_\perp$ and $E_\parallel$ polarization of the indicated surfaces of GaAs after brief exposure to an aqueous $(NH_4)_2S$ solution.

surfaces prepared by different techniques are being compared. Our results suggest that chemical passivation using the recipe we have employed (brief exposure at 25°C, water rinse and no annealing) produces a GaAs(001)-S surface structure with better order than solution passivation at elevated temperature, although not as well ordered as that generated by exposure to atomic S in a UHV environment, followed by extensive annealing.

Figure 6 depicts the structures of GaAs(111) and GaAs(001) surfaces, based on truncation of the bulk structure (no re-organization). Polarization dependent S 1s XAFS has been used to study the structure of sulfur passivated GaAs(111)A, GaAs(111)B and GaAs(001) surfaces, prepared using aqueous $(NH_4)_2S$. For the GaAs(001) surface, in addition to the polar angle study, the normal incidence spectrum was investigated with the in-plane E-vector oriented along the [011] and [0-11] azimuths. If the system adopted the postulated Ga-S-Ga adsorption site [28,31], the signal variation in these two azimuths would be as dramatic as that displayed by the σ*(Cl-E0) resonance in the Ge(111)-Cl and GaAs(111)A-Cl cases.

The S 1s XANES of GaAs(001)-S (Figure 7) does indeed show considerable polarization dependence, with the form expected in that the intensity of the first feature, corresponding to S 1s excitations into σ*(S-Ga) or σ*(S-As) orbitals, is much stronger for alignment along [011] than [0-11]. At the same time, the polarization is less than that found for the Cl passivated surfaces suggesting incomplete alignment and thus possibly multiple adsorption sites. When we compare these results to XANES of GaAs(001)-S in the literature [33,34], we find the spectra, particularly the change between the [011] and [0-11], are more similar to those of the vacuum prepared system [34] and different from those previously reported for GaAs(001)-S prepared using (NH₄)₂S with longer exposure, at higher temperature [33]. **Figure 8** is a careful comparison to the literature XANES [33,34]. As the ratio of the [011] and [0-11] signals (insert) indicates, the low temperature solution passivation is in between the high temperature solution and the UHV passivation. The reduced polarization dependence of either solution preparation indicates either poor alignment in the proposed Ga-S-Ga site, or (more likely) superposition of multiple sites with opposite polarization dependence. In addition, our GaAs(001)-S exhibits an additional peak at 2483 eV, which may be associated with partial oxidation (the white line in the S 1s spectrum of SO₂ is at ~2482 eV). The aqueous S-passivated surface has shown excellent electrical characteristics [3,4].

GaAs(001)-S **Figures 9** presents the S 1s XAFS of GaAs(001)-S with three different polarizations, with a comparison of experiment and that calculated by Feff based on the proposed Ga-S-Ga [011] bridge site (see Fig. 6). The isolated XAFS signals (compared to the Feff results) are shown in **Figure 10** while the Fourier transforms of the extracted extended fine structure are plotted in **Figure 11**. Again, the relative amplitudes of the XAFS signals are consistent with the polarization dependence of the XANES in all three directions. The signal has similar periodicity, but the amplitude envelope is quite different for $E_{(0-11)}$ than the other two. If the surface structure was uniquely the postulated Ga-S-Ga bridge site [28] then the $E_{(0-11)}$ signal should be negligible. The residual first shell $E_{(0-11)}$ signal (about 20% of that seen in $E_{(011)}$) can be considered a measure of the proportion of S atoms that are NOT in the Ga-S-Ga bridge site. Not surprisingly, attempts to fit the three XAFS spectra for S/GaAs(001) to ONLY the simple Ga-S-Ga bridge model were unsuccessful (**Figure 9, 11**), consistent with this observation. It is well-known that (NH₄)₂S as-treated surface is covered with a variety of S species including elemental S, sulfur oxide, As sulfide, as well as the most stable Ga-S bridge site. The physisorbed S species can be effectively rinsed off the surface by water in the case of InP(100) [31, 51]. In the present case, it is likely that the surface is covered with some residual disordered S and sulfur compounds which contribute to, and broaden, the XAFS. There is little evidence from our

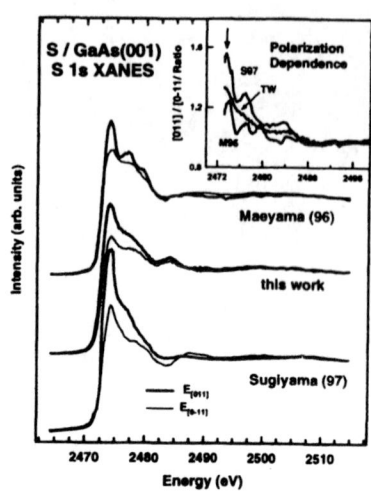

Fig. 8 S 1s XANES of GaAs(001)-S for $E_{\parallel}$ with [011] and [0-11] azimuths. Chemical preparation (Maeyaa et al. [33], this work) results in a less ordered surface than vacuum deposited S (Sugiyama et al [34]). Inset compares the ratio of [011] and [0-11] signals indicating our passivation preparation gives a surface intermediate to those of [33] and [34].

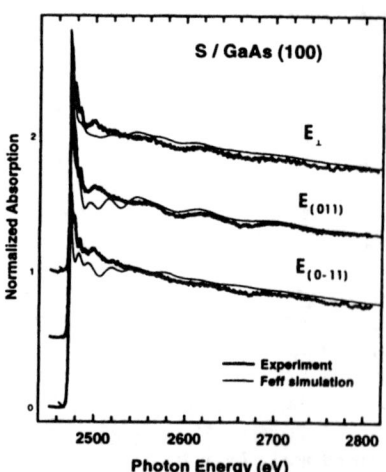

Fig. 9 S 1s XAFS of GaAs(001)-S with $E_{\perp}$ and $E_{\parallel}$ aligned along the [011] and [0-11] directions. Feff 6.01 simulations for the proposed Ga-S-Ga [011] bridge site are indicated.

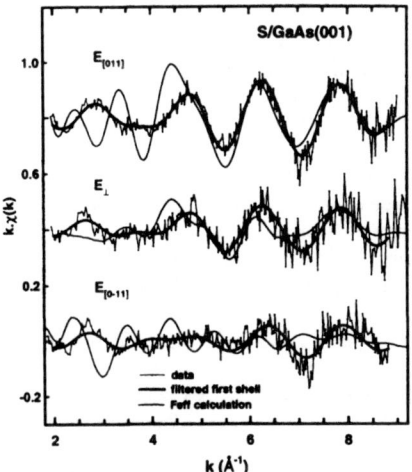

Fig. 10 First shell polarization-dependent XAFS of GaAs(001)-S compared to Feff 6.01 for the Ga-S-Ga bridge site.

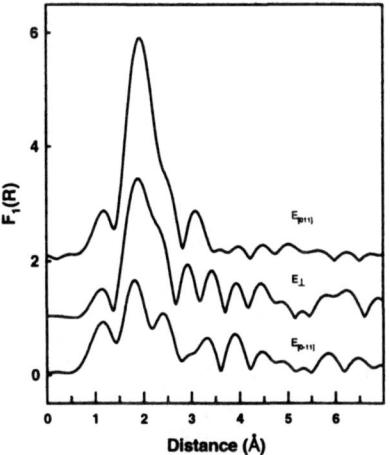

Fig. 11 Pseudo-radial distributions GaAs(001)-S for $E_\perp$ and $E_\parallel$ aligned along the [011] and [0-11] directions.

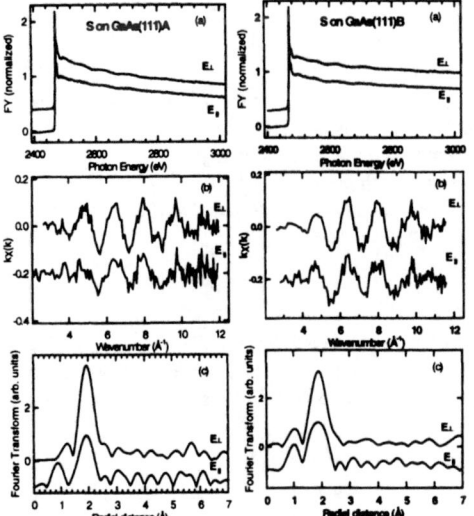

Fig. 12 S 1s XAFS of GaAs(111)A-S and GaAs(111)B-S at $E_\perp$ and $E_\parallel$ polarization.

XAFS study supporting S-S dimers, since there is no evidence for a low-lying XANES signal associated with a short Si-Si contact [52], and the amplitude envelopes of the strong first shell backscattering signal or the weak signal at ~3.2 Å in the FT are not consistent with S backscattering.

GaAs(111)-S **Figure 12** presents the extended fine structure for GaAs(111)A-S and GaAs(111)B-S systems for both $E_\parallel$ and $E_\perp$ polarization. Consistent with the polarization dependence of the XANES (**Fig. 7**), there is considerable signal for the $E_\parallel$ polarization, with essentially the same periodicity. This suggests the $E_\parallel$ and $E_\perp$ signals arise from the same bond, just projected on the different axes. The S – Ga first shell distance derived from the initial EXAFS analysis (2.29 Å) is similar to that reported by Sugiyama (2.27 Å) [53] and relatively close to that in GaS (2.33 Å). The EXAFS oscillations in the "in plane" spectrum are consistent with a relatively strong "in plane" component of the S-Ga bond. Relative to the GaAs(111)A-S system, the near edge measurements for GaAs(111)B-S show more anisotropy while the EXAFS for GaAs(111)B-S exhibits similar or perhaps less anisotropy, along with wider first shell signals. It is interesting to note that there is much less higher shell

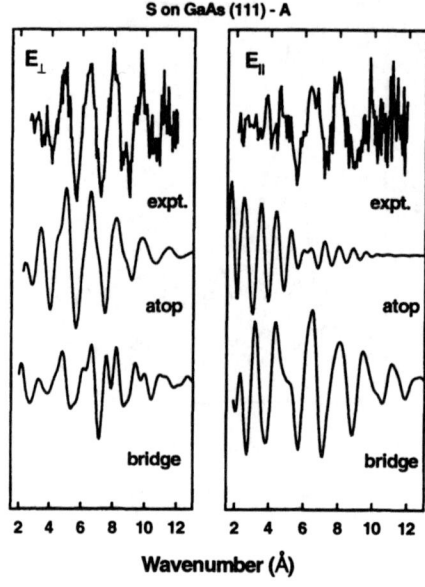

S on GaAs (111) - A

$E_\perp$

$E_\parallel$

expt.

expt.

atop

atop

bridge

bridge

2 4 6 8 10 12

2 4 6 8 10 12

Wavenumber (Å)

Fig. 13 Comparison of measured XAFS for GaAs(111)A-S for $E_\perp$ and $E_\parallel$ with that calculated for S-Ga atop and Ga-S-Ga bridge sites.

signal in both of the GaAs(111)-S systems (little asymmetry on the high-R side of the first shell peak), than found in GaAs(001)-S. This could be explained either by destructive interference from multiple sites or large amplitude bending motions. **Figure 13** compares the measured polarization dependent XAFS to Feff calculations for a S-Ga atop site (assumes that the S would also bond to a H atom) and a Ga-S-Ga bridge site. The predicted XAFS for each site disagrees with experiment, although the periodicity and amplitude in the $E_\perp$ XAFS is close to that observed, suggesting some of the atoms might be in this site. Indeed, a mixture of 70% of calculated atop and 30% calculated bridge site is in significantly better agreement with the data than either of the calculated XAFS for individual sites.

SUMMARY

Surface passivation is an important issue for the semiconductor device industry. This article has described recent fluorescence yield X-ray absorption spectroscopy studies of the structures of group IV and III-V single crystal semiconductors passivated using wet chemistry. The capability of XAFS to study local environments of elements of interest with both chemical state and physical site sensitivity, makes it an ideal tool for quantifying surface structures of as-prepared surfaces. These results in turn can assist the engineering of better passivated surfaces.

ACKNOWLEDGEMENTS

This work was supported financially by NSERC (Canada). We thank the staff of CSRF and SRC for enabling the synchrotron measurements and J. Lipkowski for generously loaning the X-ray detector. This work is based on research carried out at SRC, University of Wisconsin-Madison which is supported by NSF grant DMR 95-31009.

REFERENCES

1. E. Yablonovitch, D.L. Allara, C.C. Chang, T. Gmitter, and T.B. Bright, Phys. Rev. Lett. **57**, 249 (1986).
2. Z.H. Lu, Progress in Surface Science **50**, 335 (1995).
3. C.J. Sandroff, R.N. Nottenburg, J.C. Bischoff, and R. Bahat, Appl. Phys. Lett. **50**, 256 (1987).
4. R. Driad, Z.H. Lu, S. Charbonneau, W.R. McKinnon, S. Laframboise, P. Poole, and S.P. McAlister, Appl. Phys. Lett. **73**, 665 (1998).
5. Z.H. Lu, M.J. Graham, X.H. Feng, and B.X. Yang, Appl. Phys. Lett. **60**, 2773 (1992).
6. T. Ohno, Phys. Rev. B **44**, 6306 (1991).

7. J.M. Jin, M.W.C. Dharma-wardana, D.J. Lockwood, G.C. Aers, Z.H. Lu, and L.J. Lewis, Phys. Rev. Lett. **75**, 878 (1995).
8. X.R. Qin, Z.H. Lu, J.G. Shapter, L.L. Coatsworth, K. Griffiths, and P.R. Norton, J. Vac. Sci. Technol. A.**16**, 163 (1998).
9. M. Sugiyama, S. Maeyama, M. Oshima, Phys. Rev. B **60**, 11037 (1993).
10. P. Moriarty, B. Murphy, L. Roberts, A.A. Cafolla, G. Hughes, L. Koenders, P. Bailey, D.A. Woolf, Appl. Phys. Lett. **67**, 383 (1995).
11. G.W. Anderson and P.R. Norton, Phys. Rev. Lett. **74**, 2764 (1995).
12. J. He, Z.H. Lu, S.A. Mitchell, and D.M. Wayner, J. Am. Chem. Soc. **120**, 2660 (1998).
13. M.R. Lindford, P. Fenter, P.M. Eisenberger, and C.E.D. Chidsey, J. Am Chem. Soc. **117**, 3145 (1995).
14. Z.H. Lu, Appl. Phys. Lett. **68**, 520 (1996).
15. P.A. Lee, P.H. Citrin, P.E. Eisenberger and B.M. Kincaid, Rev. Mod. Phys. 53769 (1981).
16. D.C. Koningsberger and R. Prins, (eds) *X-ray Absorption: Principles, Applications and Techniques of XAFS, SEXAFS and XANES* (Wiley, NY, 1988)
17. H. Oyanagi, *Recent Progress in X-ray Absorption Fine Structure*, Res. Electrotechnical Laboratory, No. 966 (1994)
18. R.D. Bringans, R.I.G. Uhrberg, R.Z. Bachrach and J.E. Northrup, Phys. Rev. Lett. **55**, 533 (1985).
19. P.H. Citrin, J.E. Rowe, P. Eisenberger and F. Comin, Physica B **117**, 786 (1983); P.H. Citrin, J.E. Rowe, and P. Eisenberger, Phys. Rev. B **28**, 2299 (1983).
20. H. Oyanagi, R. Shioda, Y. Kuwahara and K. Haga, J. Synchrotron Rad. **2**, 99 (1995).
21. S. Pascarelli, F. Boscherini, C. Lamberti and S. Mobilio, Phys. Rev. B **56**, 1936 (1997).
22. J.C. Woicik, J.G. Pellegrino, B. Seiner, K.E. Miyano, S.G. Bompadre, L.B. Sorensen, T.-L. Lee and S. Khalid, Phys. Rev. Lett. **79**, 5026 (1997).
23. T. Tyliszczak and A.P. Hitchcock, Physcia B **158**, 335 (1989)
24. A.P. Hitchcock, T. Tyliszczak, P. Aebi, X.H. Feng, Z.H. Lu, J.-M. Baribeau, and T.E. Jackman, Surf. Sci. **301**, 260 (1994).
25. T. Tyliszczak, A.P. Hitchcock and T.E. Jackman, J. Vac. Sci. Tech. A **8**, 2020 (1990).
26. L. Troger, D. Arvanitis, K. Baberschke, H. Michaelis, U. Grimm and E. Zscech, Phys. Rev. B **46**, 3283 (1992).
27. Z.H. Lu, T. Tyliszczak and A.P. Hitchcock, Phys. Rev. B **58**, 13820 (1998).
28. T. Ohno and K. Shiraishi, Phys. Rev. B **42**, 11194 (1990); T. Ohno, Surf. Sci. **255**, 229 (1991).
29. H. Oigawa, J.-F. Fan, Y. Nannichi, H. Sugahara and M. Oshima, Jap. J. Appl. Phys. **30**, L322 (1991).
30. M. Sugiyama, S. Maeyama and M. Oshima, Phys. Rev. B **50**, 4905 (1994).
31. Z.H. Lu, M.J. Graham, Phys. Rev. B **48**, 4604 (1993).
32. P. Moriarity, B. Murphy, L. Roberts, A.A. Cafolla, G. Hughes, L. Koenders and P. Bailey, Phys. Rev. B **50**, 14237 (1994).
33. S. Maeyama, M. Sugiyama and M. Oshima, , Surf. Sci. **357-358**, 527 (1996).
34. M. Sugiyama and S. Maeyama, Surf. Sci. **385**, L911 (1997).
35. B.X. Yang, F. Middleton, B. Olssen, G.M. Bancroft, J.M. Chen, T.K. Sham, K.H. Tan and D. Wallace, Rev. Sci. Inst. **63** (1992) 802.
36. J.J. Rehr, J. Mustre deLeon, S.I. Zabinsky, and R.C. Albers, J. Am. Chem. Soc. **113**, 5135 (1991); J.J. Rehr, R.C. Albers and S.I. Zabinsky, Phys. Rev. Lett. **69**, 3397 (1992).
37. J.C. Phillips, *Bonds and Bands in Semiconductors*, (Academic Press, New York, 1973), p. 22.
38. W. Kohn and L. J. Sham, Phys. Rev. A **140**,1133 (1965).
39. L. Kleinan and D. M. Bylander, Phys. Rev. Lett, **48**, 1425 (1982).
40. G.B. Bachelet and M. Schlüter, Phys. Rev. B **28**, 2302, (1983).

41. Z. Tian, M.W.C. Dharma-wardana, Z.H. Lu, R. Cao, and L.J. Lewis, Phys. Rev. B **55**, 5376 (1997).
42. Z.H. Lu, F. Chartenoud, M. Dion, M.J. Graham, H.E. Ruda, I. Koutzarov, Q. Liu, C.E.J. Mitchell, I.G. Hill, and A.B. McLean, Appl. Phys. Lett. **67**, 670 (1995).
43. W.A. Harrison, *Electronic Structure and the Properties of Solids*, (Dover, New York, 1989).
44. L. Pauling, *The Nature of the Chemical Bond*, (Cornell, Ithaca, 1960), p.246.
45. G.P. Jiang and H.E. Ruda, Appl. Phys. Lett. **67**, 3334 (1995).
46. C. Dahma-wardana, Z.H. Lu, T. Tyliszczak, and A.P. Hitchcock (to be published).
47. Y. Nannichi, J.-F. Fan, H. Oigawa and A. Koma, Jap. J. Appl. Phys. **27**, L2367 (1988)
48. C.J. Sandroff, R.N. Nottenberg, T.C. Bischoff, and R. Bhat, Appl. Phys. Lett. **51**, 33 (1987).
49. H. Sugahara, M. Oshima, H. Oigawa and Y. Nannichi, Thin. Sol. Films **220**, 212 (1992).
50. S. Tsukumoto, N. Koguchi, Jap. J. Appl. Phys. **33**, L1185 (1994); Appl. Phys. Lett. **65**, 2201 (1994).
51. Y. Tao, A. Yelon, E. Sacher, Z.H. Lu, and M.J. Graham, in "Chemical Surface Preparation, Passivation and Cleaning for Semiconductor Growth and Processing", edited by R. Nemanich, C.R. Helms, M. Hirose, and G.W. Rubloff, MRS Proceedings 259, pp293 (MRS, Pittsburg, 1992).
52. J. Chauvistre et al., Chem. Phys. **223**, 293 (1997).
53. S. Maeyama, M. Sugiyama and M. Oshima, J. Electron Spectrosc. **80**, 209 (1996).

Part II

Novel Approaches for Surface Passivation and Device Processing

A NOVEL SURFACE PASSIVATION STRUCTURE FOR III-V COMPOUND SEMICONDUCTORS UTILIZING A SILICON INTERFACE CONTROL LAYER AND ITS APPLICATION

Tamotsu Hashizume, Hideki Hasegawa
Research Center for Interface Quantum Electronics and Graduate School of Electronics and Information Engineering, Hokkaido University, Sapporo, Japan, hashi@rciqe.hokudai.ac.jp

ABSTRACT

The present status of surface passivation research for III-V compound semiconductors utilizing a novel unique structure including a silicon interface control layer (Si ICL) is presented and discussed. The basic principle of passivation is to insert an ultrathin MBE-grown Si layer between the III-V compound semiconductor and a Si-based thick insulator so as to terminate the surface bonds of the III-V material with Si atoms and then to transfer Si-bonds smoothly to those of the Si based insulator. Based on the calculation of quantized levels in strained Si ICL, the passivation structure was optimized. Such a structure was realized by partial nitridation of Si ICL surface. In-situ surface characterization techniques including newly developed UHV contactless C-V technique, were used for optimization of each passivation step. Surface reconstruction of initial semiconductor surface was found to have a great influence on passivation. In the case of GaAs, the c(4x4) surface is preferable to the (2x4) surface. The novel process has realized the oxide-free surface passivation of InP with a N_{ssmin} value of 2 x 10^{10} cm^{-2} eV^{-1}. Furthermore, the novel passivation technique has been successfully applied to the fabrication of MISFETs and IGHEMTs, and the passivation of quantum structures.

INTRODUCTION

In future high-performance multi-media systems including photonics as well as wireless communication computing, low-power, high-speed and high-frequency electronic and optoelectronic devices based on III-V materials are required. For these, surafce passivation is a critical issue. As for the gate structure, a MIS gate structure is preferable than the Schottky gate structure, to realize high-speed and low-power consumption devices. III-V materials are also very important for nano-scale quantum devices, because they have mono-layer precision in epitaxy and a variety of hetero-interfaces. In these low-dimensional structures, passivation becomes even more crucial due to increase of the surface-to volume ratio.

However, control of surfaces and interfaces of compound semiconductors is much more difficult than Si, usually resulting in formation of high-density surface and interface states which cause the so-called Fermi level pinning. To overcome this serious problem, much efforts have been devoted to establishment of a suitable surface passivation process for III-V compound semiconductors. However, none of them except for one recent attempt[1] have so far been proved to be sufficiently effective.

According to the disorder-induced gap-state model (DIGS model) propesd by Hasegawa and Ohno[2], a disordered region is formed at the interface between insulator and semiconductor. This produces high-density interface states and causes a firm Fermi level pinning. Based on the DIGS model, Hasegawa and co-workers [3,4] proposed and experimentally confirmed a silicon interface-control-layer (ICL)-based passivation process for the control of surface and interfaces of compound semiconductors. The basic principle of passivation, as shown in Fig. 1(a), is to insert an ultrathin MBE-grown Si layer between the III-V compound semiconductor and a Si-based thick insulator so as to terminate the surface bonds of the III-V material with Si atoms and then transfer Si-bonds smoothly to those of the Si based insulator. This results in the reduction of surface states in III-V compound semiconductors. Such a principle works excellently for narrow-gap materials such as InGaAs after suitable process optimization[5,6]. For materials with wider enrgy gaps, one has to take account of the fact that the coherently strained Si layer has a very narrow bandgap and band states of Si ICL behave as

45

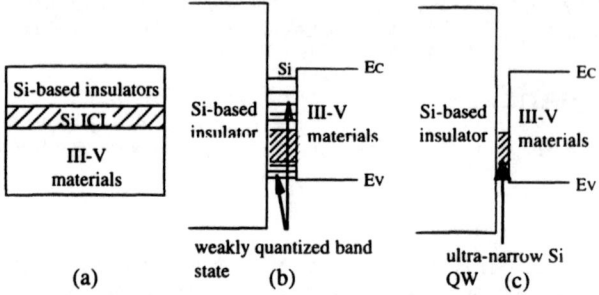

Fig.1 (a) Basic structure for the Si-ICL based passivation, (b) an example of band structure for strained Si ICL-III-V material, and (c) a band structure including ultra-narrow Si surface quantum well.

interface states within the energy gap of the III-V material, as shown in Fig.1 (b). This problem can be solved using a structure having an ultranarrow silicon surface quantum well, as shown Fig.1 (c). In this structure, the band states of the Si ICL are pushed out of the band gap region of the compound semiconductor due to the quantum confinement effect.

In this paper, a novel Si-ICL based passivation structure for III-V compound semiconductor surfaces are described and discussed. Based on the calculation of quantized levels in strained Si surface quantum well, the passivation structure is optimized. Experimentally, an optimal structure is realized by the partial nitridation of Si ICL surface, and each passivation step is characterized using various types of in-situ surface characterization techniques. It is shown that surface reconstruction of initial semiconductor surface was found to have a great influence on passivation. In the case of GaAs, the c(4x4) surface is preferable to the (2x4) surface. It is also shown that the novel process realizes the oxide-free surface passivation of InP with a N_{ssmin} value of $2 \times 10^{10} \, cm^{-2} \, eV^{-1}$, and is successfully applied to the fabrication of MISFETs and IGHEMTs and the passivation of quantum structures.

BAND LINEUPS and QUANTIZED LEVELS in Si SURFACE QUANTUM WELL

Figure 2 shows band lineups of strained Si ICL/GaAs (100) and strained Si ICL/InP (100) interfaces calculated by the model-solid theory[7]. First, the basic band alignments for a hypothetical unstrained interface were determined from the values of average valence band and the spin-orbit splitting. Then, assuming that coherent strain takes place only in the Si ICL , the lineup was modified by calculating the energy shift in valence band by taking account the hydrostatic strain and the splitting of degenerate hole bands. For the conduction band, the energy shift by the hydrostatic strain and the splitting of the indirect conduction-band minima, was calculated[7]. In both interfaces, the bandgap of Si becomes extremely narrow. Then, band states of the Si ICL will exist within the energy gap of compound semiconductors and act like interface states. However, if the Si layer is made thin enough, these quantum states can be pushed away from the quantum wells.

Figure 3 shows the calculated energy positions of the lowest subbands versus the Si layer thickness. The vertical axis gives the energy depth of the ground subband level for electrons in the conduction band and that of the lowest subband level for

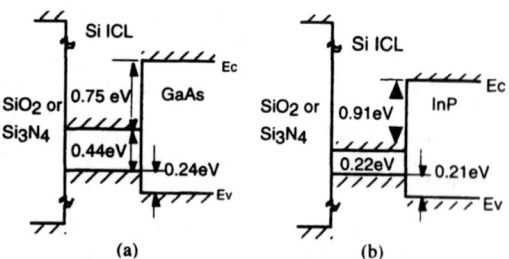

Fig.2 Band lineups for (a) strained Si ICL/InP and (b) strained Si ICL/GaAs interfaces

46

light holes in the valence band, both being measured from the conduction band edge and valence band edge of the substrate materials. The quantum well states for light holes are easily pushed away because the effective mass of the light hole is lighter and its barrier height is lower than those for electrons. From this calculation, it was found that the quantum well states of electrons can be completely removed by decreasing Si ICL thickness to less than 5Å for both GaAs and InP.

PASSIVATION PROCESS AND CHARACTERIZATION METHODS

Theoretical calculation of quantized states in a Si surface quantum well required the monolayer-level controlled formation of Si ICL. This is very difficult if we use simple direct deposition of Si-based insulators onto Si ICL. Instead, a novel structure including ultrathin SiNx can be formed by partial nitridation of a silicon layer with low-energy ECR N_2 plasma or N_2 radicals by rf-plasma excitation, as shown Fig.4.

Such sample structures were produced and characterized by a UHV-based multi-chamber system, as shown in Fig.5, installed in the super clean room facility of the Research Center for Interface Quantum Electronics (RCIQE), Hokkaido University. In this system, all the chambers are connected by a UHV transfer chamber and the samples, and the base pressure of the system was 3×10^{-10} Torr.

After some surface treatments of compound semiconductors, an ultrathin Si layer (0.2-1 nm) was grown at 250-280 °C by MBE. Subsequently, the surface of the Si ICL was partially nitrided in the same MBE chamber by the following two ways. One was to expose the Si surface to ECR N_2 plasma with an micro-wave power level of 50 W at room temperature for 10-30 s, and the other was to expose the surface to a nitrogen radical flux excited by rf-plasma (100W) at room temperature for 1-2 min. For the latter, a commercially available nitrogen radical source (Eiko Engineering ER1000) was used. Then some sample surfaces were covered by a thick Si_3N_4 film by ECR PECVD using N_2 and SiH_4.

The microscopic surface structures and atomic arrangements were investigated by a UHV-STM system (JEOL JSTM-4600). The surface chemical composition, the band bending of surface and thickness of ultrathin film were characterized by X-ray photoelectron spectroscopy (XPS) using a

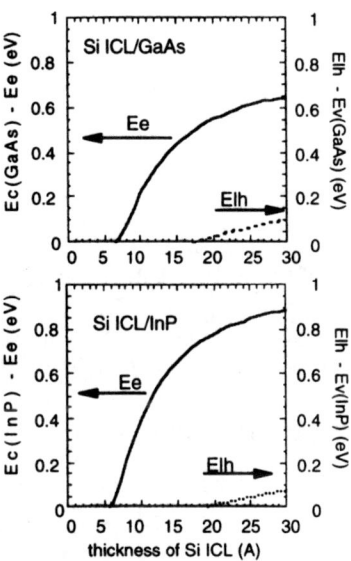

Fig.3 Quantized levels in strained Si surface QW as a function of Si ICL thickness

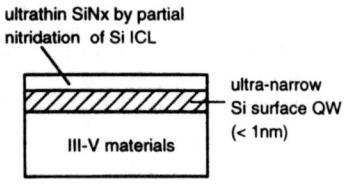

Fig.4 A novel passivation structure.

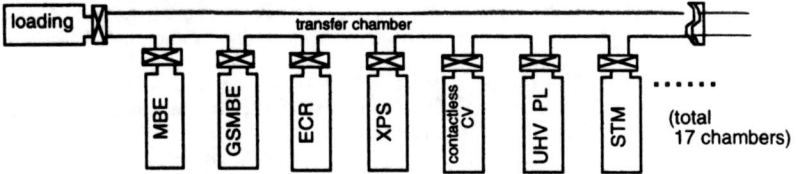

Fig.5 UHV-based growth/fabrication/characterization

spectrometer (Perkin-Elmer PHI 1600) with a spherical capacitor analyzer (SCA) and a monochromatic AlKα radiation (hv = 1486.6 eV). The binding energies of the spectra were calibrated by measuring the peak positions of Au4f$_{7/2}$, Ag3d$_{5/2}$ and Cu2p$_{3/2}$. Electronic properties of the sample surfaces were investigated by UHV-photoluminescence (PL) measurement and UHV contactless capacitance-voltage (C-V) measurement. In the UHV contactless C-V method, an ultranarrow constant and parallel "UHV gap" (100-300 nm) is realized and maintained between the field plate electrode and the sample surface, as shown in Fig.6, using a piezo electric feedback mechanism based on capacitance measurements from the three parallelism electrodes surrounding the field plate electrode. Then, C-V measurements were carried out between the field plate electrode and the sample by changing the voltage. The value of the UHV gap length was determined by an optical technique utilizing the Goos-Haenchen effect. When the UHV gap is in the

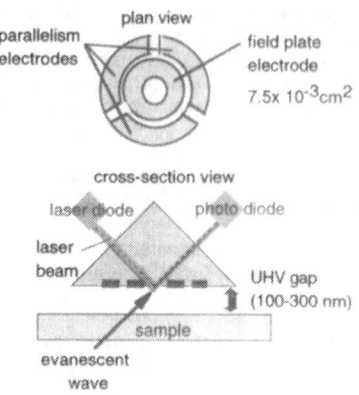

Fig.6 Schematic view of sensor head used in the UHV contactless C-V system.

range of sub-micronmeter, the reflectivity of the laser light is reduced due to the disturbance of evanescent wave. From this reflectivity change, the UHV gap length can easily be calculated. More details of the UHV contactless C-V measurement system are described elsewhere[8,9].

RESULTS AND DISCUSSION

Passivation and in-situ characterization of GaAs surfaces

The first passivation step is MBE growth of an ultrathin pseudomorphic Si layer on MBE-grown GaAs. Thus, in order to understand formation process of the ultrathin Si layer and resultant passivation mechanism, we tried to investigate the initial formation process of the ultrathin MBE Si layer on GaAs surface by using in-situ UHV STM.

Si-doped (2x4) GaAs layers with a carrier concentration of 1-5 x 10^{16} cm^{-3} were first grown on n$^+$ GaAs (100) substrates at a substrate temperature of 580 °C in the MBE chamber. Control of the initial surface reconstruction of (2x4) and c(4x4) was carried out in the following way. To maintain the initial surface reconstruction of the (2x4) pattern at the temperature (300 °C) used for subsequent growth of the Si ICL, the intensity of the As$_4$ flux was gradually reduced during cooling process. On the other hand, without reducing the intensity of the As$_4$ flux during cooling process, c(4x4) reconstructed surfaces could be obtained at about 500 °C and remained even at 300 °C.

Typical reflection high-energy electron diffraction (RHEED) patterns observed on (2x4) and c(4x4) surfaces at 300 °C are shown in Fig. 7. Clear and streaky patterns show that well-defined surface structures were established. As seen in Fig. 7(a), the initial (2x4) pattern changed to the (2x1) pattern after 2-Å Si deposition. On the other hand, the c(4x4) surface changed to an asymmetric (3x1) structure after the Si deposition, as shown in Fig. 7(b). These results agree with previous observations.[4,10] Thus, different initial surface reconstructions of GaAs led to different Si atom arrangements. However, both surfaces showed the same (1x1) RHEED pattern after deposition of a 10-Å Si layer.

Figure 8 show the STM images taken from the GaAs surfaces before and after 2-Å Si deposition. Before Si deposition, a well-known missing dimer structure was observed on the (2x4) surface, as shown in Fig. 8(a), whereas more regular featureless atom arrangements were seen on the c(4x4) surface, as shown in Fig. 8(b). After Si deposition, (2x1) and (3x1) microscopic structures were observed on the initially reconstructed (2x4) and c(4x4) surfaces, respectively. These correspond well to the macroscopic RHEED observation. Many holes were found to exist after Si deposition on the (2x4) surfaces, as seen in Fig. 8(a). In fact, the evolution of the surface structure during the initial phase of Si deposition on the (2x4) surface was found to be extremely complex, involving trench filling, hole formation and later filling by Si as well as partial Si overlayer formation, as discussed in detail previously[11]. On the other hand, after Si deposition on the initially reconstructed c(4x4) surface, such holes were not formed, and atom arrangements were more regular, as seen in Fig. 8(b). In fact, the root mean square (RMS) values of surface flatness for the initially reconstructed (2x4) and c(4x4) surfaces after Si deposition was 0.17 nm and 0.11 nm, respectively. Thus, the Si layer on the initially reconstructed c(4x4) surface was more ordered and flatter than that on the initially reconstructed (2x4) surface.

Figures 9(a) and (b) show the well-accepted models of the (2x4) and c(4x4) surfaces. The model in Fig. 9 (c) was constructed on the basis of our previous study[11], where very complex processes including filling of the missing dimer trench by Si, hole formation, and its filling by Si, as well as partial Si overlayer formation were carried out. It was thus found to be difficult to grow a two-dimensional Si layer on the (2x4) surface and Si atoms are expected to lie in three layers in a random fashion at the interface region. On the other hand, no dimer trenches and no holes existed on the initial c(4x4) surface. Thus, it is highly probable that a layer-by-layer growth of Si takes place in this case, replacing the top extra As layer with Si, as indicated in Fig. 9 (d). There may be some other processes such as the attachment of a Si atom to an extra As atom, and the attachment of a Si atom to a Ga atom after removal of two As atoms. However, this surface definitely has a

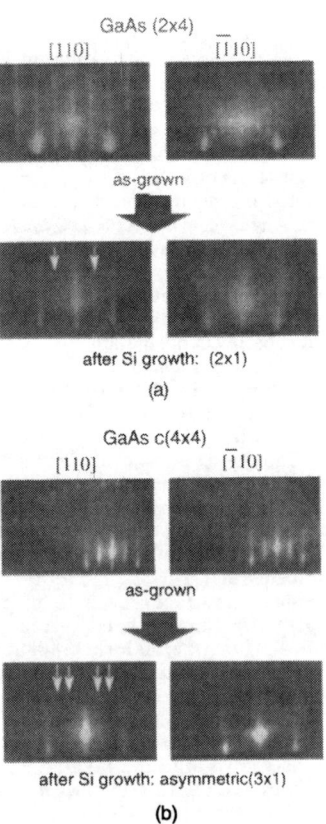

Fig. 7 RHEED patterns before and after the Si deposition on (100) GaAs

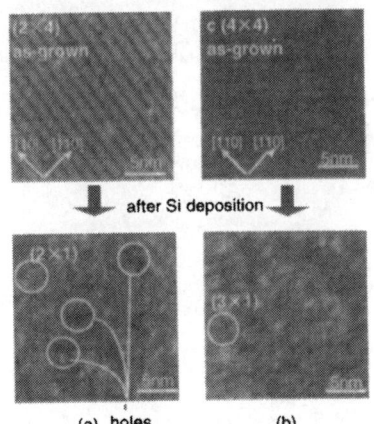

(a) holes   (b)

Fig. 8 STM images of Si ICL on GaAs (100)

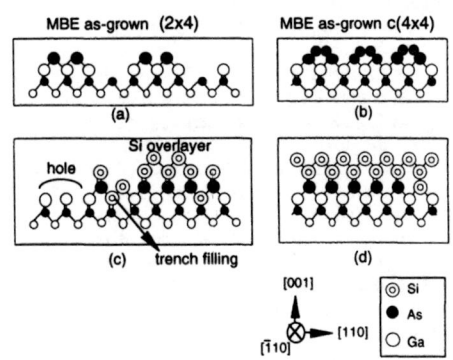

Fig. 9 Models for interface formation of Si/GaAs (100)

49

larger chance of forming a well-ordered interface.

The XPS Si2p spectra after Si layer formation and subsequent partial nitridation by ECR N_2 plasma or nitrogen radicals are shown for the initially reconstructed (2x4) surface in Fig. 10. The formation of Si-Si bonds on GaAs surfaces was confirmed by the spectrum shown in Fig. 10(a). As seen in Fig. 10(b), where the Ga3p spectrum was subtracted from the original spectra for clarity, the Si nitride component was created on the surfaces after the ECR N_2 plasma process for 10s. Angle-resolved XPS analysis clearly showed that only the upper part of Si layer was nitrided. No remarkable difference in XPS spectra existed between the initially (2x4) surface and c(4x4) surface. Furthermore, there was no XPS trace for introduction of damage to GaAs surfaces after Si deposition and subsequent nitridation.

The amount of band bending and the position of the surface Fermi level can be determined from the peak position of the GaAs core levels. Figure 11 (a) shows the Ga3d spectra obtained from the GaAs surfaces after Si ICL deposition. For the initially reconstructed (2x4) surfaces, the peak position of Ga3d is almost the same before and after the deposition of Si ICL. On the other hand, large shift of the Ga3d peak towards the higher binding energy was observed after the deposition of Si ICL for the initially reconstructed c(4x4) surfaces. This indicates significant reduction of Fermi level pinning. Change in surface Fermi level position is summarized in Fig.11 (b). Surface band bending is reduced step by step, indicating that Fermi level pinning is reduced step by step by Si ICL-based passivation. The passivation effect was found to be much larger on the initially reconstructed c(4x4) surface.

Figure 12 (a) shows the measured UHV contactless C-V curves of as-grown and passivated GaAs surfaces. On the as-grown MBE GaAs surfaces, almost flat C-V curves were obtained for both (2x4) and c(4x4) surfaces. This indicates the presence of strong Fermi level pinning on the initial surfaces. A remarkable increase in capacitance variation was obtained on the initially c(4x4) surface after passivation with a Si interlayer and partial nitridation by ECR N_2 plasma. Figure 12(b) summarizes the interface state density (Nss) distributions calculated from the measured UHV contactless C-V curves. A narrow U-shaped distribution is seen for as-grown MBE surfaces. On the other hand, for passivated surfaces using the Si ICL-based method, wider U-shaped curves have been obtained. We obtained the minimum value of the state density of 6×10^{11} cm^{-2}eV^{-1} for c(4x4) surfaces after passivation. Furthermore, the width of the

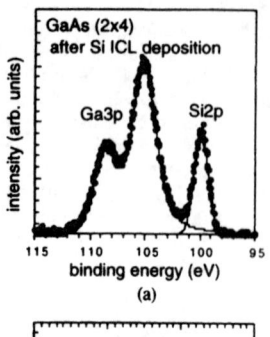

(a)

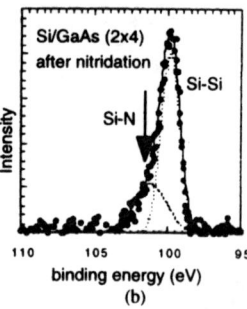

(b)

Fig.10 XPS spectra of Si/GaAs
(a) before and (b) after nitridation

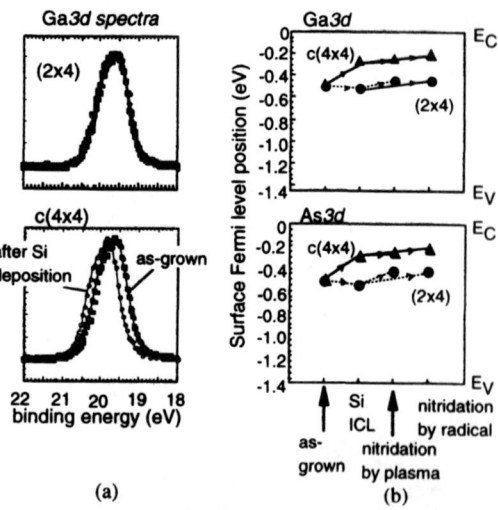

(a)

(b)

Fig.11 (a) Ga3d spectra of Si/GaAs surfaces before and after nitridation and (b) Change in surface Fermi Level position.

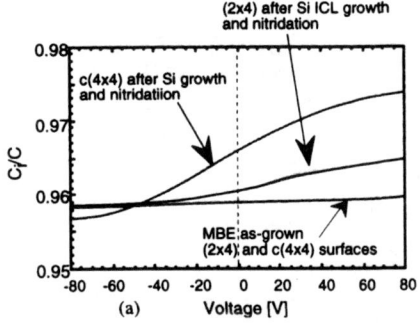
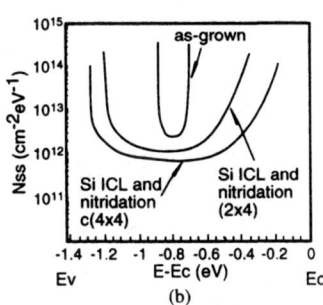

Fig.12 (a) Measured contactless C-V curves and (b) Nss distributions for as-grown
MBE GaAs surface and passivated surfaces

distribution was very much increased. This indicates that the Si-ICL based passivation process is very effective for the reduction of the "pinning" of Fermi level, particularly on initially c(4x4) surface. This seems to be consistent with the results of STM study, which clearly showed that the c(4x4) surfaces lead to a well-ordered interface after Si deposition.

Figure 13 shows the measured changes in the normalized PL intensity of the edge emission from the GaAs surface at room temperature for the as-grown and passivated surfaces, using a specially designed UHV PL chamber. It is clearly seen in Fig. 13 that the PL intensity increases step by step as the passivation procedure proceeds. However, this increase is much larger on the initially reconstructed c(4x4) surface.

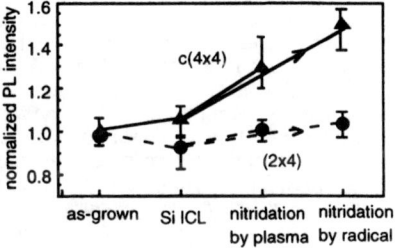

Fig.13 Change in normalized PL intensity of GaAs band-edge emission

All the techniques used here for characterization of macroscopic electronic properties including XPS band bending measurement, UHV PL measurement and UHV contactless C-V measurement, consistently indicated that the effectiveness of the Si ICL-based passivation of GaAs depends on the initial reconstruction of the GaAs surface. This is probably related to the marked difference in the initial surface structure immediately after Si deposition, between (2x4) and c(4x4) surfaces, as shown in Fig.9. From such a viewpoint, the present results indicate that a two-dimensional order at the Si/GaAs interface is extremely important for the success of the Si ICL-based passivation. This seems to support the DIGS model for Fermi level pinning[2].

Passivation of InP surfaces

We have attemted to passivate two different types of surfaces, namely on in-situ MBE-grown surface and an air-exposed surface. For in-situ passivation process, InP layers were grown by GSMBE. Then, in-situ Si ICL passivation process was applied to the fresh (2x4) InP surfaces. Air-exposed n-InP epilayers ($n=5 \times 10^{16}$ cm^{-3}) were prepared as the starting materials to comply with usual device processing requirements. They were cleaned in an organic solution followed by surface treatment in a solution of HF:H$_2$O=1:1 for 1 min. Then, they were loaded into the GSMBE chamber and heated up to 490 °C to remove natural oxides under phosphorus overpressure produced by cracking of tertialybutylphosphine (TBP). After removal of natural oxides, the RHEED pattern showed a clear (2x4) surface reconstruction. Then, a Si ICL with a thickness of 1 nm was grown by MBE at a substrate temperature of 270 °C.

Figure 14 (a) shows the Si 2p spectra at various nitridation steps of Si-covered InP

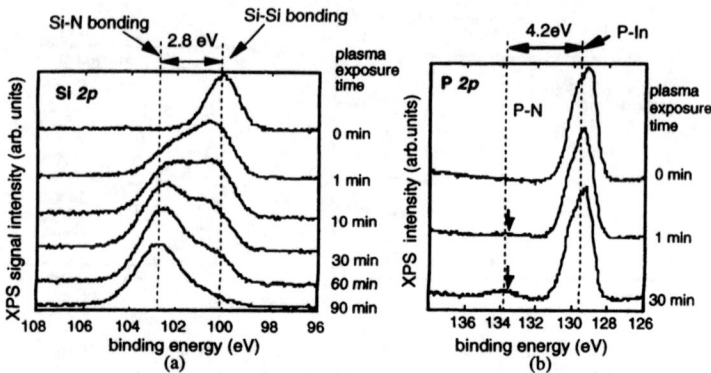

Fig.14 (a) Si2p and (b) P2p spectra of Si/InP surfaces before and after nitridation.

surfaces. Before nitridation, only Si-Si bonding was observed. After nitridation, the Si 2p spectra began to include the peak of the Si-N bonding near 102 eV in addition to the peak corresponding to the Si-Si bonding. Furthermore, a N 1s peak appeared at 377.5 eV. These situation is very similar to the case of GaAs. The nitridation process was further optimized to prevent penetration of nitrogen radicals and ions into the InP surface through the Si layer. Figure 14 (b) shows changes of the P 2p core-level spectra during nitridation of the Si layer. The spectrum started to show a small peak near 134 eV in addition to the peak corresponding to the P-In bonding after nitridation. The position of this higher energy peak is very close to that of the P2p spectrum obtained from a PNx film[12]. This indicates that P-N components were produced after prolonged ECR plasma nitridation, most probably due to the penetration of nitrogen radicals and ions into InP underneath the Si layer.

In situ UHV contactless C-V measurements were performed in order to characterize the effect of each processing step of the present surface passivation of InP. Figures 15(a) and (b) show contactless C-V curves of InP surfaces after the HF treatment and after the UHV thermal cleaning, respectively. Although no trace of the O1s XPS peak was found on the surface after UHV thermal cleaning process, both the slope and the range of capacitance variation of the C-V curve became smaller

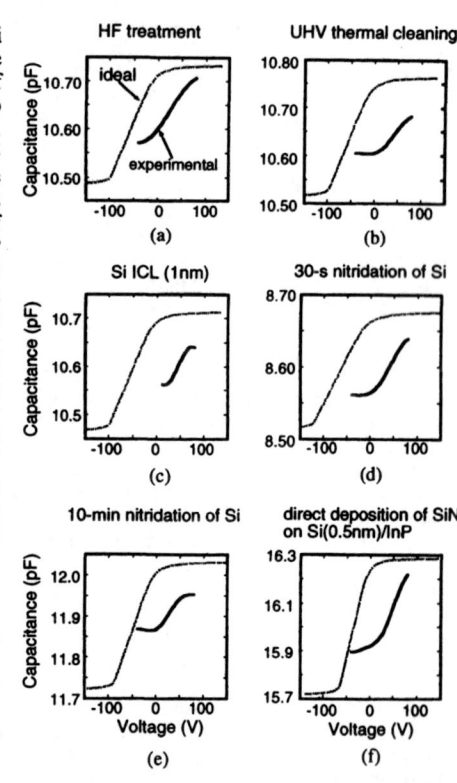

Fig.15 In-situ C-V curves of passivated InP surfaces.

than those of the HF-treated InP surface. This indicates that simple removal of the natural oxide from the surface does not directly cause a surface passivation effect, but it actually increased the surface state density. Figures 15(c) and 15(d) show C-V curves after 30-s and 10-min nitridation of the Si layer, respectively. The slope of the C-V curve became very close to that of the calculated ideal curve after 30-s nitridation of the Si layer, indicating that a significant reduction of the interface state density took place. On the other hand, after 10-min nitridation, C-V behavior became poor again. The latter change appears to be related to the surface damage introduced by a plasma process, as indicated by the XPS result in Fig. 14 (b). Figure 15 (f) shows the C-V curve after growth of a thinner Si ICL (0.5 nm) followed by direct deposition of SiNx on the Si ICL surface by ECR-excited plasma CVD process. Although a comparatively large capacitance variation was observed, the slope was not close to that of the ideal curve.

Figure 16 summarizes the distributions of the interface state density (Nss) by applying Terman's method to the measured UHV C-V curves. All the curves are U-shaped, and the energy position for minimum Nss remained more or less at the same position of Ec-0.35 eV. These features are in good agreement with the DIGS model concerning the origin of interface states [2].

After HF treatment, the surface was found to have the N_{ssmin} value of 4×10^{11} cm^{-2}eV^{-1} and slightly increased after UHV thermal treatment to 6×10^{11} cm^{-2}eV^{-1}. This is because the surface dangling bonds were properly not terminated only by UHV thermal treatment. The reason why the Nss distribution became narrower after growth of the Si ICL on the InP surface can be explained by referring to the energy band diagram of the strained Si/InP structure, as shown in Fig.2. Namely, the bandgap of coherently strained Si in the Si/InP structure becomes very narrow. Because of this, the width of the Nss distribution became narrower. After 30-s nitridation of the Si ICL, the width of the Nss became wider to 600 meV and N_{ssmin} became lower value of 2×10^{10} cm^{-2} eV^{-1}. After the nitridation process, the thickness of the Si ICL became narrower and this effectively increases the bandgap of Si surface quantum well by the quantum confinement effect. Additionally, successful termination of interface bonds of the Si ICL reduces the density of gap states. On the other hand, after longer nitridation for 10 min of the Si ICL by the ECR plasma, N_{ssmin} value increased to a large value of 5×10^{11} cm^{-2} eV^{-1}, because the prolonged nitridation damaged the interface structure, as shown in Fig. 14. In the case of direct CVD deposition of SiNx onto a thinner Si ICL, N_{ssmin} value remained to be 5×10^{11} cm^{-2} eV^{-1}. This is probably due to the damage of the InP surface introduced during direct deposition of thick SiNx film. Thus, effect of each step of the UHV-based passivation could be sensitively detected with this technique, confirming the extremely powerful nature of the present method. Furthermore, very wide Nss distribution at the InP surface was obtained after in-situ Si-ICL based passivation. Thus, the Si-ICL based passivation process is very effective for the reduction of the "pinning" of Fermi level.

MISFETs and insulated-gate HEMTs

We applied the Si-ICL based passivation process to fabrication of InGaAs and InP MISFETs. Figure 17 (a) shows a drain I-V characteristics of the fabricated InGaAs MISFETs. Good gate control of drain currents was obtained. The obtained maximum value of transconductance (gm) is 61 mS/mm. Figure 17 (b) shows normalized drain current drift after application of a gate voltage step. Extremely stable operation characteristics was obtained for the InGaAs MISFET with the Si-based passivation.

Figure 18 shows a drain I-V characteristics of the fabricated InP MISFET. Good gate controllability were obtained. The effective channel mobility is calculated by use of the following formula

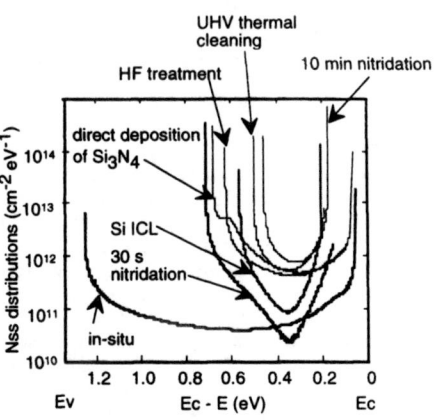

Fig.16 Distributions of surface state density for various passivated InP surfaces.

$$m_{eff} = g_m Lg/(V_{dsat}C_{ox}Wg)$$

where Lg and Wg are the channel length and width, respectively, C_{ox} is the oxide capacitance per unit area, and V_{dsat} is the drain voltage at saturation. The calculated value is 810 cm²/Vs which is very close to that obtained from Hall measurement. No observation for pinch-off behavior of the drain current is due to the non-optimized thickness of active layer. The low doping concentration (1x 10¹⁷ cm⁻³) in the channel layer has caused a high source resistance, thus resulting in high on-resistance. Further optimization of the material design and device processes will significantly improve the device performance.

InP-based HEMT's are promising for use in various future multi-media systems based on microwave/milimeterwave communication, owing to high electron mobility, high sheet carrier density and a large conduction band off-set of the InAlAs/InGaAs system. However, for power application, a conventional Schottky gate structure is not favorable because of its large gate leakage currents which as well as severely limit the rf power handling capability. This problem can be solved by applying an insulated gate (IG) structure to InAlAs/InGaAs HEMT. Our Si-ICL based passivation technique was also applied to fabrication of an insulated-gate type InAlAs/InGaAs HEMT for a novel power HEMT technology.

The fabrication process of the IGHEMT is shown in Fig.19 (a). First, a modulation-doped InGaAs/InAlAs heterostructure lattice-matched to InP was grown by MBE on a semi-insulating InP substrate. After taking out the epi-wafer from MBE chamber, the GeAu/Ni ohmic electrode formation, device isolation and gate recess formation were performed in air. Then, the sample was dipped in HF solution for 1min in N_2 ambient and again introduced to UHV system. HF treatment was found to be effective in reducing native oxides in InGaAs surface. In MBE chamber, the novel Si-ICL passivation process was applied to the InGaAs surfaces. Then, a thick SiNx film (50-70nm) was deposited on the passivated HEMT wafer surface. Finally, the Al gate electrode was formed by lift-off process.

Figure 19 (b) shows I-V characteristics of the fabricated IGHEMT. The device showed a good gate controllability of drain current without hysteresis even under large positive gate bias, resulting in very large gate swing capability. A maximum transconductance of 170mS/mm was obtained. Taking account of a large contact resistance of 1.1 Ωmm determined by TLM method in this device, the intrinsic value of the transconductance was estimated to be 220 mS/mm for LG=2 μm.

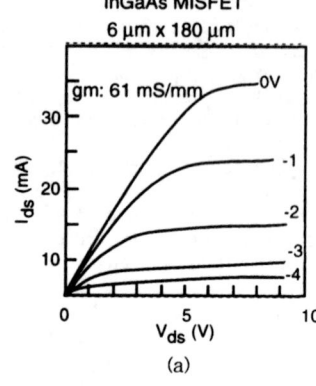

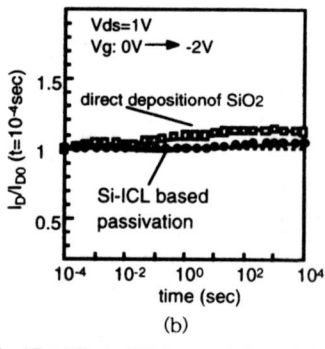

Fig. 17 (a) Drain I-V characteristics and (b) the drift characteristics of InGaAs MISFET

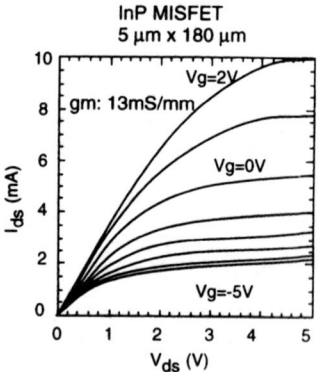

Fig. 18 Drain I-V characteristics of InP MISFET.

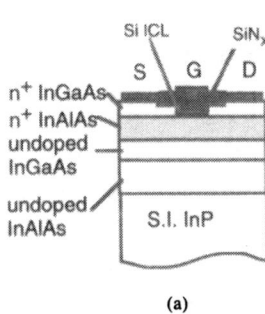

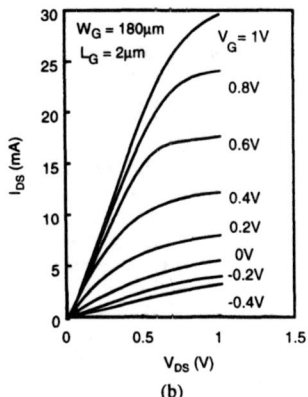

(a)

(b)

Fig.19 (a) A structure and (b) I-V characteristics of insulated-gate HEMT

Passivation of near-surface quantum wells

In future electronics and optelectronics, integrate circuits based on quantum devices will be fabricated near surface. In this case, serious problems arise from the interaction between the quantized states and the surface states.

This can be seen from the behavior of near-surface quantum well samples shown in Fig. 20. In all the samples, two AlGaAs/GaAs/AlGaAs quantum wells, QW1 and QW2, were grown on SI GaAs substrate by MBE. The surface-to-QW1 distance t_B, i.e., the thickness of top AlGaAs barrier layer was varied from 5nm to 90nm. QW2 was formed at a deeper position so as to not to be affected by the surface, and was used as the reference for ex-situ PL measurements. After the growth of such quantum well samples, some sample surfaces were passivated using the Si-ICL passivation technique, as shown in Fig.20(b).

Figure 21 summarizes the normalized PL intensity (the ratio of PL intensity of QW1 to QW2 at each t_B) obtained from the samples with and without passivation. In the samples without passivation, the PL intensity from QW1 was found to start to decrease exponentially when t_B becomes less than about 10nm, probably due to interaction between quntum confined states and surface states, while such degradation of the PL behavior was not observed in the samples with the Si-ICL based passivation. Even for t_B= 5nm, the PL intensity from QW1 in the passivated sample is almost the same with that of the samples with t_B=90nm, realizing more than a 1000 times PL intensity enhancement. This clearly indicates that the present process is very effective in passivating quantum well surfaces. Similar passivation effect was observed in InP-based quantum wires[13].

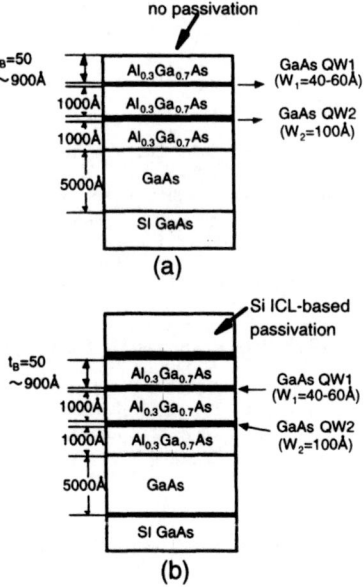

Fig.20 (a) Unpassivated and passivated near-surface quantum well structures.

55

CONCLUSIONS

The present status of surface passivation research for III-V compound semiconductors utilizing a novel unique structure including a silicon interface control layer (Si ICL) is presented and discussed. The basic principle of passivation is to insert an ultrathin MBE-grown Si layer between the III-V compound semiconductor and a Si-based thick insulator. Based on the calculation of quantized levels in strained Si ICL, the passivation structure was optimized. Such a structure was realized by partial nitridation of Si ICL surface. In-situ surface characterization techniques including newly developed UHV contactless C-V technique, were used for optimization of each passivation step. Surface reconstruction of initial semiconductor surface was found to have a great influence on passivation. In the case of

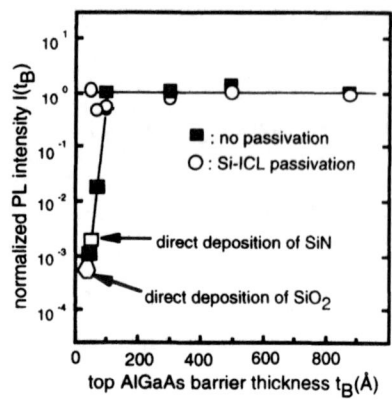

Fig.21 Normalized PL intensity of QWI as a function of thickness of top AlGaAs barrier layer.

GaAs, the c(4x4) surface is preferable to the (2x4) surface. The novel process has realized the oxide-free surface passivation of InP with a N_{ssmin} value of $2 \times 10^{10} cm^{-2} eV^{-1}$. Furthermore, the novel passivation technique has been successfully applied to the fabrication of MISFETs and IGHEMTs, and the passivation of quantum structures.

ACKNOWLEDGEMENT

The authors would like to thank Dr. S. Kodama, Dr. S. Suzuki, H. Takahashi, T. Yoshida, M. Mutoh, N. Tsurumi, and M. Yamada for fruitful discussions and technical supports. This work was partly suppoted by a Garnt-in Aid for Scientic Research of Priority Area (#08247101) and Scientic Research B (#1055098).

REFERENCES

[1] M. Passlack, E. F Schubert, W. S. Hobson, M. Hong, N. Moriya, S. N. G. Chu, K. Konstadinidis, J. P. Mannaerts, M. L. Schnoes and G. J. Zydzik, J. Appl. Phys. **77** , 686(1995).
[2] H. Hasegawa and H. Ohno: J. Vac. Sci. Technol. **4** , 1130(1986).
[3] H. Hasegawa, M. Akazawa, K. Matsuzaki, H. Ishii and H. Ohno, Jpn. J. Appl. Phys. **27**, 2265(1988) .
[4] M. Akazawa, H. Ishii and H. Hasegawa, Jpn. J. Appl. Phys. **30**, 3744 (1991) .
[5] H. Hasegawa, M. Akazawa, H. Ishii and K. Matsuzaki, J. Vac. Sci. Technol. **B7**, 870 (1989).
[6] S. Kodama, S. Koyanagi, T. Hashizume and H. Hasegawa, Jpn. J. Appl. Phys. **34**, 1143(1995).
[7] C.G. Van de Walle, Phys. Rev. B **39**, 1871(1989).
[8] T. Sakai, M. Kohno, S. Hirae, I. Nakatani, and T. Kusuda, Jpn. J. Appl. Phys., **32**, 4005(1993).
[9] T. Yoshida, H. Hasegawa and T. Sakai., Jpn.J. Appl. Phys., **38**, (1999) (in press).
[10] A. G. Taylor, A. R. Turner, M. E. Pemble and B. A. Joyce, J. Cryst. Growth **172**, 275(1997).
[11] N. Tsurumi, Y. Ishikawa, T. Fukui and H. Hasegawa: Jpn. J. Appl. Phys. **37**,1501(1998) .
[12] Y. Matsumoto, T. Suzuki, K. Haga, H. Sasaki, M. Sakuma, T. Hanajiri, T. Sugano and T. Katoda: Proc. 7th Int. Conf. Indium Phosphide and Related Materials (1995) , p. 609.
[13] H. Fujikura, S. Kodama, T. Hashizume and H. Hasegawa, J. Vac. Sci. Technol. **B14**, 2888(1996).

THE (Ga$_2$O$_3$)$_{1-x}$(Gd$_2$O$_3$)$_x$ OXIDES WITH x= 0 – 1.0 FOR GaAs PASSIVATION

J. Kwo, M. Hong, A. R. Kortan, D. W. Murphy, J. P. Mannaerts,
A. M. Sergent, Y. C. Wang, and K. C. Hsieh*
Bell Laboratories, Lucent Technologies, Murray Hill, NJ 07974, raynien@lucent.com
*Department of Electrical and Computer Engineering, Univ. Illinois, Urbana, IL 61801

ABSTRACT

Ga$_2$O$_3$(Gd$_2$O$_3$) film was previously discovered to effectively passivate GaAs surface, and the employment of this oxide as gate dielectric has led to the first demonstration of the enhancement-mode GaAs metal oxide semiconductor field effect transistors (MOSFETs) with inversion. In order to gain insight into the passivation mechanism and elucidate the role of Gd$_2$O$_3$, we have carried out a systematic study of the dependence of the structural and dielectric properties of (Ga$_2$O$_3$)$_{1-x}$(Gd$_2$O$_3$)$_x$ on the Gd (x) content. Our studies indicate that it is necessary to have the Gd addition exceeding 14% in order to form an electrically insulating dielectric with low interfacial state density. Furthermore, we found that pure Gd$_2$O$_3$ film grows epitaxially on GaAs in the single domain, (110) oriented Mn$_2$O$_3$ structure. This new crystalline dielectric has a dielectric constant ~10, and shows excellent dielectric properties with low leakage and high breakdown strength even for films as thin as 2.5 nm. We expect that epitaxial growth of the Mn$_2$O$_3$ structure can be extended to other rare earth oxides, and to other semiconductor substrates like Si.

INTRODUCTION

Modern electronic technologies to-date are predominantly based on Si CMOS circuits. However, for applications at frequencies over 20GHz, there has been increasing demand to call for GaAs based MOSFET circuits for the following reasons. The GaAs substrate is semi-insulating, and GaAs has inherently higher electron mobility when compared to Si. Furthermore, the circuits of GaAs MOSFETs are simpler, and consume less power when compared to GaAs MESFETs currently being used. The applications for GaAs MOSFETs are particularly useful as high power, linear amplifiers for wireless based stations, and low voltage, efficient power amplifiers for mobile phones.

However, successful development of GaAs MOSFETs has been hampered due to the lack of a thermodynamically stable passivating oxide for the GaAs surface. Numerous efforts over the last three decades have been made to search for a suitable dielectric to passivate GaAs. Unlike SiO$_2$/Si system, direct oxidation of GaAs by thermal, anodic, or other treatments produces, in general, amorphous mixtures of Ga$_2$O$_3$ and As$_2$O$_3$ with some inclusion of elemental As. The results tend to be poor in terms of thermal stability, chemical stability, and compatibility with other processing technologies[6]. In addition, strong evidence of the Fermi level pinning was observed in these types of approaches. Deposition of more stable insulators such as SiO$_2$, Al$_2$O$_3$, MgO, and Si$_3$N$_4$ did not unpin the Fermi level. Reviews or reports on those efforts of GaAs oxidation or deposition of thermodynamically stable insulators can be found in Refs. 6-9.

Recently we have discovered that in-situ deposition of mixed oxide Ga$_2$O$_3$(Gd$_2$O$_3$) on GaAs using electron beam evaporation from single crystal Ga$_5$Gd$_3$O$_{12}$ (GGG) has produced an oxide/GaAs interface with a remarkably low recombination velocity and a low interfacial density of states (D$_{it}$)[1,2] in the mid 10^{10} cm^{-2} eV^{-1}. Consequently, employment of this novel oxide as a

Mat. Res. Soc. Symp. Proc. Vol. 573 © 1999 Materials Research Society

gate dielectric along with an ion implantation process has led to the first demonstration of the enhancement-mode GaAs metal oxide semiconductor field effect transistors (MOSFETs) with inversion on semi-insulating GaAs substrates in both n- and p- channel configurations[3]. The n-channel enhancement-mode InGaAs MOSFETs with inversion were fabricated using $Ga_2O_3(Gd_2O_3)$ as the gate dielectric on InP semi-insulating substrates. The devices showed excellent device performance[4] with an extrinsic transconductance of 190 mS/mm, and an effective mobility of 470 cm^2/Vs. Recently, greatly improved depletion-mode GaAs MOSFETs have exhibited negligible drain current drift and hysteresis[5], an important technological advance toward the manufacturing of this class of devices. The effective passivation of GaAs using $Ga_2O_3(Gd_2O_3)$ may find more applications in other electronic and photonic devices, e.g. laser facet coating.

The fundamental cause responsible for such a remarkably low D_{it} at the $Ga_2O_3(Gd_2O_3)$/GaAs interface is not clear. Low D_{it}'s were not obtained in other oxides on GaAs, using a similar in-situ approach[1,9]. Stable, insulating dielectric layers on GaAs, as formed by electron beam evaporation of GGG were initially thought to stem from the preferential evaporation of Ga_2O_3 since the vapor pressure of Ga_2O_3 is much higher than that of Gd_2O_3[10]. It was then assumed that the deposited film consists essentially of pure Ga_2O_3, with the presence of only minute amount of (e.g., 0.1 at %) Gd. This was thought to form higher purity Ga_2O_3 film than possible from evaporation of a powder-packed Ga_2O_3 source. Furthermore, it was believed that the Gd is an undesirable dopant, and that ideally the film should be pure Ga oxide.

However, chemical analysis of our effective passive films indicated the presence of considerable amount of Gd (as high as 20-40 at.%)[11]. This raised the question of the role of Gd_2O_3 in GaAs surface passivation. As a major focus of recent investigations, we have studied the systematic dependence of the structural and dielectric properties of $(Ga_2O_3)_{1-x}(Gd_2O_3)_x$ films on the Gd (x) content. The $Ga_2O_3(Gd_2O_3)$ films are predominantly amorphous and may contain microcrystallites on the order of nano-meters in size. The interface between the $Ga_2O_3(Gd_2O_3)$ and GaAs is smooth and sharp, as studied by high-resolution transmission electron microscopy and x-ray reflectivity. As for pure Gd_2O_3 films, it is a single crystal epitaxially grown in (110) orientation on the GaAs (100) surface. Our structural studies indicated the epitaxy takes place in single domain growth over an oxide thickness from 25 Å to at least 250 Å.

The current-voltage (I-V) and capacitance-voltage (C-V) results showed that pure Ga_2O_3 films have poor electrical leakage properties, and do not passivate the GaAs surface. Additions of Gd exceeding x = 6% greatly improved the dielectric performance in terms of the leakage current density and the electrical breakdown field strength. Low D_{it} was achieved in films with x > 14%. Our results show that the inclusion of Gd_2O_3 reduces the conductivity of $(Ga_2O_3)_{1-x}(Gd_2O_3)_x$ films. Contrary to the earlier assumption[10], our findings strongly suggest the necessity of Gd_2O_3. The electrical measurements showed that pure Gd_2O_3 is an excellent dielectric with a dielectric constant of ~10. Typical electrical leakage current density J_L is less than 10^{-9}A/cm^2 at zero bias, and the breakdown strength is 4-5 MV/cm for a film 100 Å thick (a higher breakdown field for thinner films). Furthermore, both inversion and accumulation layers were observed in the Gd_2O_3/GaAs MOS diodes. Recent advances in GaAs MOSFET devices using the $Ga_2O_3(Gd_2O_3)$ as a gate dielectric are given by Y. C. Wang et al[12]. Both conduction and valence band offsets in the $Ga_2O_3(Gd_2O_3)$/GaAs films were measured by T. S. Lay et al[13]. The detailed structural analysis of single crystal Gd_2O_3 on GaAs (100) and the local bonding configuration at the interface are discussed by A. R. Kortan et al[14].

EXPERIMENTAL

The growth of mixed oxides $(Ga_2O_3)_{1-x}$ $(Gd_2O_3)_x$ (x= 0 – 1.0) on GaAs was performed in a multi-chamber UHV system under conditions described previously[1]. Reflection high-energy electron diffraction (RHEED) was used to monitor film growth of both GaAs epi-layer and oxide films. Prior to oxide deposition either an As-stabilized (2x4) or a Ga-stabilized (4x6) reconstructed surface was obtained after heating to 550°C or above. Substrate temperature was held at 450-550°C during the oxide growth.

Both methods of e-beam evaporation from a single source and co-evaporation from two-sources were employed. Various oxide charges were used including a single crystal GGG, a powder packed composite oxide consisting of $(Ga_2O_3)_{1-x}$ $(Gd_2O_3)_x$ with a composition close to $Ga_5Gd_3O_{12}$, and powder packed pure Gd_2O_3 and pure Ga_2O_3 sources. When evaporated from a single source of GGG, the Gd composition in the deposited films depended on the usage of the source and the substrate temperature. Precise control of the film composition, thus, was quite limited. Since the vapor pressure of Gd_2O_3 is much lower than that of Ga_2O_3 at a given temperature, after repeated evaporation runs Gd-rich material was left in the charge. It is easier to control the film compositions by co-evaporation from two separate sources. The evaporation rate from each source can be individually adjusted.

The film composition was determined by using Rutherford backscattering spectrometry (RBS) and Auger electron spectroscopy. Oxide film thickness was determined by ellipsometry and x-ray reflectivity[15]. The structural properties were characterized by x-ray diffraction using a rotating anode source equipped with a triple-crystal four-circle diffractometer[16]. The MOS diode structure was fabricated by evaporating Au/Pt dots 75, 100, 150 μm in diameter to the oxide film surface for electrical contacts. The C-V measurements were performed in a probe station using a HP 4284A. The I-V curves were taken using a HP 4140B with a pico-ammeter and a voltage source. The voltage was stepped in 0.2 V increment and the resultant current was measured after a 5-second settling time. The other probe for both the C-V and the I-V measurements was placed to the back of the GaAs substrate with an N dopant concentration of $10^{18}/cm^3$.

RESULTS AND DISSCUSION

(A) Structural Properties

We note that both the oxide film and the substrate contain the Ga species. In order to resolve the Ga edge of thin oxide film from that of the substrate in the RBS spectrum, the substrate plane was rotated so that the $^4He^+$ ion beam was incident at an angle $\theta \sim 80$-85° to the substrate normal. The effective film thickness is enhanced by a factor of $(1/\cos\theta)$. Figure 1 shows a RBS spectrum taken using 2.0 MeV $^4He^+$ ion beam for a $Ga_2O_3(Gd_2O_3)$ sample 160 Å thick, e-beam evaporated from GGG. The incident angle of the ion beam is set at 83.5° to the substrate normal to separate the two Ga edges. A variation of Gd composition in the oxide is detected in the depth profile. The best fit to the spectrum by the RUMP program suggests a bi-layer profile, consisting of an upper layer of approximately $Ga_9Gd_{1.6}O_{15}$ 80 Å thick and a lower layer of approximately $Ga_9Gd_2O_{15}$ 80 Å thick near the GaAs interface. The average composition for Gd, x, is ~17%. Auger measurements on our passive films of $(Ga_2O_3)_{1-x}(Gd_2O_3)_x$ indicate the presence of a considerable quantity of Gd, and the Gd to Ga composition ratio is consistent with the RBS results. Notice the difficulty in determining the exact amount of oxygen in the films using RBS.

A cross-section high-resolution TEM picture of $Ga_2O_3(Gd_2O_3)$/GaAs heterostructure shows a abrupt transition from GaAs to $Ga_2O_3(Gd_2O_3)$ (Fig. 2). The interfacial abruptness was also studied by X-ray reflectivity. The $Ga_2O_3(Gd_2O_3)$ appears to be amorphous. However, after removing the high-frequency Fourier components, the reconstructed lattice imaging exhibits granular-like features, which are distinctly different from typical amorphous films such as thermal SiO_2 or electron beam evaporated SiO_2 as prepared from our multi-chamber UHV system (not shown but was discussed in Ref. 15). The granular feature indicates the presence of short-range ordering or even microcrystallites on the order of nanometers in size. Based on X-ray reflectivity measurement, we have deduced the root-mean-square (rms) roughness of air-$Ga_2O_3(Gd_2O_3)$ interface, $Ga_2O_3(Gd_2O_3)$-GaAs interface and the oxide thickness, and the typical results are summarized in Table I. The air-oxide interfacial roughness remains around 1 nm (ranging from 0.86 to 1.2 nm), regardless of exposure time to air. The oxide-GaAs interfaces are very smooth with a roughness of at most 0.78 nm, which is slightly less than three atomic layers of GaAs. For sample MH1342, the roughness (0.33 nm) is as small as one atomic layer of GaAs.

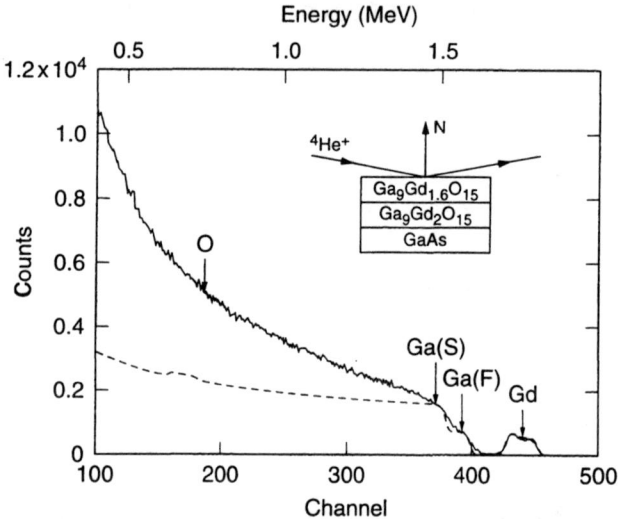

Figure 1. RBS spectrum for a $(Ga_2O_3)_{82}(Gd_2O_3)_{18}$ film 160 Å thick with the incident $^4He^+$ ion beam directed at 83.5° to the film normal. The Ga edges for the film and the substrate are denoted as Ga(F) and Ga(S), respectively.

As for x=1, pure Gd_2O_3 film is grown as an epitaxial single crystal with a crystal structure isomorphic to Mn_2O_3. In-situ RHEED studies (Fig. 3) indicated that deposition of a Gd_2O_3 film 25 Å thick on an As stabilized (2 x 4) reconstructed GaAs surface resulted in streaky diffraction patterns of two-fold symmetry as reported earlier[17]. The quality of the RHEED patterns improved for thicker films of 250 Å, suggesting the epitaxial growth continues, but is not limited to this thickness. We also notice surface reconstruction occurring on the Gd_2O_3 films. Analysis of the diffraction patterns indicates that the Gd_2O_3 film is (110) oriented and grown in single domain. The in-plane epitaxial relationship between (100) GaAs substrate and (110) Gd_2O_3 film is $[001]_{Gd2O3}$ // $[011]_{GaAs}$ and $[\bar{1}10]_{Gd2O3}$ // $[01\bar{1}]_{GaAs}$. The lattice constant mismatches between Gd_2O_3 and GaAs in these two directions are 1.9%, and 3.9%, respectively.

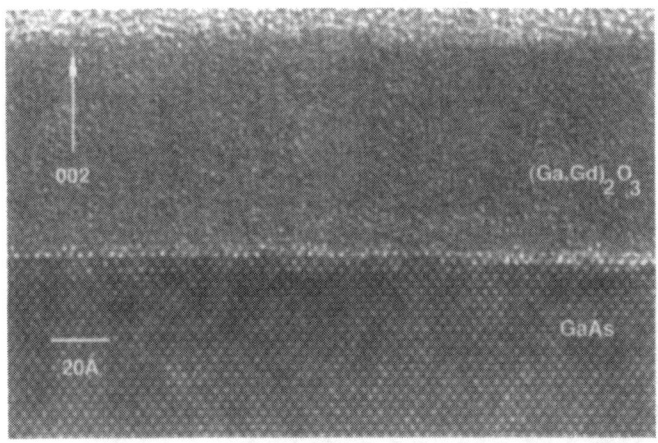

Fig. 2 Cross-sectional HRTEM of a $Ga_2O_3(Gd_2O_3)$/GaAs interface.

Table I   X-Ray Reflectivity Studies on $Ga_2O_3(Gd_2O_3)$/GaAs

Samples	Surface Roughness Air/oxide (nm)	Film Thickness (nm)	Interfacial Roughness oxide/GaAs (nm)
MH1336	1.2	16.6	0.78
MH1338	1.0	14.5	0.54
MH1342	0.86	14.1	0.33
MH1369	1.29	7.67	0.81
MH1370	0.82	8.84	0.24

X-ray diffraction studies of the Gd_2O_3 films showed that these oxide films are indeed single crystals[14,17]. The crystallographic orientation relationship between Gd_2O_3 and GaAs determined from x-ray diffraction is in agreement with the RHEED analysis. Additional single crystal scans made on the samples confirmed the bulk structure as Mn_2O_3 archived in powder diffraction file data bank[18]. The rocking scans show substantial broadening over the instrument resolution of 0.25°, indicating that films have a mosaic spread[17]. The film 25 Å thick, however, shows a very sharp component in the rocking scan, meaning that the epitaxial film at such a fine thickness elastically distorts its unit cell in order to conform to the in-plane perfect epitaxial condition. As the film grows thicker, it is energetically favorable to relax the strain possibly by generating misfit dislocations.

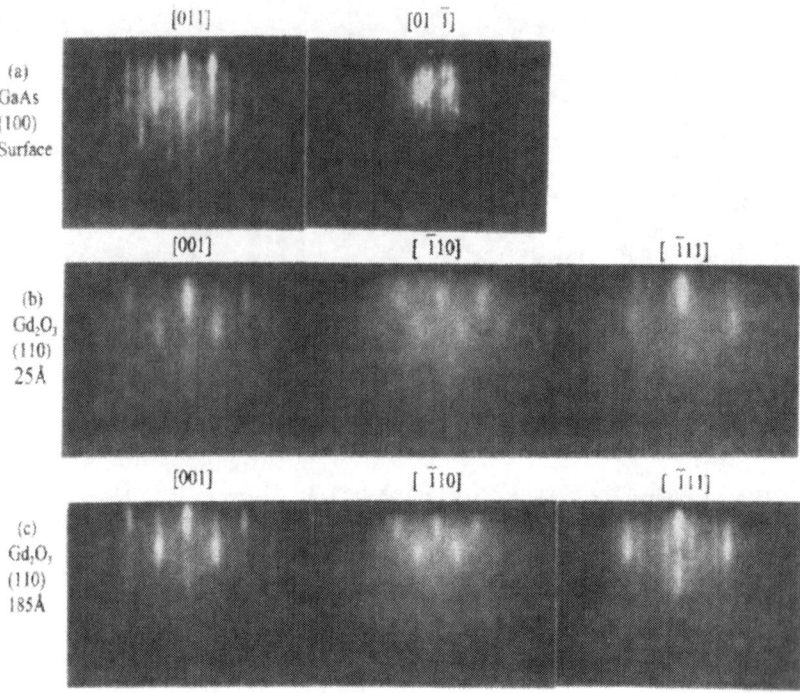

Figure 3 In-situ RHEED patterns of (a) (100) GaAs surface along [011] and [01 1̄] axes, (b) (110) Gd₂O₃ film 25Å thick along [001], [1̄10], and [1̄11] axes, and (c) (110) Gd₂O₃ film 185 Å thick along same axes.

Additional x-ray scans were taken to ascertain that Gd_2O_3 film is indeed of single domain. This was done by fixing the detector to the lattice spacing of the (222) reflection of the oxide, and then rotating the sample (ϕ angle) 360° on a cone centered to the surface normal. Observed reflections of two-fold symmetry unequivocally suggest a nearly perfect epitaxial growth of a single-domain, single crystal Gd_2O_3 film on GaAs (100)[14,17]. This is unusual considering the two-fold degeneracy of aligning the (110) Gd_2O_3 plane of a rectangular symmetry onto the square symmetric (100) GaAs surface. The attainment of single domain may be attributed to the (2x4) reconstruction occurring on the GaAs surface that removes the two-fold degeneracy, thus favors the single variant growth. Other possible mechanisms were given in Ref. 14. HRTEM is now being used to examine the details of film microstructure and interfacial structure.

(B) Dielectric Properties

Figure 4 shows J-E characteristics for a set of $(Ga_2O_3)_{1-x}(Gd_2O_3)_x$ films on GaAs with the Gd content, x, systematically increased from 0, 6, 14, 20, to 100%. All the mixed-oxide films (6, 14, and 20 %) are structurally amorphous based from RHEED studies. The pure Gd_2O_3 sample is an epitaxial, single crystalline film and the pure Ga_2O_3 is a poly-crystalline film with a preferred orientation. The film thickness is kept around 200 Å for all Gd-added samples. The positive bias means that the top metal electrode (Pt/Au) is positive with respect to GaAs. Pure Ga_2O_3 films show very high leakage current with poorly defined breakdown characteristics. Small additions of Gd with x= 6, 14 % to the dielectric films drastically reduced the leakage current density by 3 or 4 orders of magnitude and also increased the breakdown field strength. The x = 20% sample showed a low leakage current density in the range of 10^{-8} - 10^{-9} A/cm^2 prior to breakdown, and an electrical breakdown field E_{br} of 2.5 MV/cm.

Interestingly, pure Gd_2O_3 film showed even lower leakage current density by one order of magnitude ($\sim 10^{-9}$-10^{-10} A/cm^2 at zero bias) when compared to the x =20% sample, and the breakdown behavior is characteristics of a hard-breakdown with an E_{br} increased to 3.5 MV/cm. Figure 5 shows the dependence of leakage current density (J_L) on the applied voltage (V) for a set of Gd_2O_3 samples with the oxide thickness (t_{ox}) systematically reduced from 260 Å to 25 Å. As t_{ox} is decreased from 260 Å to 45 Å, the respective breakdown field E_{br} increases systematically from 3 to 10 MV/cm, yet J_L at a fixed field of 1.5 MV/cm increases merely by one order of magnitude. The fact that the low electrical leakage remains intact even for films as thin as 25 Å suggests that a high degree of structural integrity is sustained through epitaxy. Based on our J-E results, we conclude that Gd_2O_3 is necessary for attaining electrically insulating films

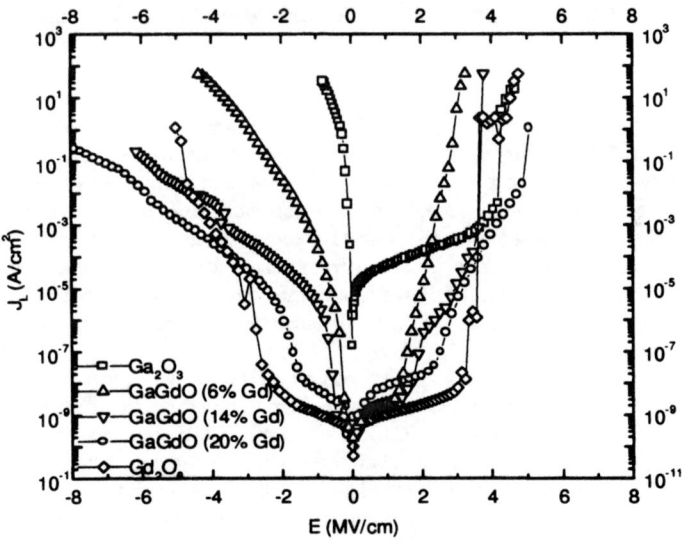

Figure 4 Dependence of leakage current density (J) on the electric field (E) for a set of $(Ga_2O_3)_{1-x}(Gd_2O_3)$ films with x varying from 0, 6, 14, 20, to 100 %.

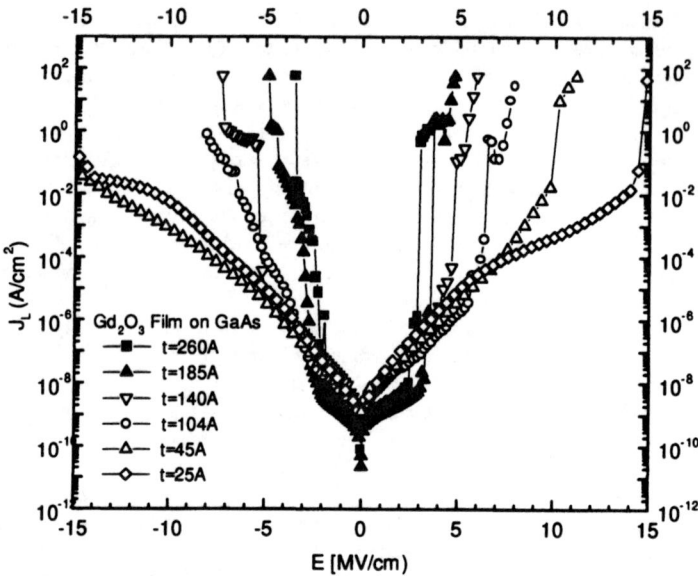

Figure 5 Leakage current density J_L vs applied voltage V for Gd_2O_3 films with decreasing thickness of 260, 185, 140, 104, 45, and 25 Å.

Figures 6(a) to (d) show the results of C-V measurements of MOS diodes for passive films of x = 6, 14, 20, and 100%, respectively. The operation in accumulation, depletion, and inversion is evident for the samples with x =14, 20 and 100 % measured at frequencies varying from 50Hz to 1MHz. In contrast, pure Ga_2O_3 samples are too resistively leaky to give meaningful C-V data. The x = 6 % sample did not show the C-V curves expected for an unpinned oxide/GaAs interface. For pure Gd_2O_3 film 185Å thick, a transition from accumulation to depletion modes occurs at ~2V. The inversion carriers (holes) follow the a.c. signal up to a frequency of 10 kHz, and do not respond to frequencies greater than 100 kHz. Typical dielectric constant is about 10 for Gd_2O_3 film, and is higher about 14 for $Ga_2O_3(Gd_2O_3)$[1,2].

The C-V characteristics can be understood by taking the conductance (G) into account[19]. The finite value of G, mainly arising from the tunneling current through the ultra-thin oxide layer, is in parallel with the oxide capacitance (C_{ox}). This simple equivalent circuit model explains that the total capacitance increases as the modulation frequency decreases. After re-plotting the C-V curves by subtracting the contributions from G and C_{ox}, the interface density of states (D_{it}), which responses only to the low-frequency measurements, is then deduced from the high- and low-frequency curves. This analysis gives a D_{it} at the midgap for Gd_2O_3 is ~ 10^{11} cm^{-2} eV^{-1}, which is comparable to that of $(Ga,Gd)_2O_3$/GaAs interface, and slightly higher than those of the best SiO_2/Si interfaces.

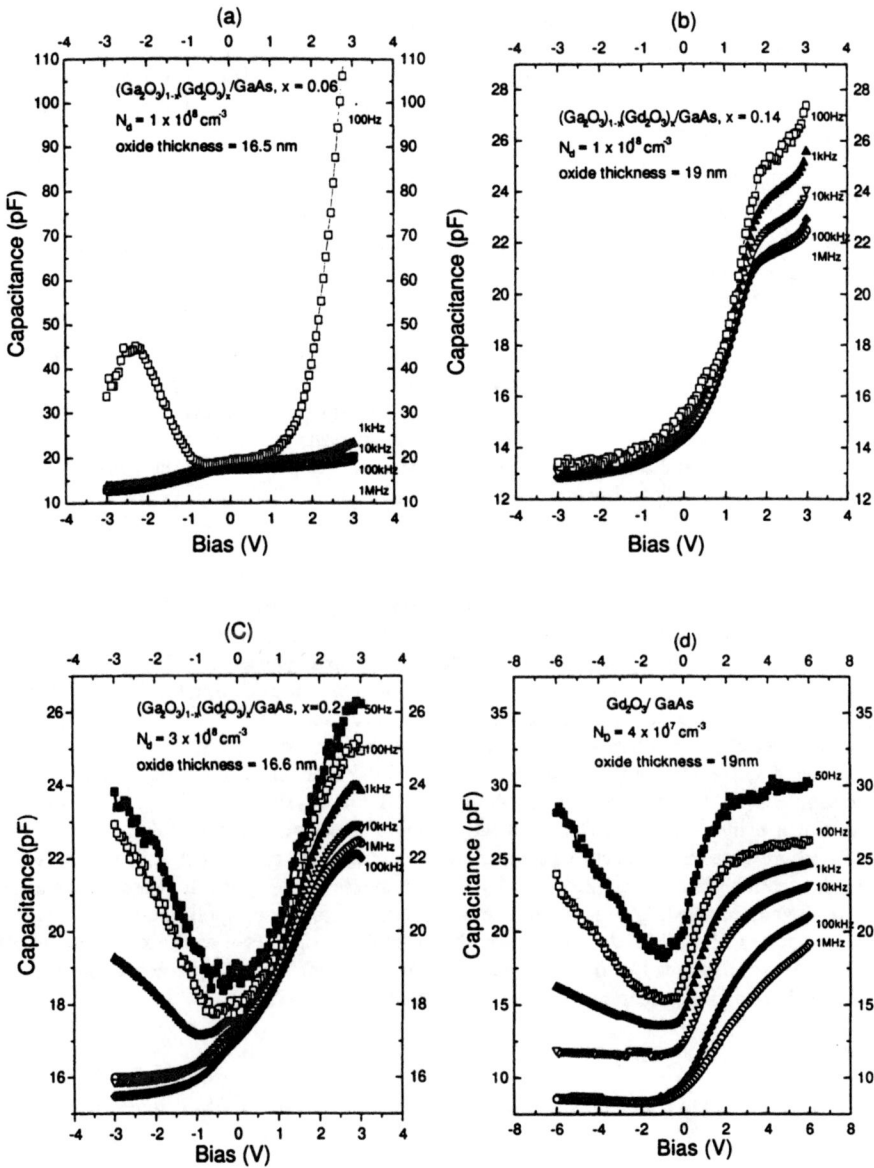

Figure 6 C-V curves as a function of frequency from 50Hz to 1MHz for $(Ga_2O_3)_{1-x}(Gd_2O_3)_x$ films with (a) x = 6%, (b) 14%, (c) 20%, and (d) 100%.

Our electrical data suggest that contrary to previous thinking that Gd was a problematic dopant[23], Gd is beneficial. This result can be understood by the following consideration. Bulk Ga oxides are known to contain[20] essentially only Ga^{3+}. However, the number of oxygen vacancies, and hence the electrical conductivity may depend strongly on the preparation environment[20-22]. E-beam evaporated $(Ga,Gd)_2O_3$ films likely contain oxygen vacancies and/or a small amount of reduced Ga in the film. Traps due to oxygen vacancies may cause the high electrical leakage in films of pure gallium oxide. By analogy with the fact that electropositive elements stabilize high oxidation states for metals with multiple oxidation states in ternary phases (e.g., $BaPbO_3$, $YBa_2Cu_3O_7$, $KMnO_4$, and $SrFeO_3$), we suggest that Gd (Pauling electronegativity 1.1) minimizes oxygen vacancies (this is equivalent to maximizing the Ga oxidation state). Furthermore, films containing other co-deposited electropositive metal oxides may also give passive films. Of the possible electropositive elements, rare earth and alkaline earth elements are preferred over alkali metals because of stability to moisture and compatibility with other processing considerations.

One remarkable feature of the epitaxial crystalline Gd_2O_3 dielectric is the thermodynamic stability of its structure and the interface with GaAs when subject to rapid thermal annealing (RTA) to 850°C for post growth processing. The I-V data for the annealed samples after RTA showed little change, and the C-V measurements still showed accumulation and inversion, indicating the intact of the oxides and the interfaces under such severe thermal stress. This is one significant advantage over other amorphous dielectrics.

CONCLUSION

Investigations in this work have shown clearly that passivating dielectric for GaAs, the $Ga_2O_3(Gd_2O_3)$ film is not pure gallium oxide as thought previously, but instead, contains a considerable amount of Gd_2O_3 exceeding 14%. Our results suggest that Gd_2O_3 is necessary in either reducing the oxygen vacancies or stabilizing gallium oxide to be in the 3^+ state. Based on our findings, we propose the passivating oxide to have the general composition of $Ga_{1-x}A_xO_z$, where A is an electropositive stabilizing element adapted for stabilizing Ga in the 3^+ oxidation state, and could preferably be the rare earth and alkaline earth elements. We also have demonstrated the epitaxial growth of single crystal Gd_2O_3 films on GaAs achieving a low interfacial density of states. The atomically smooth interface and single domain structure of Gd_2O_3 ensure excellent dielectric properties with low leakage conduction and high breakdown strength even for films of only 25 Å thick. The robustness of the epitaxial dielectric against post high temperature annealing adds another attractive feature for device application. We expect that epitaxial growth of the Mn_2O_3 structure can be extended to other rare earth oxides, and to other semiconductor substrates like Si. Our findings from this work thus suggest new opportunities of producing high ε gate dielectrics for Si and GaAs based MOSFETs.

ACKNOWLEDGMENTS

We would like to thank valuable suggestions from A. Y. Cho, R. M. Fleming, C. T. Liu, and Y. H. Wong.

REFERENCES

1. M. Hong, M. Passlack, J. P. Mannaerts, J. Kwo, S. N. G. Chu, N. Moriya, S. Y. Hou, and V. J. Fratello, *J. Vac. Sci. Technol.* **B 14 (3)**, 2297, (1996).

2. M. Passlack, M. Hong, J. P. Mannaerts, J. Kwo, R. L. Opila, S. N. G. Chu, N. Moriya, and F. Ren, *IEEE Transaction of Electron Devices*, **44 (2)**, 214, (1997).

3. F. Ren, M. Hong, W. S. Hobson, J. M. Kuo, J. R. Lothian, J. P. Mannaerts, J. Kwo, Y. K. Chen, and A. Y. Cho, *IEEE IEDM Technical Digest p.*943, (1996) and *Solid State Electronics*, **41** (11), p.1751, (1997).

4. F. Ren, J. M. Kuo, M. Hong, W. S. Hobson, J. R. Lothian, J. Lin, W. S. Tseng, J. P. Mannaerts, J. Kwo, S. N. G. Chu, Y. K. Chen, and A. Y. Cho, *IEEE Electron Device Letters*, **19 (8)**, 309, (1998).

5. Y. C. Wang, M. Hong, J. M. Kuo, J. P. Mannaerts, J. Kwo, H. S. Tsai, J. J. Krajewski, Y. K. Chen, and A. Y. Cho, *IEEE IEDM Technical Digest p.*67, (1998).

6. H. Hasegawa. "Properties of Gallium Arsenide", 3rd Edition, p. 447, Ed. By M. R. Brozel and G. E. Stillman, Published by INSPEC, The Institution of Electrical Engineers, London, UK, 1996.

7. A review can be found in "Physics and Chemistry of III-V Compound Semiconductor Interfaces", Ed. C. W. Wilmsen, Plenum, New York, 1985.

8. M. Hong, C. T. Liu, H. Reese, and J. Kwo, "Semiconductor-Insulator Interfaces" in Encyclopedia of Electrical and Electronics Eng., John Wiley & Sons, New York, 1999.

9. M. Passlack, M. Hong, J. P. Mannaerts, and L. W. Tu, *Appl. Phys. Lett.* **68(25)**, 3605, (1996).

10. N. K. Dutta, R. J. Fischer, N. E. J. Hunt, M. Passlack, E. F. Schubert, and G. J. Zydzik, "Gallium oxide coatings for optoelectronic devices using electron beam evaporation of a high purity single crystal $Ga_5Gd_3O_{12}$ source", US Patent 5,550,089.

11. J. Kwo, D. W. Murphy, M. Hong, J. P. Mannaerts, R. L. Opila, R. L. Masaitis, and A. M. Sergent, presented at *17th North American MBE Conf. Oct.4-7, 1998 at Penn State Univ. and to be published in JVST* (1999).

12. Y. C. Wang, M. Hong, J. M. Kuo, J. P. Mannaerts, J. Kwo, H. S. Tsai, J. J. Krajewski, J. S. Weiner, Y. K. Chen, and A. Y. Cho, paper Z5.5 in this Symposium.

13. T. S. Lay, M. Hong, J. Kwo, J. P. Mannaerts, W. H. Hung, D. J. Huang, paper Z3.7, in this Symposium.

14. A. R. Kortan, M. Hong, J. Kwo, J. P. Mannaerts, and N. Kopylov, paper Z1.7, in this Symposium.

15. M. Hong, J. P. Mannaerts, M. A. Marcus, J. Kwo, A. M. Sergent, L. J. Chou, K. C. Hsieh, and K. Y. Cheng, *J. Vac. Sci. Technol.* **B16 (3)**, 1395, (1998).

16. A. R. Kortan, A. Erbil, R. J. Birgeneau., M. S. Dresselhaus, *Phys. Rev. Lett.* **47**, 1206, (1981).

17. M. Hong, J. Kwo, A. R. Kortan, J. P. Manaerts, and A. M. Sergent, *Science*, 283, pp.1897-1900, (1999).

18. S. Geller, *Acta Cryst.* **B27**, 821, (1971).

19. W. E. Dahlke and S. M. Sze, *Solid State Electronics*, **10**, 865-873 (1967).

20. M. Fleischer and H. Meixner, J. Appl. Phys. **74**, 300, (1993).

21. T. Harwig and J. Schoonman, J. Solid State Chem. **23**, 205, (1978).

22. L. N. Cojocaru and I. D.Alecu, Z. Phys. Chem. **84**, 325, (1973).

23. M. Passlack, and J. K. Abrokwah, *"Method of Forming a Ga_2O_3 layer"*, US patent 5,597,768

DEVELOPMENT OF LOW TEMPERATURE SILICON NITRIDE AND SILICON DIOXIDE FILMS BY INDUCTIVELY-COUPLED PLASMA CHEMICAL VAPOR DEPOSITION

J. W. Lee*, K. D. Mackenzie*, D. Johnson*, S. J. Pearton**, F. Ren*** and J. N. Sasserath*
* Plasma-Therm Inc., St. Petersburg, FL 33716
** Dept. of Materials Sci. Eng., University of Florida, Gainesville FL 32611
***Dept. of Chemical Eng., University of Florida, Gainesville FL 32611

ABSTRACT

High-density plasma technology is becoming increasingly attractive for the deposition of dielectric films such as silicon nitride and silicon dioxide. In particular, inductively-coupled plasma chemical vapor deposition (ICPCVD) offers a great advantage for low temperature processing over plasma-enhanced chemical vapor deposition (PECVD) for a range of devices including compound semiconductors. In this paper, the development of low temperature (< 200 °C) silicon nitride and silicon dioxide films utilizing ICP technology will be discussed. The material properties of these films have been investigated as a function of ICP source power, rf chuck power, chamber pressure, gas chemistry, and temperature. The ICPCVD films will be compared to PECVD films in terms of wet etch rate, stress, and other film characteristics. Two different gas chemistries, $SiH_4/N_2/Ar$ and $SiH_4/NH_3/He$, were explored for the deposition of ICPCVD silicon nitride. The ICPCVD silicon dioxide films were prepared from $SiH_4/O_2/Ar$. The wet etch rates of both silicon nitride and silicon dioxide films are significantly lower than films prepared by conventional PECVD. This implies that ICPCVD films prepared at these low temperatures are of higher quality. The advanced ICPCVD technology can also be used for efficient void-free filling of high aspect ratio (3:1) sub-micron trenches.

INTRODUCTION

There is a growing interest in high-density plasma processing in both the semiconductor and the magnetic thin film head industry [1-7]. In particular, much research has been conducted on dry etching with inductively coupled plasma (ICP) sources because they may provide advanced processes for pattern transfer [8-10]. A lot of research has been reported for dielectric film deposition using remote or high-density plasmas [11-14]. However, there is relatively little work on deposition technology for dielectric materials using ICP [15,16]. Using an ICP source, we explored high-density plasma (>10^{11} cm^{-3}) chemical vapor deposition (HDPCVD) of SiN_x and SiO_2. In comparing HDPCVD technology with conventional PECVD, some potential advantages are lower hydrogen content films, higher quality films at lower process temperatures (< 200 °C), void-free gap filling of high aspect ratio features, and self-planarization. Low temperature SiN_x film deposition of low hydrogen content by HDPCVD is of special interest for cap and capacitor layers in III-V semiconductor devices [17-19]. Due to the relatively low dissociation efficiency of N_2, typical process recipes for SiN_x deposition by PECVD use NH_3 as the source of nitrogen. Therefore, some portion of hydrogen incorporation from NH_3 in deposited SiN_x films is inevitable. However, HDPCVD technology enables us to deposit SiN_x with a NH_3-free recipe because high-density plasma sources have typically one order of magnitude higher ion dissociation efficiency (i.e. ~0.1 % for PECVD and ~1 % for HDPCVD).

Some advantages of an ICP source over other types of high-density sources include easier scale up, advanced automatic tuning for the source, and lower cost of ownership. In addition,

69

with a hybrid ICP configuration, such as used in this work, it is possible through the addition of rf power to the wafer chuck, to control ion flux and ion energy independently. This expands the applications for dielectric film deposition by ICP. For example, gap-filling techniques require simultaneous high ion bombardment by an inert gas, such as Ar, during deposition to prevent void formation. To achieve high ion bombardment, high ion energy can be induced by controlling the rf power on the wafer chuck in HDPCVD without change to the ion density in the source. In PECVD, that approach is not feasible and alternating sequences of deposition and sputter etch are usually required. This requires extra steps and more process time. For damage sensitive devices, such as the high electron mobility transistor (HEMT), it is essential to use a very low ion energy process because ion energy is a major factor causing ion damage to the device [20]. With HDPCVD, the ion energy can be reduced to minimize damage of the devices. The process pressures for PECVD and HDPCVD are quite different. More than 500 mTorr is common for PECVD. The pressure range of HDPCVD is 1 to 30 mTorr. We will report on the effect of rf chuck power and pressure for SiN$_x$ films deposited by HDPCVD with an ICP source.

EXPERIMENTAL

Both SiN$_x$ and SiO$_2$ films were deposited in a Plasma-Therm HDPCVD Versalock system on 4" Si wafers. Figure 1 shows a schematic of the ICP chamber configuration. The load module of the system can handle up to about 50 wafers at a time. For the SiN$_x$ deposition, two different gas chemistries, SiH$_4$/N$_2$/Ar and SiH$_4$/NH$_3$/He, were explored. A gas chemistry of SiH$_4$/O$_2$/Ar was used for the SiO$_2$ film deposition. Electronic mass flow controllers regulated all gas flows. The SiH$_4$ was fed through a lower gas ring (gas inlet 2). All other gases entered the chamber via a showerhead in the ICP source (gas inlet 1). The process chamber is a hybrid configuration consisting of an inductively coupled plasma (ICP) source and an rf powered wafer chuck. An oil recirculating heat exchanger connected to the chuck controls the wafer temperature. Helium backside cooling of the wafer is used for efficient heat transfer. For this work, the ICP source power and chuck temperature were fixed at 800 W and 150 °C respectively. The rf chuck power and chamber pressure were varied from 25 to 150 W and 1 to 20 mTorr, respectively.

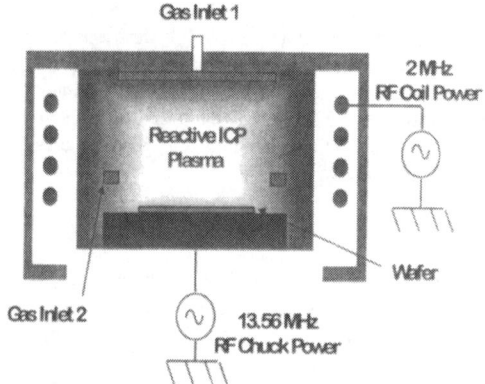

Figure 1. Schematic of the ICP chamber.

Langmuir probe measurements of ICP Ar plasmas were used to determine the ion density characteristics of the process chamber. In-situ optical emission spectroscopy (OES) was also used to characterize the ICP SiH$_4$/N$_2$/Ar plasmas.

Deposition rate and uniformity were determined by a NanoSpec model 4150 metrology system. Refractive index measurements were made on a Gaertner model L116D-PC ellipsometer. Film stress measurements were done with a Tencor model P-2 profilometer. A buffered oxide etch (BOE) solution of 7:1 NH$_4$: HF was used for the wet-etch rate measurements.

RESULTS AND DISCUSSION

ICPCVD Plasma Characterization

Figure 2 shows ion density and negatively induced dc bias as a function of ICP source power and rf chuck power for Ar plasmas at a fixed pressure of 10 mTorr. These data were obtained from a design of experiment simulation, which we developed with experimental data obtained on the ICP process chamber. As shown in Figure 2 (a), the ion density of the Ar plasma can be controlled by ICP source power and is almost independent of rf chuck power. The Ar ion density is about 2×10^{11} cm^{-3} with 800 W ICP power, 20 sccm Ar, 10 mTorr chamber pressure, and 100 W rf chuck power. As illustrated in Figure 2 (b), it was also found that the dc bias on the chuck was a strong function of ICP source power and rf chuck power. The dc bias, in other words, ion energy increased with rf chuck power and decreased with ICP source power.

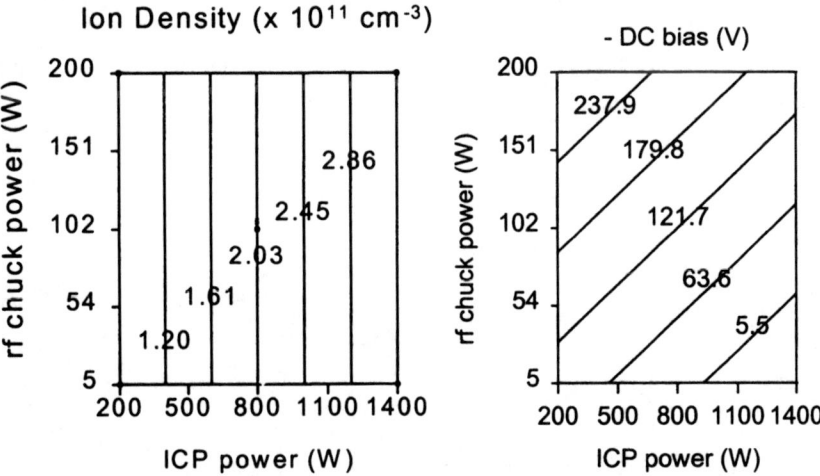

Figure 2. Ion density (a) and negatively induced dc bias (b) as a function of rf chuck power in an ICP chamber for Ar plasmas.

Silicon Nitride: Optical Emission Spectra

In Figure 3, a typical optical emission spectrum (OES) is presented for an ICP SiH$_4$/N$_2$/Ar plasma. This spectrum was recorded during a SiN$_x$ deposition. All N$_2$, SiH, H$_2$, and Ar related peaks were identified. Figure 4 shows more detailed OES data taken at different rf powers. Many N$_2$ related peaks were detected between 300 and 400 nm. A N$_2^+$ peak is also noticed at 391.4 nm. A SiH and two H$_2$ peaks are observed at 414.2, 486.1 and 656.2 nm, respectively. The peak intensities of both SiH and H$_2$ increased with increasing rf power.

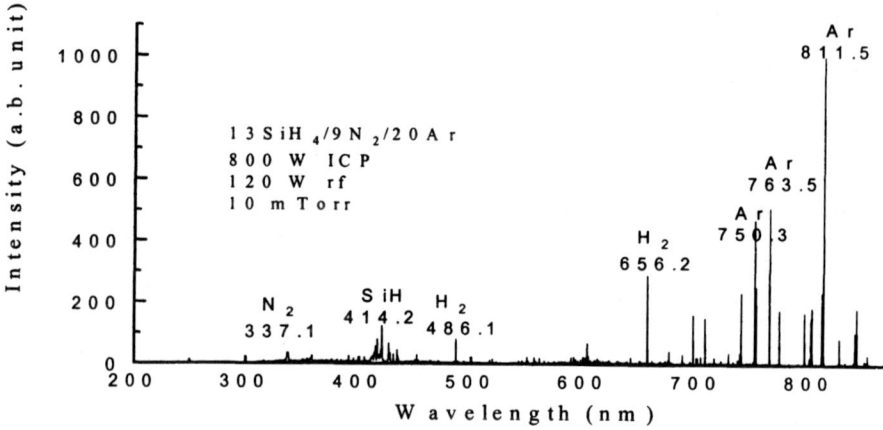

Figure 3. Optical emission spectrum of an ICP SiH$_4$/N$_2$/Ar plasma at 800 W ICP, 120 W rf, and 10 mTorr.

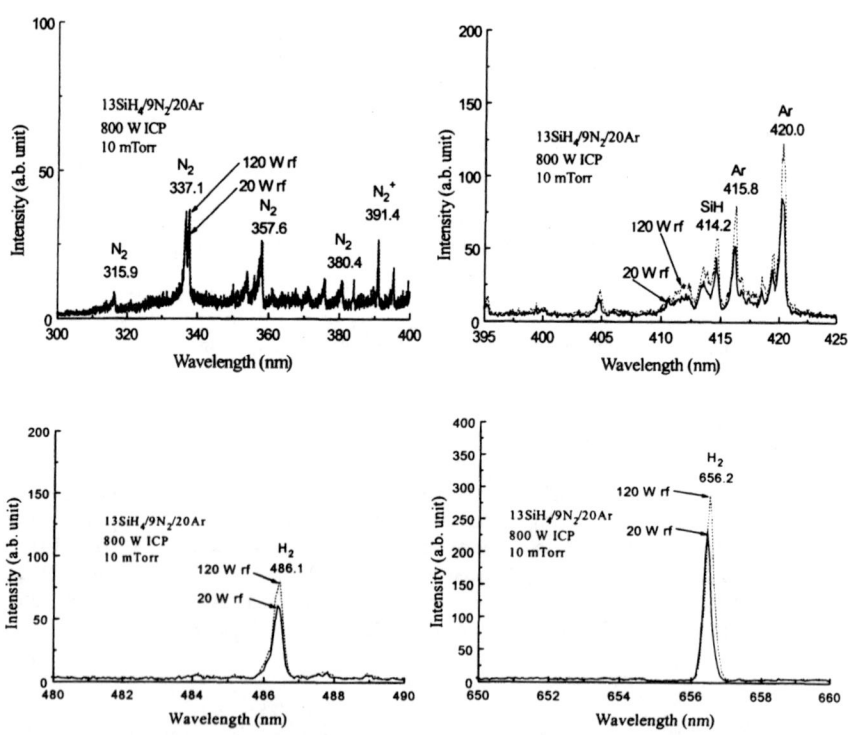

Figure 4. Detailed optical emission spectra for an ICP SiH$_4$/N$_2$/Ar plasma for rf chuck powers of 20 and 120 W. The ICP power and pressure were 800 W ICP and 10 mTorr, respectively.

72

Silicon Nitride: Film Properties

1. Results for SiH$_4$/N$_2$/Ar

In Figure 5, the dependence of deposition rate on rf chuck power is shown. The deposition rate decreases with increasing rf chuck power. At high rf power, it is expected that more ion bombardment during deposition will occur. High rf power increases the dc bias and consequently the ion energies of the bombarding species, e.g. Ar$^+$ and N$_2^+$, will increase. This will enhance the sputtering of Si and N reactive neutrals before and during the formation of the SiN$_x$ film. This accounts for the observed reduction in deposition rate with rf power. The typical deposition rate was in the range 400 to 600 Å/min.

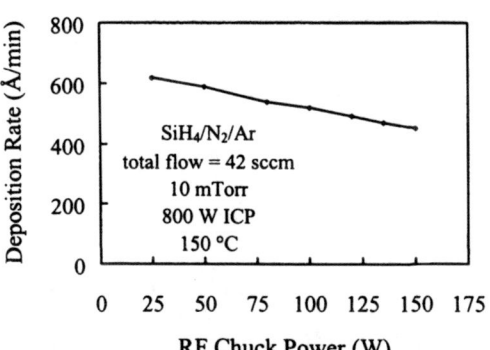

Figure 5. SiN$_x$ deposition rate as a function of rf chuck power.

Figures 6 and 7 show the variation of refractive index and BOE rate with rf chuck power. At 25 W rf chuck power, a refractive index of 1.95 was achieved. The combination of decrease of deposition rate and BOE rate, and increase of refractive index with rf chuck power indicates that deposited films become Si-rich and densified as the rf chuck power is increased. The density, ρ may be calculated from the following equation:

$$\rho = (28[Si] + 14[N])/A \tag{1}$$

where [Si] is atomic concentration of Si, [N] is atomic concentration of N, and A is Avogadro's number.

One possible explanation for the increase of refractive index with rf power is that the

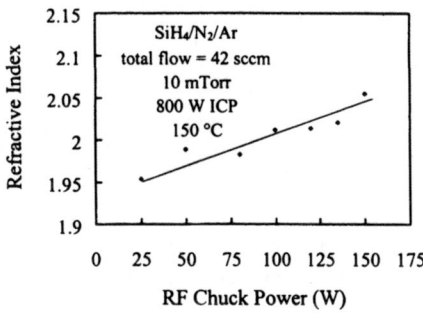

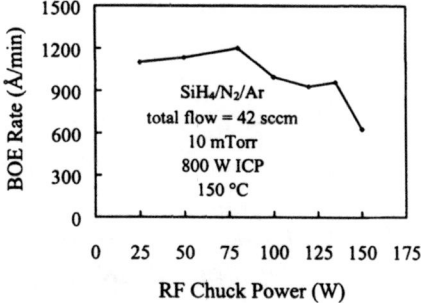

Figure 6. Refractive index of ICPCVD SiN$_x$ as a function of rf chuck power.

Figure 7. BOE rate of ICPCVD SiN$_x$ as a function of rf chuck power.

dissociation of SiH_4 into reactive neutrals in the $SiH_4/N_2/Ar$ plasma increased relative to that of N_2 with rf chuck power (in other words, ion energy). Another possible explanation is that high rf power brings more Si-related species than N to participate in the film growth. Table I shows the dissociation energy of related diatomic molecules for SiN_x deposition. Notice that the N-N bond has a much higher bond strength (945.33 kJ/mol) than that of Si-H ($\leq$ 299.2 kJ/mol). Dissociation of N_2 requires more energy than that of SiH_4. All of this data suggests that more SiH_4 species are brought to the substrate than N_2 with increase of rf chuck power.

Table I. Dissociation energies of diatomic molecules.

Bond	Dissociation Energy at RT (kJ/mol)
Si-H	$\leq$ 299.2
N-H	$\leq$ 339
N-N	945.33
H-H	435.99

It is worthwhile commenting further on the BOE data presented in Figure 7. Considering the low deposition temperature of 150 °C, the typical BOE rate of 1000 Å/min is very low. At 150 °C, the BOE rate of PECVD SiN_x is greater than 5,000 Å/min. This implies that the ICPCVD SiN_x films are denser and contain less hydrogen [21].

As indicated in Figure 8, the stress of all ICPCVD SiN_x films prepared from $SiH_4/N_2/Ar$ was compressive. At 25 W rf chuck power, the film stress was as high as 900 MPa. Raising the rf chuck power lowers the film stress. A stress of 300 MPa is achieved at 120 W. Film stress of less than 500 MPa is acceptable for many applications. We speculate that incorporated hydrogen relieves the stress in the SiN_x film, although the correlation with rf chuck power is not fully understood. In PECVD, the film stress increases with rf power regardless of whether it is compressive or tensile. This is diametrically opposite to the current observations with ICPCVD.

Figure 9 shows deposition uniformity versus rf chuck power. The typical uniformity is less than 4 %. The uniformity becomes somewhat worse as the rf chuck power is increased. It was observed that the center of the deposited films was thicker than edge of the wafer.

As presented in Figures 10 and 11, raising the chamber pressure increased both the deposition rate and refractive index. The deposition rate increases from 350 Å/min at 1mTorr to 600 Å/min at 20 mTorr. Over this pressure range, the refractive index increases from 1.92 to 2.32. As shown in Figure 12, the BOE rate decreases from 1200 to 400 Å/min as the pressure is

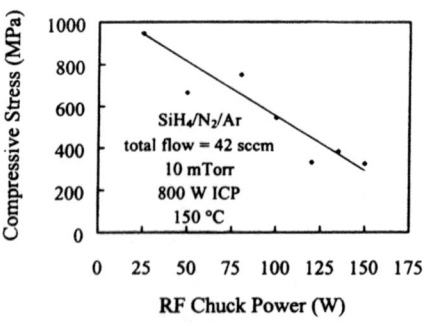

Figure 8. Compressive stress of ICPCVD SiN_x as a function of rf chuck power.

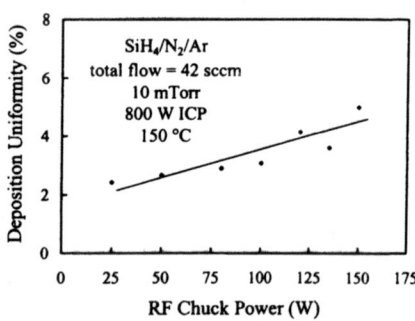

Figure 9. Deposition uniformity of ICPCVD SiN_x as a function of rf chuck power.

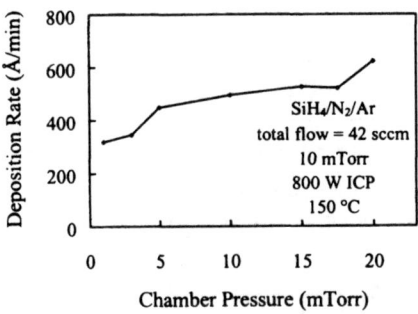

Figure 10. Deposition rate of ICPCVD SiN$_x$ as a function of chamber pressure.

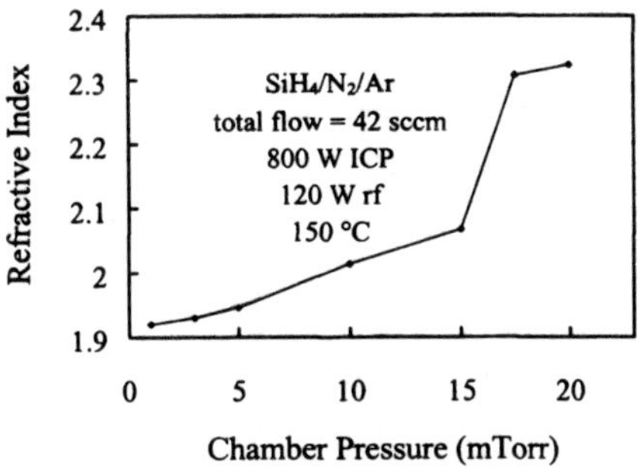

Figure 11. Refractive index of ICPCVD SiN$_x$ as a function of chamber pressure.

increased. At high pressure, the residence time of gas species increases by the equation:

$$\tau = PV/Q \qquad (2)$$

where τ, P, V, and Q are the residence time, pressure, volume, and throughput, respectively. With longer residence time and higher Ar ion density at high chamber pressure, more SiH$_4$ may be dissociated into neutrals, which will be incorporated in the films and cause a Si-rich composition at high pressure. It is well known that increased refractive index with Si-rich composition reduces the BOE rate of the films.

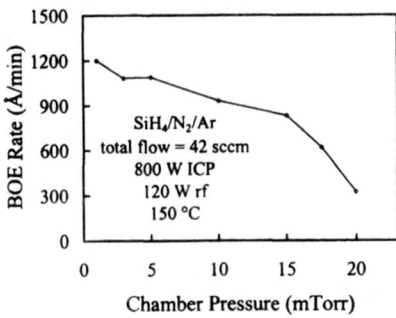
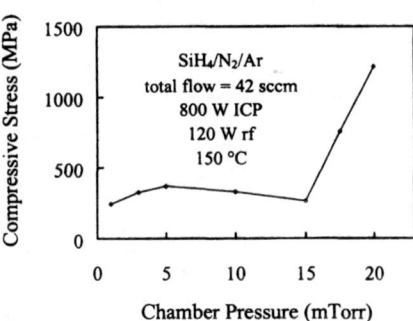

Figure 12. BOE rate of ICPCVD SiN$_x$ as a function of chamber pressure.

Figure 13. Compressive stress of ICPCVD SiN$_x$ as a function of chamber pressure.

As shown in Figure13, the stress of the deposited films is constant at about 400 MPa, compressive over the pressure range of 1 to 15 mTorr. However, at higher pressure (> 15 mTorr), stress increased rapidly and reached 1300 MPa at 20 mTorr.

2. Results for SiH$_4$/NH$_3$/He

ICPCVD SiN$_x$ deposition from SiH$_4$/NH$_3$/He was also investigated. Table II summarizes some of data obtained. The process conditions were 800 W ICP power, 30 W rf chuck power, 10 mTorr, and 150°C.

Table II. Characteristics of ICPCVD SiN$_x$ with SiH$_4$/NH$_3$/He.

Refractive Index	2.0	BOE Rate	1200 Å/min
Deposition Rate	300 Å/min	Uniformity	+/- 2.5 %
Film Stress	200 MPa, (compressive)		

Deposition rate was not a strong function of rf chuck power (Figure 14) while stress was controlled by rf chuck power (Figure 15). The film stress was compressive at low rf chuck power and it became tensile with more rf chuck power.

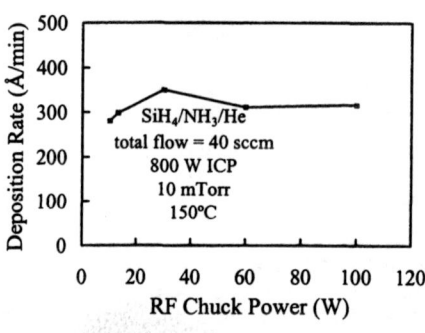 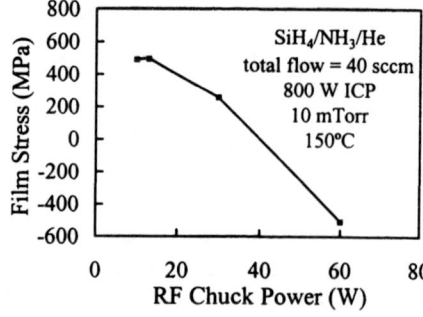

Figure 14. Deposition rate of ICPCVD SiN$_x$ as a function of rf chuck power.

Figure 15. Compressive stress of ICPCVD SiN$_x$ as a function of rf chuck power.

Silicon Dioxide: Gap Filling

The application of ICPCVD SiO$_2$ for void-free filling of 1 μm gap features was investigated. Table III summarizes some of the main properties of these films.

Table III. Characteristics of ICPCVD SiO$_2$ film.

Gas Chemistry	SiH$_4$/O$_2$/Ar
Deposition Rate	~1000 Å/min
Film Stress	300 MPa, (compressive)
Refractive Index	1.46

Figure 16 shows an SEM micrograph of a gap-filled Si trench. The aspect ratio of the trench was about 3:1. Notice that there is no void after gap filling is completed. A similar technique was applied on an electroplated metal which had an irregular shaped sidewall. Figure 17 shows self-planarization after gap filling. Note that there were small voids after the process was completed. The origin of the void is from the irregularity of sidewall at the bottom of the trench. As illustrated in Figure 18, with modified process conditions to compensate for the sidewall problem of the control samples, the trench can be filled with ICPCVD SiO$_2$ without any void. Further processing would bring self-planarization after gap filling.

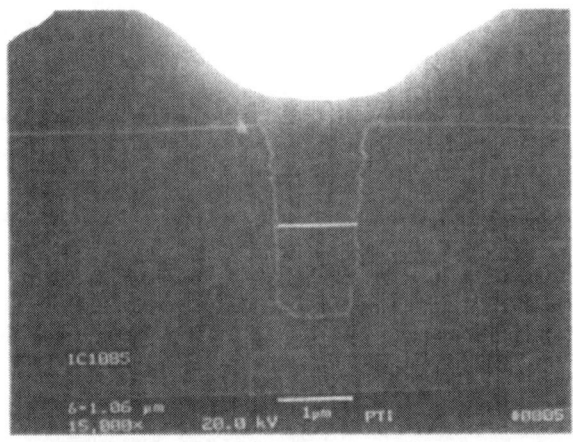

Figure 16. SEM micrograph showing a Si trench filled with ICPCVD SiO$_2$.

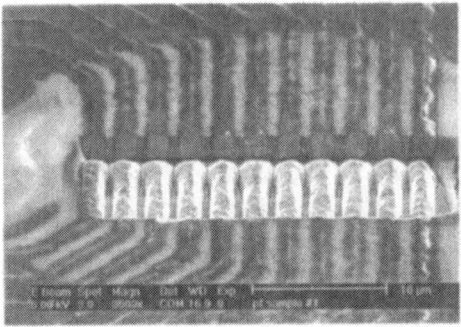

Figure 17. Gap-filled and self-planarized SiO$_2$ on metal coils by ICPCVD.

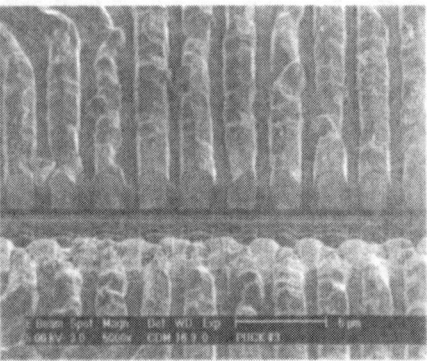

Figure 18. Partially gap-filled metal coils by ICPCVD SiO$_2$. No voids are evident. The SiO$_2$ film is planarized.

SUMMARY AND CONCLUSIONS

We investigated low temperature deposition (150 °C) of ICPCVD SiN$_x$ prepared from SiH$_4$/N$_2$/Ar and SiH$_4$/NH$_3$/He. For SiN$_x$ films prepared from SiH$_4$/N$_2$/Ar, rf chuck power increased refractive index and deposition uniformity while it decreased deposition rate, stress, and BOE rate. Increasing rf chuck power may bring more Si species than N species to the wafer surface leading to Si-rich films. Raising the chamber pressure resulted in higher deposition rate, refractive index, and film stress. The BOE rate of ICPCVD SiN$_x$, prepared from either SiH$_4$/N$_2$/Ar or SiH$_4$/NH$_3$/He, is much lower compared to PECVD SiN$_x$ deposited at the same low temperature. This implies that the ICPCVD SiN$_x$ films are denser and contain less hydrogen. An

ICPCVD SiN$_x$ process for adjustable film stress was demonstrated using SiH$_4$/NH$_3$/He.

ICPCVD SiO$_2$ films were prepared from SiH$_4$/O$_2$/Ar. Void-free gap filling of high aspect ratio trench was demonstrated with ICPCVD SiO$_2$.

In conclusion, high-density ICPCVD technology clearly has many advantages compared to PECVD for dielectric film deposition in a broad range of advanced processes for electronic device fabrication, especially at low temperatures.

ACKNOWLEDGMENTS

The authors appreciate Dr. J. Donohue, Mr. R. Westerman, and Mr. Wei Pen for fruitful discussions and assistance. We also thank Mr. R. McAfee and Mr. L. Heckerd for their technical support. The work at UF is partially supported by a DOD MURI, monitored by AFOSR (H. C. DeLong), contract no. F49620-96-1-0026.

REFERENCES

1. C. Constantine, C. Barratt, S. J. Pearton, F. Ren and J. R. Lothian, Appl. Phys. Lett. 61, 2899 (1992).

2. J. W. Lee, H. Cho, C. Hays, C. R. Abernathy, S. J. Pearton, R. J. Shul and G. A. Vawter and H. Han, IEEE J. of Selected Topics in Quantum Electronics, 4, 557 (1998).

3. F. Ren, J. W. Lee, C. R. Abernathy, S. J. Pearton, C. Constantine, C. Barratt and R. J. Shul, J. Vac. Sci. Technol. B 15, 983 (1997).

4. S. J. Pearton, J. W. Lee, E. S. Lambers, J. R. Mileham, C. R. Abernathy, W. S. Hobson and F. Ren, J. Vac. Sci. Technol. B 14, 118 (1996).

5. S. Thomas III, K. K. Ko and S. W. Pang, J. Vac. Sci. Technol. A 13, 894 (1994).

6. K. B. Jung, J. Hong, H. Cho, S. Onishi, D. Johnson, Y. D. Park, J. R. Childress, and S. J. Pearton, J. Vac. Sci. Technol. A 17, 535 (1999).

7. R. J. Shul, G. B. McClellan, R. D. Briggs, D. J. Rieger, S. J. Pearton, C. R. Abernathy, J. W. Lee, C. Constantine and C. Barratt, J. Vac. Sci. Technol. A 15, 633 (1997).

8. J. W. Lee, E. S. Lambers, C. R. Abernathy, S. J. Pearton, R. J. Shul, F. Ren, W. S. Hobson and C. Constantine, Solid-State Electron. 42 A65-A73 (1998).

9. R. J. Shul, C. G. Wilson, M. M. Bridges, J. Han, J. W. Lee, S. J. Pearton, C. R. Abernathy, J. D. MacKenzie, L. Zhang and L. F. Lester, J. Vac. Sci. Technol. A 16, 1621 (1998).

10. F. Ren, J. W. Lee, C. R. Abernathy, R. J. Shul, C. Constantine and C. Barratt, Semicond. Sci. Tech. 12, 1154 (1997).

11. G. Lucovsky, P. D. Richard, D. V. Tsu, S. Y. Lin and R. J. Markunas, J. Vac. Sci. Technol. A 4, 681 (1986).

12. C. B. Labelle, S. J. Limb and K. K. Gleason, J. Appl. Phys. 82, 1784 (1997).

13. S. J. Limb, C. B. Labelle, K. K. Gleason, D. J. Edell and E. F. Gleason Appl. Phys. Lett. **68**, 2810 (996).

14. M. Lapeyrade, M. P. Besland, C. Meva'a, A. Sibai and G. Hollinger, J. Vac. Sci. Technol. A **17**, 433 (1999).

15. Y. B. Hahn, J. W. Lee, K. D. Mackenzie, D. Johnson, S. J. Pearton and F. Ren, Solid-State Electron. **42**, 2017 (1998).

16. Y. B. Hahn, J. W. Lee, K. D. Mackenzie, D. Johnson, D. Hays, C. R. Abernathy and S. J. Pearton, Electrochemical and Solid-State Lett. **1**, 230 (1998).

17. J. W. Lee, K. D. Mackenzie, D. Johnson, R. J. Shul, S. J. Pearton, C. R. Abernathy and F. Ren, Solid-State Electron. **42**, 1031 (1998).

18. J. W. Lee, C. R. Abernathy, S. J. Pearton, F. Ren, R. J. Shul, C. Constantine and C. Barratt, Solid-State Electron. **41**, 829 (1997).

19. F. Ren, J. W. Lee, C. R. Abernathy; S. J. Pearton, R. J. Shul, C. Constantine and C. Barratt, Semicond. Sci. Technol. **12**, 1154 (1997).

20. J. W. Lee, K. D. Mackenzie, D. Johnson, R. J. Shul, S. J. Pearton, C. R. Abernathy and F. Ren, Solid-State Electron. **42**, 1021 (1998).

21. J. Kanicki, Mat. Res. Soc. Symp. Proc., **118**, 671 (1988).

SELECTIVE OXIDATION TO FORM DIELECTRIC APERTURES FOR LOW THRESHOLD VCSELS AND MICROCAVITY SPONTANEOUS LIGHT EMITTERS

D.G. DEPPE, D.L. HUFFAKER, L.A. GRAHAM, Z. ZOU, and S. CSUTAK
Microelectronics Research Center, Department of Electrical and Computer Engineering,
The University of Texas at Austin, Austin, Texas 78712, deppe@mail.utexas.edu

ABSTRACT

Selective oxidation of AlAs (or AlGaAs) can be used to form buried, low refractive index apertures within high Q Fabry-Perot microcavities. These apertures provide electrical and optical confinement, and for vertical-cavity surface-emitting lasers (VCSELs) have resulted in ultra-low threshold room temperature lasing with threshold currents under 25 μA. When used with quantum dot light emitters, the oxide-apertured microcavity can also be used to control the spontaneous lifetime. We describe the microcavity fabrication based on high Q Fabry-Perot microcavities and selective oxidation, and design and cavity Q constraints for apertured microcavities for quantum well and quantum dot VCSELs and microcavity LEDs. Threshold current densities of quantum well VCSELs are as low as 98 A/cm^2, while ground state lasing is also obtained for quantum dot VCSELs. Our initial experiments on microcavities with very small apertures and quantum dot emitters demonstrate up to a factor of 2.3 increase in the spontaneous emission rate.

INTRODUCTION

Oxide-confined vertical-cavity surface-emitting lasers (VCSELs) [1] have demonstrated impressive device performance when used with quantum well (QW) active regions. Multiple QWs allow high output coupling along with moderately heavy doping to reduce electrical resistance. High wall-plug efficiency can then be obtained for output powers of several milliwatts [2], [3], and the VCSEL performance appears close to ideal. In contrast, single QW active regions combined with higher Q cavities have allowed very low threshold currents to be obtained, but with the wall-plug efficiency reduced due to internal cavity losses that compete with the output coupling [4] - [6]. As the size of the aperture is reduced, additional losses due to the lack of electronic confinement in QWs also impact the device performance. However, there may be some applications for very low threshold VCSELs with maximum output powers in the sub-100 μW range, such as high density arrays used for free space interconnects, that benefit from low loss cavities that can also achieve high efficiency. In addition, quantum dots have been shown capable of laser action in edge-emitters in the 1.0 μm to 1.3 μm wavelength range [7], and QD VCSELs will likely require high Q cavities. One of the sources of optical losses that must be controlled in high Q VCSELs is due to absorption that accompanies the p and n type doping.

As we also show below, the oxide-confined Fabry-Perot microcavity is not only of interest for low power VCSELs, but may also be useful for high speed, high efficiency microcavity light emitting diodes (LEDs). Controlling the spontaneous emission of semiconductor light emitters has become a popular topic over the last ten years or so. Early experiments on planar microcavities have shown that while the angular spontaneous emission is easily changed using distributed Bragg reflectors (DBRs) or metal reflectors, modifying the spontaneous lifetime presents a bigger challenge [8] - [14]. For very small oxide apertures used for either VCSELs or microcavity LEDs, the confinement effects of the electromagnetic field that lead to modified spontaneous emission due to the Purcell effect [15] are unavoidable.

Below we describe the basic principles of mode confinement along with experimental results for high Q QW and QD oxide-apertured VCSELs and lower Q but small area microcavities. We show that the oxide-apertured microcavity can possess a mode volume that is sufficiently small with a Q that is sufficiently high to enter a regime in which the spontaneous lifetime can be controlled. This is most readily demonstrated with QD light emitters.

THEORY AND BACKGROUND

Active Region - Quantum Dots vs. Quantum Wells

One of the challenges in realizing very low threshold VCSELs and high efficiency, high speed microcavity LEDs based on aperture confinement is in obtaining small optical mode sizes while retaining high Q. The QD active regions are highly desirable if ground state operation can be achieved, because lateral electronic confinement becomes more critical as the aperture size is reduced. With QWs the carriers injected through the aperture spread laterally through diffusion. In our laboratory we have measured diffusion coefficients of ~10 cm^2/sec even at 77 K, and if we estimate a 3 nsec lifetime for reasonable pump levels this gives a minimum pumped diameter of ~3 to 4 μm independent of the aperture size. Although carrier thermalization from the QDs is a problem at room temperature, there is hope that placing large, closely spaced barriers around the QDs will eliminate this problem and lead to electron confinement within apertures of sub-micron dimensions.

The difficulty with QD active regions is that they have a reduced density of electronic states as compared to QWs. In the limit of an inhomogeneously broadened system with a Gaussian distribution of ground state energy levels, the QD density of states takes on the degeneracy of the two-dimensional harmonic oscillator, and can be expressed analytically as [16]

$$\rho_{red,QD} = 4\sqrt{\frac{\ln(2)}{\pi}} \frac{n_{QD}}{\hbar\Delta\omega} \sum_{m_{x,y}}^{\infty} (m_{x,y} + 1) \exp\{-\ln(2)\frac{[\omega - (\omega_g + m_{x,y}\omega_o)]^2}{(\Delta\omega/2)^2}\} \qquad (1)$$

where we use $m_{x,y}$ as an integer characterizing each energy level of a 2-dimensional harmonic oscillator, ω is the transition frequency, $\hbar\omega_o$ is the energy level separation where $\hbar$ is Planck's constant divided by 2π, $\hbar\omega_g$ is the transition energy for the ground state emission, $\Delta\omega$ is the full-width at half-maximum, and n_{QD} is the quantum dot density per unit area. The factor of $m_{x,y}+1$ in the summation of Eq. (1) is due to the two-dimensional nature of the confinement, assuming a larger QD diameter than height. For comparison, the first sub-band of the ideal QW reduced density of states is given by $\rho_{red,QW} = m_{red}/(\pi\hbar^2)$, where m_{red} is the reduced electron-hole mass. For $m_{red} \approx 0.058\ m_o$, the QW density of states is 2.4x10^{13} (eV·cm^2)$^{-1}$. If we consider typical values for a QD ensemble of $n_{QD} \approx 5$x10^{10} cm^{-2} and $\hbar\Delta\omega \approx 60$ meV, the reduced density of states for this QD ensemble is $\rho_{red,QD} \approx 1.6$x10^{12} (eV·cm^2)$^{-1}$.

The factor of >10 reduction in the density of states for the QD active region versus the QW active region has important implications for the optical loss that must be achieved in the QD VCSEL. First is that the QD VCSEL will necessarily require high reflectivity mirrors to achieve the Q sufficient for lasing, especially for the ground state. Although stacking multiple QD layers can increase the ground state gain, some of the attractive benefits of lateral electronic confinement are then lost. On the other hand, the factor of >10 reduction in the density of states yields a reduction in the transparency current (the current for which optical gain can be achieved). The transparency current of a high quality QW is ~50 A/cm^2, and this sets a lower limit on the lasing threshold current. The transparency current density of the typical QD ensemble considered above, with a spontaneous lifetime of ~1 nsec, is ~8 A/cm^2. For edge-emitting lasers, the very low QD ensemble transparency current density has already resulted in low temperature threshold current densities of ~12 A/cm^2 [7]. A 10 μm diameter QW VCSEL therefore has a minimum threshold current of ~40 μA, while the 10 μm diameter QD VCSEL has a minimum threshold of ~6 μA. Of course, the cavity Q required to reach the transparency current threshold is much higher than what is obtained with present day VCSELs. However, as we show below, the latest VCSEL results have approached within a factor of two (~100 A/cm^2) of the transparency value for QWs. Further advances require identifying and eliminating all non-essential optical losses from the cavity.

Mode Confinement

For microcavity LEDs the optical mode size must be small in order to control the spontaneous lifetime. The apertured microcavity is an "open" cavity in the sense that significant coupling occurs to waveguide modes with coupling rates that are nearly unchanged as compared to that without the aperture. However, with moderate Q's on the order of 1000 and mode diameters of ≤ 1 μm, the spontaneous lifetime in the apertured-microcavity can be increased for resonant narrow band emitters. We describe this as well more fully below, but emphasize that as the coupling to the aperture-confined optical mode increases so does the fraction of spontaneous emission that is coupled to this mode, and both the efficiency and the speed of the device are increases. In our own experiments we have measured a spontaneous lifetime increase of a factor of 2.3 [17], and another recent report on an etched-pillar semiconductor microcavity has reported a factor of 5 increase [18]. In order to obtain reasonable power levels, microarrays of microcavity LEDs will be required.

At room temperature, these electromagnetic effects due to the cavity confinement start to become noticeable with oxide aperture sizes of less than ~3 μm diameter, appearing first as a narrowing of the spectrally integrated, angular spontaneous radiation pattern when the emitter is resonant with the lowest order transverse cavity mode [19] - [21]. The angular narrowing results from the elimination of the higher order transverse modes in the spontaneous emission from the aperture. Once the aperture is small enough so that the spontaneous emission into the cavity normal is coupled to only one transverse aperture mode, further decrease in aperture size spreads the angular radiation pattern in the manner expected from the Gaussian mode approximation [21]. The aperture size for which this mode competition becomes observable also depends on the spectral linewidth of the spontaneous emission, so that for a QW active region the angular narrowing is observed at larger aperture sizes for lower temperatures.

The mechanism of the aperture confinement has been discussed by Hadley [22], who described it as a refractive index change due to the shift in the longitudinal resonances between the oxidized and unoxidized regions. For VCSELs, the oxide aperture reduces the diffraction loss that occurs when an otherwise planar microcavity optical mode is reduced below a certain value that depends on its Q and effective cavity length. Hadley's identification of the importance of the vertical resonance has prompted a further description of the mode confinement for which the minimum lateral size of the mode, and thus the mode volume, is approximately found from the vertical resonance shift [23].

Figure 1 illustrates the optical modes in the apertured-microcavity. The spontaneous emission can be separated into modes that are directed into the aperture and radiated with angles

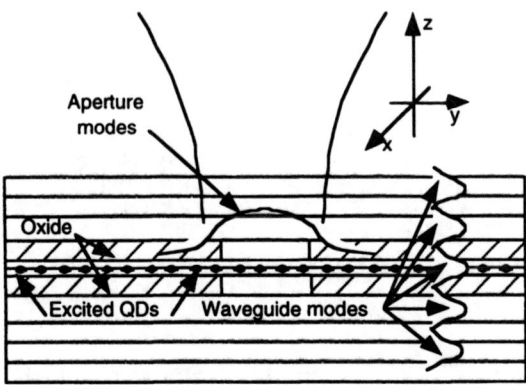

Fig. 1 Schematic illustration of an apertured-microcavity using semiconductor DBRs and two oxide apertures. Each high index layer of the microcavity forms waveguide modes.

close to the z-axis, and into modes that are confined to propagate along the high index layers of the cavity, including the DBRs [24]. Without the aperture nearly all of the spontaneous emission (approximately 80 to 90%) is constrained to radiate into waveguide modes, and the optical mode size is limited by the effective cavity length and mirror reflectivity [15]. In Fig. 1 QDs are considered as the active emitters. If optical pumping is used the QDs may be excited uniformly along the active region even between the oxide aperture. For electrical injection only the QDs within the aperture are excited, to within some lateral carrier diffusion length.

As described in [23], the aperture is effective in confining the optical mode to small areas through cut-off of some of the waveguide modes formed by the planar microcavity, and its effectiveness in generating a small mode size depends on both the vertical resonance shift and the cavity length. In Fig. 2 an ideal wavelength long cavity is considered, and the blue-shifted resonance of the oxide region cuts off all waveguide modes except the $m_z = 1$ and $m_z = 0$ modes. The spike below the $m_z = 2$ resonance (not labeled) is due to the frequency of the mode confined by the aperture. This frequency is set by both the vertical resonance in the unoxidized region and the lateral size of the optical mode. For an aperture with cylindrical symmetry, the transverse mode is given by Bessel functions. Since the mode has the same frequency in the oxidized region, the frequency is related to wavevector components in either region by

$$\frac{\omega_o}{c} = \sqrt{\frac{(4.810)^2}{\varepsilon w_o^2} + k_z^2} = \sqrt{k_{\rho,ox}^2 + k_{z,ox}^2} \qquad (2)$$

where ω_o is the mode frequency, c is the speed of light, ε is an average refractive index in the unoxidized region, w_o is the lowest order transverse mode size, k_z is the vertical wavevector component in the unoxidized region, $k_{\rho,ox}$ is the transverse wavevector component in the oxidized (aperture) region, and $k_{z,ox}$ is the vertical wavevector in the oxidized (aperture) region. The factor

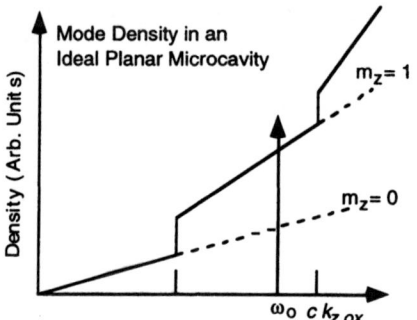

Fig. 2 Mode density plot for a wavelength long ideal cavity, in which the aperture confined mode has a frequency just below the $m_z = 2$ step.

of 4.810 in Eq. (2) comes from the Bessel function assumption. The plane wave modes associated with the lower order resonances of $m_z = 1$ and 0 are not cut off from aperture confined mode, and can give diffraction loss depending on how they interact with the confined mode [4].

As the cavity length increases, the amount of resonance shift for a fixed aperture index and thickness decreases. In [23] the optical mode size is estimated from the vertical resonances to be

$$w_{o,min} = \frac{4.810}{\sqrt{\varepsilon}\sqrt{k_{z,ox}^2 - k_z^2}} \qquad (3)$$

so that a large resonance shift is needed in order to obtain a small optical mode size. As the optical mode size is reduced there is an inevitable increase in the optical loss of the confined mode. In some cases this may partly be caused by scattering due to imperfect interfaces. However, even with perfect interfaces diffraction loss will increase for reasons similar to that of the problem in the purely planar microcavity [25], [26]. As the size of the optical mode is reduced a faster "walk-off" of the mode occurs due to propagation of the field back and forth in the cavity. In other words, from Fig. 1, some coupling always occurs to the modes that are confined as waveguide modes.

EXPERIMENTAL RESULTS

QW VCSELs With Low Loss Cavities

The QW VCSELs we study are grown by molecular beam epitaxy. In an attempt to reduce the optical loss to a minimum we have used intracavity contacts for both the p and n-side electrical connections, with an undoped lower AlAs/GaAs DBR and an upper post-growth deposited dielectric DBR. The VCSEL structure is shown in Fig. 3. The lower mirror consists of 35 AlAs/GaAs pairs, and only the two mirror pairs adjacent the half-wave cavity are Si doped n-type at a level of 5×10^{17} cm^{-3}. For the upper DBR a single p-type layer of $Al_{0.2}Ga_{0.8}As$ is used along with 5 to 7 pairs of MgF/ZnSe quarter-wave layers. The half-wave cavity consists of n-type $Al_{0.9}Ga_{0.1}As$ in the lower quarter-wave layer and p-type $Al_{0.6}Ga_{0.4}As$/AlAs in upper quarter-wave layer, as shown in Fig. 3. At the center of the cavity is a single 60 Å $In_{0.2}Ga_{0.8}As$ QW.

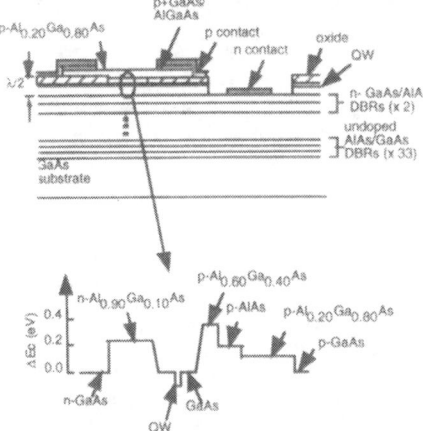

Fig. 3 Schematic illustration of the QW VCSEL with intracavity contacts for elimination of much of the doping from the cavity mirrors. A conduction band structure is also illustrated, ignoring band bending due to doping. The upper MgF/ZnSe DBRs are not shown.

The QW VCSELs are fabricated with oxide-confining apertures [1] using the wet oxidation process discovered by Holonyak and coworkers [27], [28]. The oxide-apertures are rectangular with sizes ranging from 11 μm x 12 μm down to 2 μm x 2 μm. Moderate sized QW VCSELs with 5 μm x 5 μm apertures have threshold currents of ~70 μA (J_{th} = 280 A/cm^2) with differential slope efficiencies of ~60% with a 5 pair upper MgF/ZnSe DBR. When the upper DBR reflectivity is increased by going to 7 pairs the threshold reduces to ~40 μA (J_{th} = 160 A/cm^2), but the differential slope efficiency also decreases to ~20%. This shows that with 5 pairs the optical loss is mainly due to transmission loss through the upper MgF/ZnSe DBR, but with 7 pairs internal losses become significant. Current spreading in the QW active region increases the threshold current density when normalized by the aperture, but we note that the 5 μm x 5 μm sized device still shows very low threshold current density. Continuous-wave light vs. current curves for the largest and smallest device sizes are shown in Figs. 4 and 5. For 11 μm x 12 μm aperture and 7 pairs the threshold current density is only 98 A/cm^2, or about a factor of 2 times the expected transparency value. The 2 μm x 2 μm sized aperture device has a threshold current of only 23 μA as shown in Fig. 5. The threshold current density is J_{th} = 575 A/cm^2, but this assumes that there is no carrier spreading outside the oxide aperture. Current spreading can readily account for a factor of 2 to 4 increase over the current density set by the area of the smaller aperture.

To our knowledge, these threshold current densities are the lowest yet achieved for these device sizes, and show that the VCSEL cavity Q is not too far from the value for which transparency values can be obtained. Therefore, QD active regions become interesting for the VCSEL.

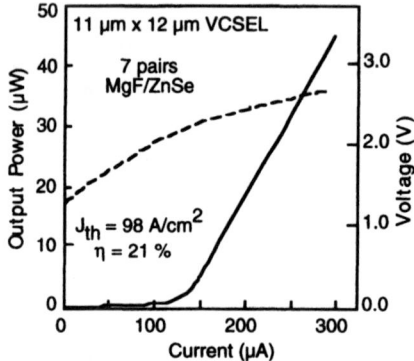

Fig. 4 Room temperature, continuous wave characteristics of the 11 µm x 12 µm VCSEL.

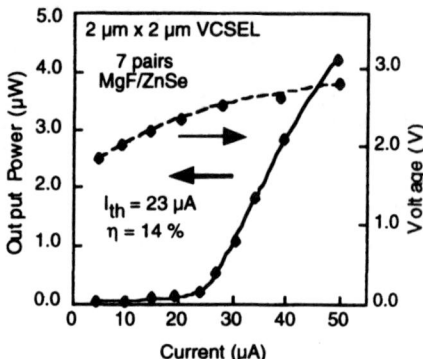

Fig. 5 Room temperature, continuous wave characteristics of the 2 µm x 2 µm VCSEL.

QD VCSELs With Low Loss Cavities

An advantage for VCSELs is that the InGaAs/GaAs QDs are compatible with AlAs/GaAs DBRs and selective oxidation. Since the optical gain from a single layer QD ensemble is limited as compared to that of a QW, in part due to an inhomogeneously broadened gain spectrum, ground state lasing operation has been troublesome. Narrow spectrum QD ensembles are of interest for VCSELs. But since the gain spectrum must also overlap the VCSEL's cavity resonance and the gain peak shifts with temperature at ~2.5 Å/K while the cavity resonance shifts with ~0.5 Å/K, a certain minimum gain bandwidth is required to obtain lasing over a useful temperature range.

Using Eq. (1) and the usual requirement of the round trip gain balancing the mirror loss, a threshold condition for the ground state is given by [16]

$$\ln(\frac{1}{\sqrt{R_T R_B}}) = \sqrt{\frac{\ln(2)}{\pi}} \frac{8\pi^2 c^2}{n^2 \omega_g^2 \tau_{sp}} \frac{n_{QD}}{\Delta\omega}, \tag{4}$$

where c/n is the speed of light in the cavity, R_T is the upper mirror reflectivity, and R_B is the bottom mirror reflectivity. For the QD layers studied below $n_{QD} \approx 5\times10^{10}$ cm^{-2}, $\omega_g = 1.76\times10^{15}$ rad/sec, $\hbar\Delta\omega \approx 76$ meV, and τ_{sp} is the spontaneous lifetime. We have measured τ_{sp} under non-resonant excitation at low temperature to be ~800 psec. For a three-stack QD active layer, and assuming $\tau_{sp} = 800$ psec and $c/n = 9\times10^7$ m/sec, we find that $\sqrt{R_T R_B} \geq 0.9983$ must be achieved for lasing threshold. However, some unknowns exist as to the actual sizes of parameters in Eq. (4), especially for τ_{sp}. Consistent with our own measurements, Wang et al. also reported a spontaneous lifetime of ~880 psec for nonresonantly pumped In$_{0.5}$Ga$_{0.5}$As/GaAs QDs [29]. However, a spontaneous lifetime of ~380 psec has been reported for resonantly pumped InP/In$_{0.48}$Ga$_{0.52}$P QDs [30]. Although the different measurement results may be due to different materials, the spontaneous lifetime may also be shorter due to resonant excitation [30]. For a spontaneous lifetime of 400 psec the mirror reflectivity product required for threshold reduces to a more easily achieved 0.9966.

The QD VCSEL structure is very similar to that of the QW VCSEL shown in Fig. 3, but with the QD active region centered within an undoped full-wave cavity spacer formed of average Al$_{0.13}$Ga$_{0.87}$As composition formed from short period superlattice. The bottom distributed

Bragg reflector (DBR) again consists of 35 AlAs/GaAs pairs, while a single upper AlGaAs/GaAs DBR pair is formed from a quarter-wave of AlGaAs (with graded Al composition), $Al_{0.97}Ga_{0.03}As$, and a quarter-wave of GaAs. The first two pairs of the lower AlAs/GaAs DBR are doped n-type, and the upper DBR layers are doped p-type. Heavily doped AlGaAs/GaAs contact layers are grown on top of the p-type quarter-wave layers, and selectively etched from the VCSEL cavity region. The QDs are formed from five monolayers of alternating InAs and GaAs. Each of the InGaAs QD layers are separated by 150 Å of GaAs with 300 Å of GaAs grown on either side of the three-stack QD active region. Selective oxidation is again used to form aperture sizes ranging from 10 μm to 2 μm in diameter. Top DBR mirrors are completed with a six pair MgF/ZnSe DBR. Broad-area edge-emitters are also fabricated with 60 μm wide stripes to study the QD spontaneous emission at low excitation level. Figure 6 shows the spontaneous electroluminescence spectrum from edge-emission at a low current of 5 A/cm². Two VCSEL resonances, one at 1.027 μm and a smaller resonance at 1.041 μm, are also collected from the edge and can be seen in the edge-emission. The QD ground-state emission is centered at 1.07 μm and has a spectral width of 76 meV due to inhomogenous broadening.

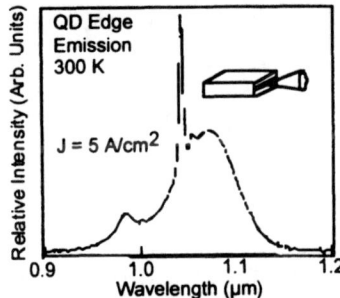

Fig. 6 Edge emission from the QD VCSEL wafer showing ground state emission at 1.07 μm wavelength. Vertical resonances can also be seen at 1.041 μm and 1.027 μm.

Figure 7 shows the QD lasing characteristics for a 10 μm diameter aperture for increasing temperatures. An inset shows the lasing spectrum above threshold consisting of three transverse modes. An uncalibrated Ge detector is used to the measure the output intensity. As the heat sink temperature is increased the lasing threshold increases from 703 μA (895 A/cm²) at 300 K, to 833 μA at 310 K, to 1037 μA at 320 K, and 1080 μA at 327K (our maximum possible measurement temperature). The peak lasing wavelength just above threshold changes from 10672 Å at 300 K, to 10679 Å at 310 K, to 10685 Å at 320 K, to 10687 Å at 327 K, giving the temperature dependence of the cavity resonance of ~0.5 Å/K typical for VCSELs. The threshold voltage is ~2.6 V, and for a thermal resistance of ~2.5 K/mW (measured on similar VCSEL designs) the threshold power at 300 K gives a temperature rise of only ~5 K. Therefore, each of the three transverse modes are within the 76 meV ground state spectral width of the low level QD spontaneous emission of Fig. 6. Lasing is also obtained in smaller apertures. However, the threshold does not scale consistently with aperture size due to some variation in cavity Q, discussed further below. Apertures with 6 μm diameter from a different processed piece of the same wafer yields ~1 mA threshold currents, but with the similar lasing wavelength to that in Fig. 7.

Figure 8 shows the lasing characteristics from a much smaller 2 μm x 2 μm square aperture. The lasing wavelength of 10705 Å is slightly longer than that for the 10 μm diameter aperture. The threshold current is 268 μA and the threshold voltage is again 2.6 V, so that heating is low enough that lasing still occurs at the QD ground state. The threshold current density of the smaller aperture is quite high at ~6.7 kA/cm². Although the high threshold current may in part be due to higher optical loss associated with the small aperture size, we speculate that it also has a significant contribution from electronic (nonradiative) losses.

We have calculated mirror reflectivities based on an idealized cavity structure that assumes only planar mirror transmission losses, and have compared the calculated linewidth to that measured for the 10 μm and 2 μm aperture sizes. The calculated mirror reflectivity of the lower 35 pair AlAs/GaAs DBR is $R_B = 0.99997$, and that of the top single AlGaAs/GaAs pair combined with the 6 pair MgF/ZnSe DBR is $R_T = 0.99958$. The calculated mirror reflectivity product is then

$\sqrt{R_T R_B} = 0.99978$, well above that needed from the estimate of the maximum QD gain assuming $\tau_{sp} = 800$ psec. However, the calculated spectral full-width at half-maximum is 0.2 Å, and gives an extremely high quality factor for the cavity of $Q = 50,000$.

The lowest order transverse mode for each aperture has two closely spaced orthogonal polarizations separated by ~3 Å, but the longer wavelength polarization is more intense for each aperture size. We estimate the linewidth of the single modes at low current density to be ~2.9 Å for the 10 µm and ~2.2 Å for the 2 µm aperture sizes. However, we cannot rule out stimulated emission narrowing in the 2 µm aperture, as the lowest current density for which the linewidth is accurately measured is 250 A/cm². The resulting Q's are ~3660 for the 10 µm and ~4800 for the 2 µm apertures, and we believe these Q's are limited in part by the upper MgF/ZnSe DBRs. This

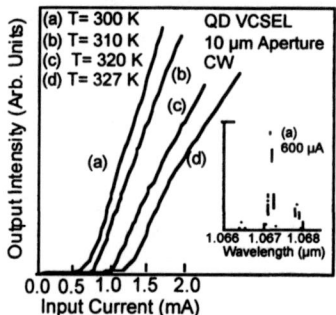

Fig. 7 Continuous wave light vs. current for a 10 µm diameter QD VCSEL with laser operation on the ground state.

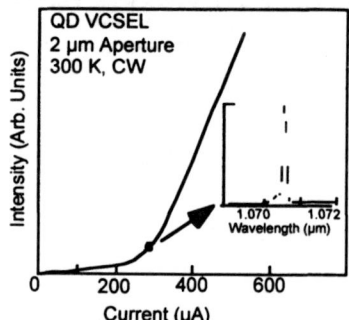

Fig. 8 Continuous wave light vs. current for a 2 µm diameter QD VCSEL, again with laser operation on the ground state.

may explain the higher Q value for the 2 µm aperture as due to variation in the upper DBR quality. If we approximate that the experimental effective cavity lengths are the same as that of the idealized cavity, the experimental reflectivity products can be estimated to be $\sqrt{R_T R_B} \approx 0.9970$ for the 10 µm and 0.9977 for the 2 µm aperture sizes. If these mirror reflectivity products are correct, from Eq. (1) we can estimate that the spontaneous lifetime must be in the range of $\tau_{sp} \approx 370$ to 550 psec for threshold. These results show the potential for low threshold VCSELs based on QD active regions, and for obtaining ground state lasing operation, but also emphasize the importance of knowing the true spontaneous lifetime of QDs. With further improvements in cavity Q and QD material quality, the QD VCSEL may represent an attractive alternative to QW VCSELs, especially for wavelengths beyond 1.0 µm.

Controlled Spontaneous Lifetime in QD Microcavities with Small Apertures

For spontaneous emission, the QD ground state exciton can be modeled as a 2-level emitter that radiates into a collection of modes, given by

$$\frac{dN_2}{dt} = -\frac{2q^2\omega_d^2}{\hbar^2}\sum_m |\mathbf{d}\cdot\mathbf{A}_m(\mathbf{r}_d)|^2 \frac{(\gamma_d + \frac{\omega_m}{2Q_m})}{(\omega_d - \omega_m)^2 + (\gamma_d + \frac{\omega_m}{2Q_m})^2} N_2(t). \tag{5}$$

In Eq. (1), N_2 is the upper level population, m labels each cavity mode, q is the electronic charge, ω_d is the resonant frequency of the 2-level emitter, ω_m is the resonant frequency of mode m, $\mathbf{d}$ is the dipole vector strength, $\mathbf{A}_m(\mathbf{r}_d)$ is the normalized vector strength of the cavity field at the QD position, $\dfrac{\omega_m}{Q_m}$ is the photon loss rate from mode m, and γ_d is the QD dephasing rate. The vector strength of the cavity field is normalized such that $\int_V d^3 r \varepsilon(\mathbf{r})|\mathbf{A}_m(\mathbf{r})|^2 = \hbar/(2\omega_m)$, where $\varepsilon(\mathbf{r})$ is the material permittivity at $\mathbf{r}$, and V is the normalized mode volume.

With optical pumping the QDs distributed as in Fig. 1 are excited uniformly in the plane between the two apertures. However, light collection is such that only QDs that couple to the microcavity apertured-modes are measured. The spontaneous photon number in these lower-order modes satisfy the rate equation (ignoring stimulated emission)

$$\frac{dn_o}{dt} = -\frac{\omega_o}{Q}n_o + \sum_{n=1}^{N_{QD}} \frac{2q^2\omega_n^2}{\hbar^2}\left|\mathbf{d}_n \cdot \mathbf{A}_o(\mathbf{r}_n)\right|^2 \frac{(\gamma_n + \frac{\omega_o}{2Q})N_{2,n}(t)}{(\omega_n - \omega_o)^2 + (\gamma_n + \frac{\omega_o}{2Q})^2} \qquad (6)$$

where the subscript n labels each emitter. We approximate the apertured-modes as Gaussian, and the QD emitters as having randomly oriented dipole moments. The coupling strength for these modes [Eqs. (5) or (6)] then becomes $\left|\mathbf{d}_n \cdot \mathbf{A}_o(\mathbf{r}_n)\right|^2 = \dfrac{|d|^2 \hbar e^{-(x_n^2 + y_n^2)/w_o^2}}{3\varepsilon_o \pi w_o^2 L_z}$, with a mode volume given by $V = \pi w_o^2 L_z$ where w_o is the mode radius and L_z is an effective cavity length. Reference [24] estimates that 80 to 90% of the spontaneous emission in a planar DBR cavity is coupled to waveguide modes (see Fig. 1), and we assume that this emission is independent of the aperture. In terms of the dipole moment, the spontaneous emission rate of the QD embedded in bulk material is $A_{sp}^B = \dfrac{q^2 \omega_o^3 n^3 |d|^2}{3\pi\varepsilon\hbar c^3}$, where n is the refractive index and c is the speed of light in vacuum. For frequencies close to that of the lowest order apertured-mode, the position dependent spontaneous emission rate is given from Eq. (5) by

$$A_{sp}(x_n, y_n, \omega_n) \approx A_{sp}^B\left[\frac{A_{WG}}{A_{sp}^B} + \frac{4c^3/n^3}{\omega_o \omega_n w_o^2 L_z} \frac{(\gamma_n + \frac{\omega_o}{2Q})e^{-(x_n^2 + y_n^2)/w_o^2}}{(\omega_n - \omega_o)^2 + (\gamma_n + \frac{\omega_o}{2Q})^2}\right], \qquad (7)$$

where A_{WG} is the spontaneous emission rate into all waveguide modes. On resonance and with $\gamma_n \ll \dfrac{\omega_o}{2Q}$ and $(x_n, y_n) = (0,0)$, the second term in brackets becomes $\dfrac{(\lambda_o^3/n^3)Q}{\pi^2(\pi w_o^2 L_z)}$ which is the enhancement described by Purcell [15].

Aside from constant factors, the cavity field intensity decay measured at frequency ω_n after a short-pulse excitation is given by

$$\frac{\omega_o}{Q}n_o(\omega_n, t) \propto \int_0^\infty d\rho_n \rho_n \frac{(\gamma_n + \frac{\omega_o}{2Q})e^{-\rho_n^2/w_o^2}e^{-A_{sp}(\rho_n, \omega_n)(t-t_o)}}{(\omega_n - \omega_o)^2 + (\gamma_n + \frac{\omega_o}{2Q})^2}. \qquad (8)$$

We have characterized the QD spontaneous emission rate into small oxide-apertured microcavities. The QD heterostructure consists of an 18 pair GaAs/AlAs DBR, an AlGaAs $\lambda/2$ cavity spacer, and a single upper $\lambda/4$ GaAs layer. The cavity spacer is $Al_{0.97}Ga_{0.03}As$ at the center of which is grown a single $In_{0.50}Ga_{0.33}Al_{0.17}As$ QD active region with 100 Å GaAs layers and 135 Å grading layers immediately adjacent on either side. The apertured-microcavity is fabricated by pattering 5 µm squares in photoresist and reactive ion etching to a depth of 2900 Å to form mesas exposing both $Al_{0.97}Ga_{0.03}As$ layers of the cavity spacer. Lateral oxidation is performed at 450 °C, after which a 5 pair $ZnSe/MgF_2$ DBR is deposited to complete the microcavity. An inset of the cavity structure is shown in Fig. 9, and the cavity spacer is similar to that illustrated in Fig. 1. At the 10 K measurement temperature, the QD ground state emission wavelength is ~9700 Å with a spectral width of ~600 Å. For time resolved measurements a mode-locked Ti-Sapphire laser beam (pulse rate reduced to 5 MHz) is focused with a microscope objective to a 5 μm diameter spot on the microcavity. Photoluminescence is collected through the same objective and time resolved using a grating spectrometer and photon counting module with rise and fall times of ~300 ps. Although InGaAs/GaAs QDs grown in our laboratory show short lifetimes of ~800 ps, the InGaAlAs/GaAs QDs have longer lifetimes of ~2 ns and are used so as not to exceed the system resolution. The data presented is for an aperture of approximately 1 μm diameter.

Figure 9 shows photoluminescence decay for wavelengths around the cavity resonance for the 1 μm diameter apertured-microcavity. Curves (a) and (c) show off-resonance decays taken at 9800 and 9950 Å, while curve (b) shows the on-resonance decay for the lowest order mode at 9860 Å. There is a factor of ~2.3 increase in the emission rate at 9860 Å as compared to off-resonance wavelengths. The fact that wavelengths both shorter and longer than resonance show similar slower decays clearly indicates that the increased rate at resonance is due to the microcavity. Considering the first 2 ns of decay, the on-resonance lifetime [curve (b) in Fig. 9] is 0.9 ns, compared with the off-resonance lifetimes of 2.2 ns at 9950 Å and 1.9 ns at 9800 Å. The off-resonance lifetimes are close to the 2.1 ns lifetime measured for the epitaxial sample before processing. Therefore, the spontaneous emission rate is increased for the spatially averaged emitter positions by a factor of ~2.3, with little inhibition off-resonance.

Figure 10 shows spontaneous spectra and decay rates plotted versus wavelength for the 1 μm apertured-microcavity. The inset shows the emission over a greater wavelength range. The longer wavelength spectral peaks at 9860 Å and 9730 Å correspond to apertured-microcavity modes. The larger emission peaks starting at 9687 Å and visible to 9600 Å are due to emission from the oxide

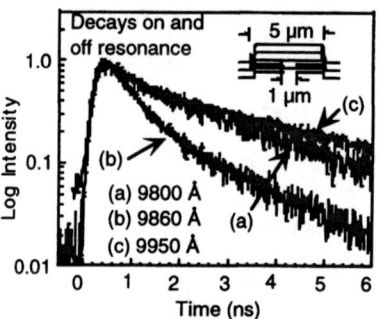

Fig. 9 Spontaneous decay curves for a 1 μm apertured microcavity, both on and off the lowest order resonance.

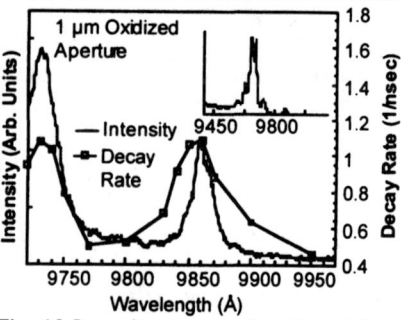

Fig. 10 Intensity vs. wavelength and decay rate vs. wavelength for the 1 μm apertured microcavity. The inset shows the intensity vs. wavelength over a wide wavelength range.

region (blue-shifted as compared to apertured modes) within the 5 μm mesa. Figure 10 shows that the spontaneous decay rate is enhanced at the lowest and next lowest order apertured-mode wavelengths of 9860 Å and 9730 Å. The spectral separation between the lowest and next lowest order modes depends on the mode area. The aperture-mode frequencies are approximately set by $\omega_m = (c/n)\sqrt{(\zeta_m/w)^2 + k_z^2}$, where the wavevector component k_z is fixed by the mirrors and w is the optical mode diameter [23]. Bessel function modes satisfy the cylindrical symmetry of the cavity, so that we take $\zeta_o = 4.810$ and $\zeta_1 = 7.664$ for Bessel functions of the first kind. From Fig. 10, $\omega_o = 1.912 \times 10^{15}$ rad/s and $\omega_1 = 1.937 \times 10^{15}$ rad/s, and the mode diameter is estimated to be $w = (c/n)\sqrt{(\zeta_1^2 - \zeta_0^2)/(\omega_1^2 - \omega_0^2)} \approx 1.8$ μm. The linewidth of the lowest order mode under continuous wave excitation is $\Delta\lambda = 15$ Å which gives a $Q = \lambda_o/\Delta\lambda = \omega_o/(2\gamma_o) \approx 650$.

Measurements to date suggest that electronic dephasing rates in the QDs can be $<10^{11}$ s^{-1}, so that we assume $\gamma_d < \dfrac{\omega_o}{Q}$ is satisfied.

Calculated decay curves using Eq. (8) are compared with the experimental data for the 1 μm apertured-microcavity. We find that values of $\dfrac{A_{WG}}{A_{sp}^{B}} \approx 1$, $A_{sp}^{B} \approx 4 \times 10^{8}$ s^{-1}, and $\dfrac{\lambda_o^3 Q}{\pi^3 n^3 w_o^2 L_z} \approx 3.1$ provide good matches between the calculated spontaneous emission rates and the experimental data in Fig. 9. Given $\lambda_o = 0.986$ μm, $Q = 650$, and an assumed value of n= 2.95 for the refractive index of the cavity spacer, the calculated curves use an effective cavity length of $L_z = 0.75$ μm and $w_o = 0.7$ μm. This Gaussian mode diameter of $2w_o = 1.4$ μm is in rough agreement with the 1.8 μm diameter estimated from the spectral separation of the transverse modes assuming Bessel functions. From the calculations, the spontaneous emission rate enhancement for emitters placed at the center of the optical mode is a factor of 3.1 compared to the factor of 2.3 for the emission rate averaged over emitter positions. This enhanced spontaneous emission rate at the mode center leads to a slight spatial hole burning of the emitters at the mode center.

Further advances for this type of microcavity may lead to practical spontaneous microcavity light emitters that provide high speed and high efficiency at low power. By combining these cavity effects with longer wavelength QDs that have deeper confinement potentials, the control of the spontaneous lifetime may be extended to room temperature.

SUMMARY

We have presented experimental results on very low threshold oxide-confined QW and QD VCSELs. For VCSELs, the cavity Q is increased by removing doping everywhere from the cavity except in thin layers directly adjacent to the active region. The result of the lower optical loss is very low threshold current and current density for the QW VCSEL, and clear ground state operation for the QD VCSEL. We have also discussed the gain limitations in QD VCSELs, and emphasize that for very high cavity Q the QDs can represent an interesting alternative to QWs due to their lower transparency current and longer wavelength operation. For very small apertures electronic confinement becomes a limitation of device performance, and can be improved with QDs. However, very small apertures (≤ 1 μm diameter) combined with QD light emitters enter the regime for which the Purcell effect controls the spontaneous lifetime. This regime of the aperturedmicrocavity may provide a route to high speed, high efficiency spontaneous light emitters.

ACKNOWLEDGMENTS

Various aspects of this work have been supported in part by the DARPA Ultraphotonics Program, the DARPA Sponsored Univ. of New Mexico OptoCenter, the ARO MURI Program, and the Texas Advanced Technology Program.

REFERENCES

1. D.L. Huffaker, D.G. Deppe, K. Kumar, and T.J. Rogers, Appl. Phys. Lett. **65**, 97 (1994).
2. K.L. Lear, K.D. Choquette, R.P. Schneider, Jr., S.P. Kilcoyne, and K.M. Geib, Electron. Lett. **31**, 208 (1995).
3. R. Jager, M. Grabherr, C. Jung, R. Michalzik, G. Reiner, B. Wiegl, and K. Ebeling, Electron. Lett. **33**, 330 (1997).
4. D.G. Deppe, D.L. Huffaker, Q.-H. Oh, H. Deng, and Q. Deng, IEEE J. Sel. Top. Quant. Electron. **3**, 893 (1997).
5. G.M. Yang, M.H. MacDougal, and P.D. Dapkus, Electron. Lett. **31**, 560 (1995).
6. D.L. Huffaker and D.G. Deppe, Appl. Phys. Lett. **71**, 1449 (1997).
7. D.L. Huffaker, G. Park, Z. Zou, O.B. Shchekin, and D.G. Deppe, Appl. Phys. Lett. **73**, 2564 (1998).

8. K.H. Drexhage in *Progress in Optics*, edited by E. Wolf (North-Holland, Amsterdam, 1974), Vol. XII, Chap. IV.
9. D.G. Deppe, J.C. Campbell, R. Kuchibhotla, T.J. Rogers, and B.G. Streetman, Electron. Lett. **26**, 1665 (1990).
10. T.J. Rogers, D.G. Deppe, and B.G. Streetman, Appl. Phys. Lett. **57**, 1858 (1990).
11. H. Yokoyama, K. Nishi, T. Anan, H. Yamada, S.D. Brorson, and E. Ippen, Appl. Phys. Lett. **57**, 2814 (1990).
12. N. Ochi, T. Shiotani, M. Yamanishi, Y. Honda, and I. Suemune, Appl. Phys. Lett. **58**, 2735 (1991).
13. T. Yamauchi, Y. Arakawa, and M. Nishioka, Appl. Phys. Lett. **59**, 2339 (1991).
14. D.L. Huffaker, C. Lei, D.G. Deppe, C.J. Pinzone, J.G. Neff, and R.D. Dupuis, Appl. Phys. Lett. **60**, 3203 (1992).
15. E.M. Purcell, Phys. Rev. **69**, 681 (1946).
16. Z. Zou, O.B. Shchekin, G. Park, D.L. Huffaker, and D.G. Deppe, IEEE Phot. Tech. Lett. **10**, 1673 (1998).
17. L.A. Graham, D.L. Huffaker, and D.G. Deppe, Appl. Phys. Lett. **74**, (26 April, 199).
18. J.M. Gerard, B. Sermage, B. Gayral, B. Legrand, E. Costard, and V. Thierry-Mieg, Phys. Rev. Lett. **81**, 1110 (1998).
19. D.G. Deppe, D.L. Huffaker, J. Shin, and Q. Deng, IEEE Phot. Tech. Lett. **7**, 965 (1995).
20. D.L. Huffaker and D.G. Deppe, Appl. Phys. Lett. **67**, 2594 (1995).
21. Q. Deng and D.G. Deppe, Opt. Express **2**, 157 (1998).
22. G.R. Hadley, Opt. Lett. **20**, 1483 (1995).
23. D.G. Deppe, T.-H. Oh, and D.L. Huffaker, IEEE Phot. Tech. Lett. **9**, 713 (1997).
24. C.C. Lin, D.G. Deppe, and C. Lei, IEEE J. Quant. Electron. **30**, 2304 (1994).
25. K. Ujihara, Jpn. J. Appl. Phys. **30**, L901 (1991).
26. Q. Deng and D.G. Deppe, Phys. Rev. A **53**, 1036 (1996).
27. J.M. Dallesasse, N. Holonyak, Jr., A.R. Sugg, T.A. Richard, and N. El-Zein, Appl. Phys. Lett. **57**, 2844 (1990).
28. S.A. Maranowski, A.R. Sugg, E.I. Chen, and N. Holonyak, Jr., Appl. Phys. Lett. **63**, 1660 (1993).
29. G. Wang, S. Fafard, D. Leonard, J.E. Bowers, J.L. Merz, and P.M. Petroff, Appl. Phys. Lett. **64**, 2815 (1994).
30. A. Kurtenbach, W.W. Ruhle, and K. Eberl, Solid State Comm. **96**, 265 (1995).

Part III

Oxides—Structural, Transport and Optical Properties

PASSIVATION OF INTERFACES IN HIGH-EFFICIENCY PHOTOVOLTAIC DEVICES

Sarah R. KURTZ, J. M. OLSON, D. J. FRIEDMAN, J. F. GEISZ, K. A. BERTNESS*, and A. E. KIBBLER
National Renewable Energy Laboratory (NREL), 1617 Cole Blvd., Golden, CO 80401, Sarah_Kurtz@nrel.gov
*NIST, 325 Broadway, Boulder, CO, 80303

ABSTRACT

Solar cells made from III-V materials have achieved efficiencies greater than 30%. Effectively ideal passivation plays an important role in achieving these high efficiencies. Standard modeling techniques are applied to $Ga_{0.5}In_{0.5}P$ solar cells to show the effects of passivation. Accurate knowledge of the absorption coefficient is essential (see appendix). Although ultralow (<2 cm/s) interface recombination velocities have been reported, in practice, it is difficult to achieve such low recombination velocities in solar cells because the doping levels are high and because of accidental incorporation of impurities and dopant diffusion. Examples are given of how dopant diffusion can both help and hinder interface passivation, and of how incorporation of oxygen or hydrogen can cause problems.

INTRODUCTION

III-V solar cells can achieve very high efficiencies (80–90% of theoretical efficiency) because the material quality and interface passivation are almost ideal. The $Ga_{0.5}In_{0.5}P$/GaAs two-junction solar cell, invented and developed at NREL, has achieved high efficiencies of around 30% [1-3] and is in large-scale production [4,5]. We have shown that very low (as low as 1.5 cm/s [6]) interface recombination velocities (IRV) can be achieved for the $Ga_{0.5}In_{0.5}P$/GaAs interface. However, in actual device structures, much higher IRVs are often observed. A detailed understanding of the structures and growth of solar cells is necessary to obtain effectively ideal interfaces (i.e., ones with IRVs less than 10,000 cm/s.) In this paper, we calculate the effects of passivation of the front and back of $Ga_{0.5}In_{0.5}P$ (hereafter, GaInP) solar cells on both the photocurrent and photovoltage, and give examples of how and why ideal passivation is not always achieved.

THEORETICAL APPROACH

We present here a brief description of the importance of interface passivation to photovoltaic device operation, using the GaInP cell as an example. Thorough treatments of photovoltaic devices can be found elsewhere [7,8]. A schematic for a solar cell is shown in Fig. 1. For convenience, we discuss an n-on-p structure, but this discussion also applies to a p-on-n structure with appropriate redefinition of terms. High-efficiency III-V solar cells typically use a thin (~0.1 μm) n layer, commonly referred to as the emitter of the device (Fig. 1). The solar spectrum, striking the front of the cell, includes ultraviolet, visible, and infrared light. The absorption coefficient for short-wavelength light (referred to hereafter as "blue" light) is quite large, and most of the blue light is absorbed very close to the front of the cell—generating photocarriers in the emitter layer. Light with energy close to, but above the band edge (referred to hereafter as "red" light), is weakly absorbed (generates photocarriers) throughout the cell. Sub-band-gap light is not absorbed and passes through the cell. The photocarriers generated by the super-band-gap light diffuse inside the cell until they are either collected at the p-n junction or recombine with a majority carrier, either by bulk or interface recombination. The efficiency of the solar cell is increased when all the photocarriers are collected at the junction instead of recombining elsewhere. Thus, effective passivation of the front and back of the cell improves the efficiency of the cell.

An ideally passivated interface "reflects" minority carriers, but passes majority carriers. Examples of two such interfaces are included in Fig. 1. The passivating layer at the front of the cell is often referred to as the "window" layer because it must be transparent if the solar cell is to have a high efficiency. The back of the cell is passivated by a structure referred to as a "back-

95

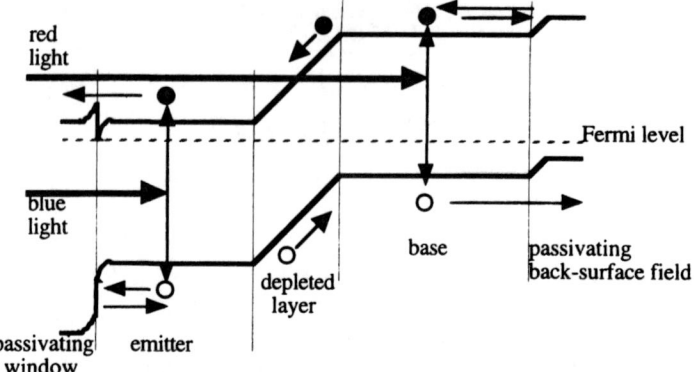

red
light

Fermi level

blue
light

base

passivating
back-surface field

depleted
layer

passivating   emitter
window

Fig. 1. Schematic of an n-on-p solar cell. Blue light is strongly absorbed close to the front of the cell, generating photocarriers in the emitter. Red light is absorbed throughout the cell. The photo-generated minority carriers diffuse within the emitter and base, may be "reflected" by the passivating layers, and are collected at the junction by the field in the depleted layer. The resulting majority carriers then pass through the passivating layers and are used in an external circuit.

surface field." This layer does not need to have a higher band gap as long as it provides an adequate barrier to minority carriers. This can sometimes be provided primarily by increased doping in the passivating layer (for n-type GaAs and for GaInP), as shown in Fig. 1, but a higher band gap material may work more reliably, especially when doping causes a reduction in the band gap, as in the case of p-type GaAs.

The current-voltage (IV) curve for a solar cell (Fig. 2) in the dark is given by the standard diode equation. If superposition holds (and the photocurrent is independent of the voltage), the light IV curve has the same shape as the dark IV curve, but is shifted downward by the short-circuit current (J_{sc}). The open-circuit voltage (V_{oc}) is the voltage at which the magnitude of the dark current equals the J_{sc}. The maximum power output of the solar cell is given by the product of the J_{sc}, the V_{oc}, and the fill factor (a measure of the "squareness" of the IV curve). Passivation of the front and back of the solar cell increases both the J_{sc} and the V_{oc}. We first describe the effect of the passivation on the J_{sc}, and then on the V_{oc}.

If certain assumptions are made, including a device structure with uniform layers similar to that shown in Fig. 1, low injection, and ideal semiconductor material, the transport equations can be solved in closed form [7,8]. The quantum efficiency (QE), defined as the probability of

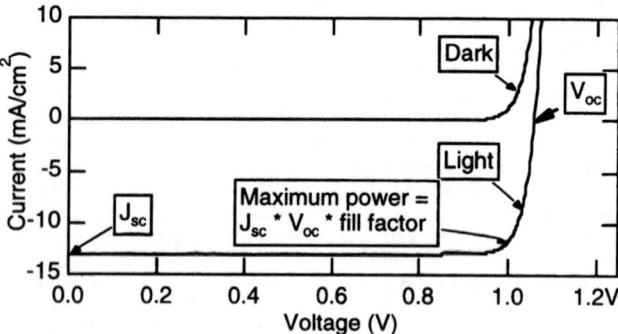

Fig. 2. Dark and light IV curves for a solar cell. The short-circuit current, open-circuit voltage, and maximum power point are labeled.

collecting a photocarrier for each photon, is given for the emitter region by

$$QE_E=\left[\frac{\alpha L_p}{\left(\alpha^2 L_p^2-1\right)}\right]\left[\frac{\left(\dfrac{S_p L_p}{D_p}+\alpha L_p\right)-\exp(-\alpha x_E)\left(\dfrac{S_p L_p}{D_p}\cosh\dfrac{x_E}{L_p}+\sinh\dfrac{x_E}{L_p}\right)}{\dfrac{S_p L_p}{D_p}\sinh\dfrac{x_E}{L_p}+\cosh\dfrac{x_E}{L_p}}-\alpha L_p\exp(-\alpha x_E)\right],(1)$$

where the variables are defined in Table 1.
Similarly, the probability of collecting a photocarrier from the base is given by

$$QE_B=\frac{\alpha L_n}{\alpha^2 L_n^2-1}\exp\!\left[-\alpha(x_E+W_D)\right]\left[\alpha L_n-\frac{\dfrac{S_n L_n}{D_n}\left(\cosh\dfrac{x_B}{L_n}-\exp(-\alpha x_B)\right)+\sinh\dfrac{x_B}{L_n}+\alpha L_n\exp(-\alpha x_B)}{\dfrac{S_n L_n}{D_n}\sinh\dfrac{x_B}{L_n}+\cosh\dfrac{x_B}{L_n}}\right].(2)$$

Table 1. Definitions of symbols and values (for GaInP) used for the fit in Fig. 3. The values specified in parentheses were used for calculations presented in other figures. Correction for absorption in a 0.025-μm-thick $Al_{0.5}In_{0.5}P$ window layer was also included.

Symbol	Significance	Value
α	absorption coefficient	see appendix
L_p	hole diffusion length in n-type emitter material	0.5 μm
L_n	electron diffusion length in p-type base material	3 μm
S_p	surface recombination velocity at window	10^4 (10^4-10^7) cm/s
S_n	surface recombination velocity at back-surface field	1.7×10^6 (10^4-10^7) cm/s
D_p	diffusion coefficient of holes in n-type material	5 cm^2/s
D_n	diffusion coefficient of electrons in p-type material	100 cm^2/s
x_E	emitter thickness (flat-band)	0.1 μm
W_D	depletion width	0.1 μm
x_B	base thickness (flat-band)	0.45 (3) μm

In the depleted layer, the electric field aids the collection of photocarriers, resulting in collection of every photocarrier generated there, as calculated by:

$$QE_D=\exp(-\alpha x_E)\left[1-\exp(-\alpha W_d)\right].$$

(3)

The total internal QE is obtained by summing the contributions of the emitter, base, and depleted layers (eqs. 1-3). The QE is a function of wavelength, λ, because of the λ-dependence of the absorption coefficient, α. In practice, the external QE and the reflectivity are measured, and the internal QE calculated from the external QE(λ) divided by [1-R(λ)], where R(λ) is the reflectivity as a function of λ. The J_{sc} is obtained from the integral of the product of the QE with the spectrum of interest. Fig. 3 shows a fit to data, including the contributions from the three regions. For all of the curves presented in this paper, the QE was also reduced to account for absorption in a 0.025-μm-thick $Al_{0.5}In_{0.5}P$ window layer. In practice, it is very difficult to differentiate low L_n from high S_n. An equally good fit was obtained by using a $S_n<10^4$ cm/s and L_n of about 0.5 μm. We emphasize that the quality of the fit is very dependent on our knowledge of $\alpha(\lambda)$ (see appendix).
Figs. 4 and 5 show the total internal QEs calculated from eqs. 1-3 using the parameters in Table 1, and varying the values for S_p and S_n. For large absorption coefficients, a high S_p causes the blue response to decrease dramatically (Fig. 4). However, a high S_p also causes a reduction in the red response as well. In contrast, high S_n causes a reduction only in the red response, with almost no measurable effect in the blue response for a thick cell (Fig. 5). If the cell

97

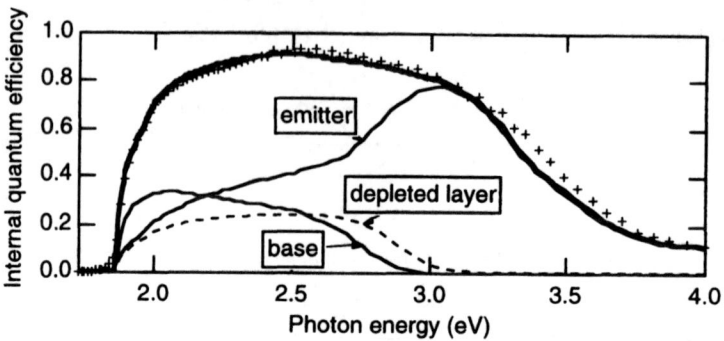

Fig. 3. Measured (+) and modeled (lines) internal quantum efficiency for a GaInP cell. The contributions of the three regions are shown in addition to their sum (heavy line). The calculated curves used Eqs. 1-3 with the values specified in Table 1, including $S_n = 1.7 \times 10^6$ cm/s.

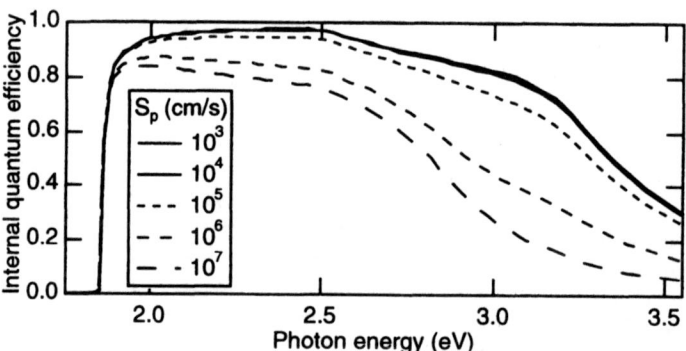

Fig. 4 Internal QE calculated using $x_B = 3$ μm, $S_n = 1000$ cm/s, and a variable value of S_p.

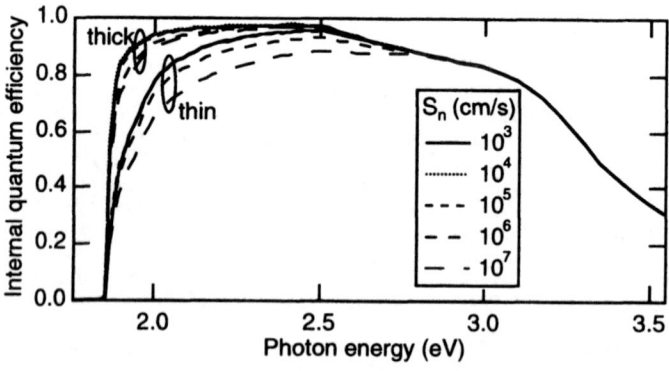

Fig. 5. Internal QE calculated using a variable value of S_n. The lower set of curves used $x_B = 0.45$ μm; for the upper set, $x_B = 3$ μm.

is thinned so that more light penetrates near the back surface, then an increase in S_n causes a decrease of both the red and the blue response, with the red response decreasing even more dramatically. The most significant result here is that the QE is increased when S_n or S_p is reduced from 10^7 to about 10^4 cm/s, but further reduction of S_n or S_p has a negligible effect on the QE.

Similar calculations for p-on-n cells show that the effect of passivation of the back of the cell is less than for the n-on-p cell. In practice, a p-on-n cell usually uses a thicker emitter and thinner base. The differences observed between the n-on-p and p-on-n cell are a result of the poor minority hole transport properties compared with those of electrons. Poor hole collection in a 0.1-μm-thick, n-type emitter is not a serious problem in the n-on-p cell, but is much more of a problem in a 3-μm-thick, n-type base of a p-on-n cell. However, in practice, because most GaInP cells are grown thin to help match the currents of the GaInP and GaAs cells [9], both cells require passivation of both the front and the back of the cells. Thick GaAs p-on-n cells benefit negligibly from back-surface passivation because the base thickness $\geq L_p$.

Passivation of the solar cell also decreases the dark current of the solar cell, and, therefore, increases the V_{oc}. Assuming ideal material, the transport equations can be solved to give the dark current associated with the base and emitter regions (first two terms in eq. 4) [7,8]. An estimation of the dark current in the depleted layer is included as the third term [8] and is discussed below.

$$J_d(V) = q\left[\frac{D_n}{L_n}\right]\left[\frac{n_i^2}{N_A}\right]\left[\frac{\frac{S_n L_n}{D_n}\left(\cosh\frac{x_B}{L_n}\right)\cosh\frac{x_B}{L_n} + \sinh\frac{x_B}{L_n}}{\frac{S_n L_n}{D_n}\sinh\frac{x_B}{L_n} + \cosh\frac{x_B}{L_n}}\right]\left[\exp\frac{qV}{kT} - 1\right]$$

$$+ q\left[\frac{D_p}{L_p}\right]\left[\frac{n_i^2}{N_D}\right]\left[\frac{\frac{S_p L_p}{D_p}\left(\cosh\frac{x_E}{L_p}\right)\cosh\frac{x_E}{L_p} + \sinh\frac{x_E}{L_p}}{\frac{S_p L_p}{D_p}\sinh\frac{x_E}{L_p} + \cosh\frac{x_E}{L_p}}\right]\left[\exp\frac{qV}{kT} - 1\right] + \left[\frac{n_i W_d kT}{2(V_d - V)\tau}\right]\left[\exp\frac{qV}{2kT} - 1\right]$$

(4)

where n_i is the intrinsic carrier concentration, N_A and N_D are the concentrations of acceptors and donors, V_d is the built-in voltage, and τ is the nonradiative carrier lifetime, given by $1/N_T \sigma v_{th}$, the reciprocal of the product of the trap density, the capture cross section, and the thermal carrier velocity [8].

The base and emitter terms are the diode injection current and are affected both by bulk recombination and by interface recombination. The impact of the two types of recombination on the relative V_{oc} can be expressed as a function of two dimensionless variables [9]. The bulk recombination is described by the ratio of the layer thickness to the diffusion length, x/L, and the ratio of the interface recombination to bulk recombination by SL/D (Fig. 6). The same results are applicable for both n- and p-type material, so the subscripts have been removed in Fig. 6. For thick layers or short diffusion lengths (layer thickness > diffusion length), the bulk recombination dominates and the V_{oc} is independent of SL/D. For thin layers, the interface recombination strongly affects the V_{oc}. Just as the QE is not affected by passivation of the back of a GaAs p-on-n cell (for which $x_B \geq L_p$), the V_{oc} is also not affected.

The third term in eq. 4 estimates recombination at defects in the depleted layer. This Shockley-Read-Hall (SRH) recombination is strongest in the part of the depleted layer for which the intrinsic level is approximately midway between the electron and hole quasi-Fermi levels. The equations describing the SRH recombination cannot be solved in closed form. In practice, the SRH recombination is difficult to quantify because the trap state density, energy, and capture cross section are usually not well known. Also, errors may be introduced because of approximating the third term in eq. 4 [10].

The terms in eq. 4 can be differentiated by their voltage dependencies. The first two terms show an exponential dependence on qV/nkT with n (diode quality factor) is equal to unity. In contrast, the third term shows an exponential dependence on V with n=2 and an additional V dependence in the preexponential. Therefore, the depleted layer dark current is sometimes referred to as "n=2" current. In practice, the voltage at the maximum-power-point on the IV curves of *small*

99

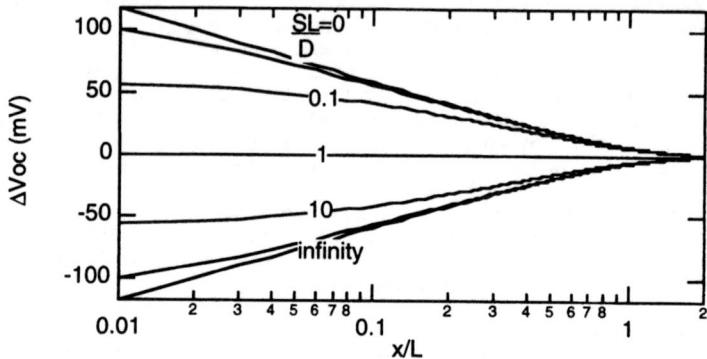

Fig. 6. Change in V_{oc} with the dimensionless thickness (x/L) as the ratio of the interface/bulk recombination (SL/D) is varied from 0 to infinity. The values apply to either p-type or n-type material.

high-efficiency cells may be dominated by an n=2 dark current that comes from recombination at the perimeter of the cells. This perimeter recombination can be reduced by passivation of the III-V exposed junction surfaces [11]. Despite the qV/2kT term in the exponential, SRH recombination in the depleted region may show n<2 because of the V dependence of the preexponential. In extreme cases, n may approach unity even when the dark current is dominated by SRH recombination [8].

When a solar cell is inefficient, the cause may be poor passivation, but it may also be something else. When the V_{oc} is low, a diode quality factor of 2 implies that the problem is in the depleted layer. An example of an extreme case of this is shown in Fig. 7. Both the n=2 nature of the dark current and the micrograph imply that recombination at defects in the depleted layer is the cause of the high dark current. When the V_{oc} is poor and the diode quality factor is unity, the cause may be either bulk or interface recombination. These, theoretically, can be differentiated by varying the thickness of the layer. If the V_{oc} improves as the cell is thinned, then the problem is likely to be bulk recombination. If the V_{oc} is reduced by thinning, then poor interface passivation is suspected. If the V_{oc} is unchanged, then, either x>L or the bulk and interface recombinations are of equal importance.

Fig. 7. Transmission electron micrograph of GaInP cell showing very high n=2 dark current.

EXPERIMENTAL DATA

Table II summarizes a few interface recombination velocities measured on double heterostructures. All (except for one) of these values are low compared with the values used in calculating the data in Figs. 3-5, implying that it should be possible to grow solar cells with essentially no losses from poor surface passivation. However, the data in Table II demonstrate that the IRV increases with both the doping of the active layer and with aluminum content of the passivating layer. Substitution of aluminum for gallium in most III-V materials increases the band gap without significantly changing the lattice constant. Thus, theoretically, aluminum-containing III-V materials make ideal passivating layers. However, practically, the use of aluminum is much more difficult because of the gettering of oxygen that comes with increased aluminum content, and because of the tendency of aluminum-containing alloys to form DX centers. The one IRV in Table II that is greater than 10^4 cm/s is the passivation of p-type GaInP with p-type AlGaInP. In the GaInP/GaAs solar cell the p-type GaInP is the most challenging layer to passivate because p-type AlInP and AlGaInP are so sensitive to low levels of oxygen.

One very interesting result shown in Table II is the use of disordered GaInP to passivate GaInP. The band gap of GaInP (at a fixed composition) varies with growth conditions because of ordering of the Ga and In atoms on the group III sublattice [12,13]. Thus, a high-band-gap (disordered) GaInP layer can be used to passivate a low-band-gap (partially ordered) GaInP layer. Higher doping also tends to create a passivating layer. (This works well for n-type GaAs, but is less effective for p-type GaAs because of heavy doping effects [14]). One of the examples cited in Table II and several of the results reported below use a high-band-gap, high-zinc-doped GaInP layer to passivate a second GaInP layer with lower band gap and lower zinc doping. The movement of zinc in a GaInP layer has been shown to disorder an ordered GaInP layer [15]. Thus, a passivating GaInP layer can be formed by diffusing zinc into it. An example of this will be discussed below.

Table II. Interface recombination velocities reported in the literature for GaAs or GaInP layers sandwiched between two barrier layers.

GaAs doping(cm^{-3})	Barrier	S (cm/s)	Reference
$n<10^{16}$	$Al_{0.3}Ga_{0.7}As$	18	[16]
$n<10^{15}$	$Ga_{0.5}In_{0.5}P$	<1.5	[6]
$n=1-3X10^{18}$	$Ga_{0.5}In_{0.5}P$	1300	[17]
$n=1.3 X 10^{17}$	$Al_{0.3}Ga_{0.7}As$	<12	[18]
$p=5X10^{15}$	$Al_{0.5}Ga_{0.5}As$	300	[19]
$p=3X10^{16}$	$Al_{0.5}Ga_{0.5}As$	350	[19]
$p=1.7X10^{17}$	$Al_{0.5}Ga_{0.5}As$	500	[19]
$Ga_{0.5}In_{0.5}P$ doping(cm^{-3})			
$p=9X10^{16}$	$Al_{0.25} Ga_{0.25}In_{0.5}P$	140,000	[17]
$n<10^{15}$	disordered $Ga_{0.5}In_{0.5}P$	<2	[20]
$n<10^{15}$	$Al_{0.25} Ga_{0.25}In_{0.5}P$	7	
$n<10^{15}$	$Al_{0.5}In_{0.5}P$	85	
$n<10^{15}$	$Al_{0.5}Ga_{0.5}As$	180	
$n<10^{15}$	$Al_{0.85}Ga_{0.15}As$	>5000	

Although aluminum-containing alloys are attractive candidates for passivating layers, they often cause incorporation of oxygen, leading to enhanced recombination at interfaces. Such an effect of poor passivation because of oxygen contamination in an $Al_{0.5}In_{0.5}P$ window layer is shown in Fig. 8 [21]. Oxygen incorporation can also lead to majority-carrier transport problems.

One challenge of achieving the low IRVs noted in Table II in solar cells is that growth of devices with p-n junctions sometimes causes dopant diffusion. The dopant diffusion may either help or hinder the passivating layers. Specifically, growth of n-type layers on top of Zn-doped layers has been shown to cause diffusion of the Zn. Deppe put forth an explanation for this effect that explains a significant fraction of the data [22]. He hypothesized that Fermi-level pinning at the growth surface controls the concentration of point defects. Although an equilibrium concentration of Ga interstitials is obtained at the growing surface, when that layer is covered by subsequent growth layers and the Fermi level shifts back to its normal value, there is then an excess concentration of Ga interstitials. These move into the underlying zinc-doped layers where they "kick out" zinc atoms, freeing them to diffuse through the material [22]. Although it is difficult to

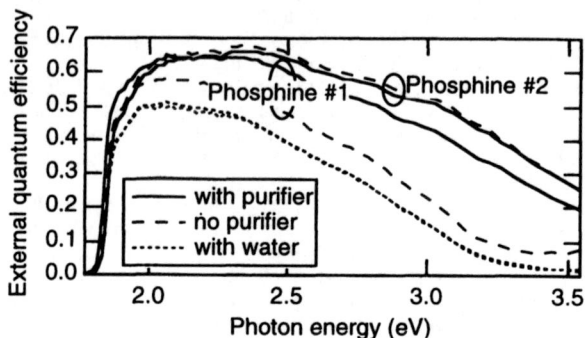

Fig. 8. QE of GaInP cells with and without a phosphine purifier for two different phosphine sources and when water was added (two different runs) intentionally by passing 0.1 sccm of hydrogen through a bubbler containing water at 3°C. The with-purifier performances differ for the two phosphine sources because the device structures were different. The top four curves were measured on two-junction (GaInP/GaAs) devices.

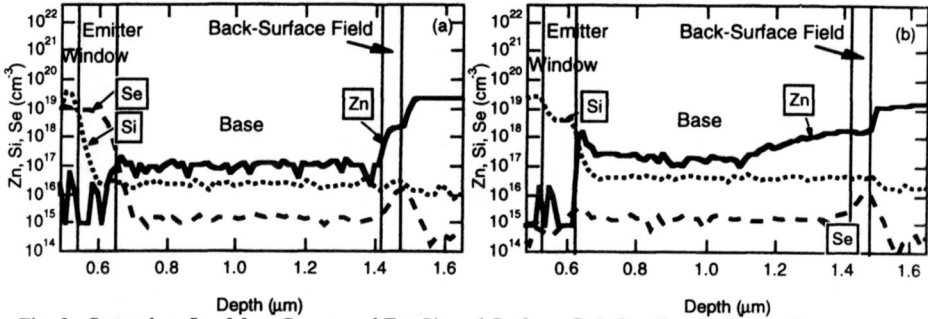

Fig. 9. Secondary Ion Mass Spectra of Zn, Si, and Se for a GaInP cell. The two cells were grown identically except for the Se (a) and Si (b) doping of the emitter and cap layers. Both used Si doping in the window. In (b) the Zn is observed to diffuse out of the back-surface field.

prove or disprove this model, the dopant diffusion can easily be observed (Fig. 9) [23]. The measured QEs of these samples are shown in Fig. 10. The sample with the degraded back-surface field also shows degraded red response (around 2 eV), as would be predicted (compare to Fig. 5). The change of blue response is less surprising because we calculate that changing the emitter will change the blue response. As pointed out below, diffusion of Zn at the back of the GaInP cell does not necessarily result in a degraded back-surface field. The situation can be more complex than only an increase in S_n, since the diffusion of Zn causes the GaInP to become disordered and higher in band gap, which also reduces the red response.

The data in Figs. 8 and 9 show how growth-induced dopant diffusion can harm intentionally added passivating layers, but the dopant diffusion can also create an unintentional passivating layer. For example, Table III compares the V_{oc}s of three GaInP cells grown identically except for the choice of the back-surface field. In this case, dopant diffusion from the Zn-doped GaAs buffer layer caused the back-most part of the GaInP base to become disordered and highly doped: almost an ideal back-surface field.

This Zn diffusion is often avoided in transistor structures by using carbon as the p-type dopant instead of Zn. However, C doping of GaInP does not work well, and the solution for GaInP cells is more difficult, especially when Si is the n-type dopant. However, Zn diffusion can be avoided, even in Si-doped GaInP cells, if the back of the n-on-p GaInP cell is passivated by using a clean,

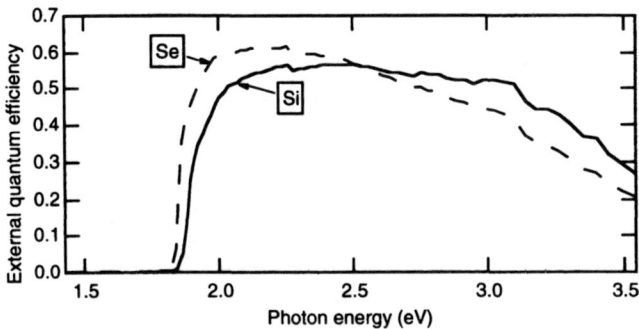

Fig. 10. External quantum efficiency of cells grown like those documented in Fig. 9, showing the poor back-surface passivation (lower QE at 2 eV) when zinc diffusion occurs. The change in blue response with change of emitter dopant is not surprising.

Table III. Open-circuit voltages of n-on-p GaInP cells, showing how unintentional Zn diffusion from a GaAs buffer can increase the V_{oc} almost as much as the intentional back-surface field design.

Back-surface-field design	V_{oc} (V)
Intentional high-E_g GaInP	1.348
Zn:GaAs buffer (no intentional back-surface field)	1.335
C:GaAs buffer (no intentional back-surface field)	1.275

p-type AlInP layer. The p-type AlInP passivation leads to lower V_{oc}s and increases the series resistance if the AlInP layer is not clean [24].

An ideal passivating interface (window or back-surface-field shown in Fig. 1) must not only reflect minority carriers, but must pass majority carriers. For example, problems are encountered if a conduction-band spike impedes electron transport past an interface between two n-type layers. This can happen for a number of reasons. We report next a problem with hole transport across a p-type AlInP/GaInP interface for a p-on-n GaInP solar cell.

Hydrogen passivation of zinc (and other) acceptors results in a reduced hole concentration. After chemical vapor deposition of GaAs and related materials, the sample is traditionally cooled in an arsine (or phosphine)-containing gas ambient to prevent loss of arsenic (or phosphorus) from the surface. The arsine is known to decompose on the surface, creating atomic hydrogen that can diffuse into the semiconductor material. In p-type material, atomic hydrogen is positively charged, and the resulting proton moves very rapidly. In n-type material, the proton picks up an electron and is much less mobile. Thus, atomic hydrogen diffusing into a p-on-n structure is concentrated in the p-type layer, dramatically reducing the hole concentration in the p-type layer. The hole concentration may be reduced to a value where the majority hole transport is impeded. This is especially a problem with p-type AlInP for which oxygen impurities compensate zinc acceptors[21]. In Fig. 11 we compare the IV curves of two GaInP p-on-n cells that were grown under identical conditions including a final GaAs cap that is used as a contacting layer, but removed from the active areas of the cells. Although the growth conditions were identical, one cell was cooled under arsine while the other was cooled with no arsine (both used molecular hydrogen as the primary gas ambient, but the molecular hydrogen does not decompose into atomic hydrogen under these conditions). Fig. 11 shows that the sample cooled in arsine shows a non-ohmic series resistance associated with the GaInP-AlInP-GaAs junction at the front of the cell. This series resistance reduces the fill factor, and, therefore, the efficiency of the solar cell.

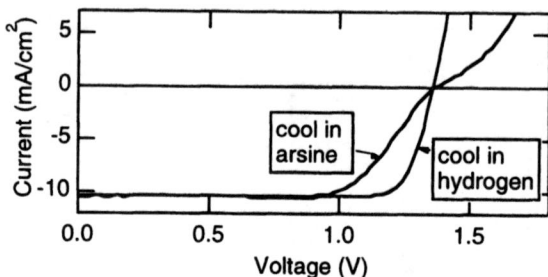

Fig. 11. IV curves for two GaInP cells grown identically, but cooled differently.

SUMMARY

We calculated the QE for GaInP cells with front and back-surface interface recombination velocities between 1000 and 10^7 cm/s. These curves showed that IRVs $<10^4$ cm/s are needed for effectively ideal passivation. Poor passivation of the front of the cell reduces the photocurrent (primarily by loss of blue response) as well as the photovoltage. Passivation of the back of a thick cell affects the red response and photovoltage of the cell, but has a negligible effect on a thick p-on-n cell. Thin cells (often needed for multijunction stacks) show an improved photovoltage and photocurrent if well passivated, even for the p-on-n design.

Although very low IRVs are reported in the literature, because of the need to dope the cells and the complexities of cell growth, much higher IRVs are frequently observed in solar cells. Dopant diffusion during growth can help or hinder passivating schemes, while accidental or intentional addition of oxygen and/or hydrogen can reduce passivation and increase barriers to majority-carrier flow. Low photovoltage from high IRV can be differentiated from low photovoltage from high defect densities by looking at the diode ideality factor.

APPENDIX—ABSORPTION COEFFICIENT OF GaInP

Although ellipsometric studies of the optical properties of GaInP and AlInP [25,26] have been reported in the literature, accurate modeling of GaInP solar cells requires more detailed knowledge of the absorption coefficient, α. A complication of evaluating α is that the band gap and optical properties of GaInP vary with growth conditions [13,27]. Also, α varies with doping. A detailed investigation of α as a function of both ordering and doping is lacking, but we find that we can model GaInP cells reasonably well using the approach presented below.

The ellipsometric studies measure the dielectric constants, $\varepsilon 1$ and $\varepsilon 2$, accurately in regions of high absorption, but give fairly inaccurate results near the band edge. Kato, et al. presented data and empirical equations fitting the data [25]. However, their empirical equations are inaccurate near the band edge (see Fig. 12) and include a tail extending well below the band gap. The curves presented in Fig. 13 of reference [25] were modified at the band edge to improve the accuracy, but the modification procedure was not described [28]. We have measured the absorption near the band edge using a simple transmission measurement and find that the near-band-edge data can be fit by

$$\alpha(E) = 5.5\sqrt{(E - E_g)} + 1.5\sqrt{(E - E_g - 0.1)}$$

where E is the photon energy and E_g is the fundamental band gap, both in eV, and α in $1/\mu m$.

This equation assumes a parabolic direct gap and is accurate near the band edge, but not at higher energies (see Fig. 12, "Direct gap only" trace). The absorption at the band edge is much weaker than the absorption associated with the E_1 transition. Omission of this higher absorption (especially in the AlInP window) prevents useful modeling of GaInP solar cells. Kato's data are accurate for photon energy >3 eV, but not near the band edge. To obtain the fits presented in Fig.

3, we have used the band-edge data for E<2.7 eV and Kato's empirical fit for E>3.1, with a smooth connection between the two (see dotted line in Fig. 12).

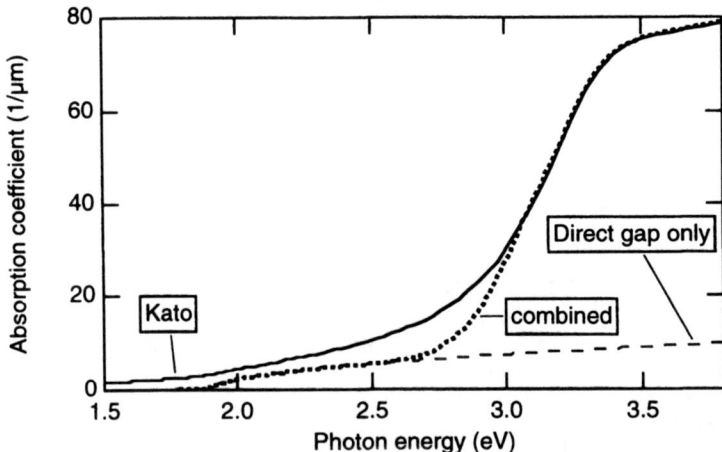

Fig. 12. Compares GaInP absorption coefficient using Kato's published empirical fit [25], direct-gap absorption near the band edge, and a combination of the two. The "combined" curve was used for the calculations presented in this paper. Use of either of the two other curves gave unsatisfactory fits.

ACKNOWLEDGMENTS

We would like to thank C. Kramer for preparing the samples, K. Jones for the TEM micrograph, R. Reedy for the SIMS measurements, and R. King for useful conversations. This work was supported under DOE contract DE-AC36-98-G010337.

REFERENCES

1. K. A. Bertness, S. R. Kurtz, D. J. Friedman, A. E. Kibbler, C. Kramer, and J. M. Olson, *Appl. Phys. Lett.* **65,** 989-991 (1994).
2. D. J. Friedman, S. R. Kurtz, K. A. Bertness, A. E. Kibbler, C. Kramer, J. M. Olson, D. L. King, B. R. Hansen, and J. K. Snyder, *Progress in Photovoltaics: Research and Applications* 3, 47-50 (1995).
3. T. Takamoto, E. Ikeda, H. Kurita, M. Ohmori, M. Yamaguchi, and M. J. Yang, *Jpn. J. Appl. Phys.* **36,** 6215-6220 (1997).
4. Y. C. M.Yeh, F. F. Ho, C. L. Chu, and P. K. Chiang, "Advance in Production of Cascade Solar Cells for Space," in *Proceedings of the 26th IEEE Photovoltaic Specialists Conference*, 1997, 827-830.
5. B. T. Cavicchi, J. H. Ermer, D. D. Krut, D. E. Joslin, M. S. Gillanders, and D. K. Zemmrich, "250,000 Watts of GaInP2/GaAs/Ge Dual Junction Production," in *Proceedings of the 2nd World Conference on PV Energy Conversion*, 1998, 3515-3519.
6. J. M. Olson, R. K. Ahrenkiel, D. J. Dunlavy, B. Keyes, and A. E. Kibbler, *Appl. Phys. Lett.* **55,** 1208 (1989).
7. H. J. Hovel, *Solar Cells*, (Academic Press, New York, 1975).
8. A. L. Fahrenbruch and R. H. Bube, *Fundamentals of Solar Cells Photovoltaic Solar Energy Conversion*, (Academic Press, New York, 1983).
9. S. R. Kurtz, P. Faine, and J. M. Olson, *J. Appl. Phys.* **68,** 1890 (1990).
10. S. M. Sze, *Physics of Semiconductor Devices*, (Wiley, New York, 1969).

11. M. S. Carpenter, M. R. Melloch, M. S. Lundstrom, and S. P. Tobin, *Appl. Phys. Lett.* **52**, 2157-2159 (1988).

12. A. Gomyo, T. Suzuki, K. Kobayashi, S. Kawata, I. Hino, and T. Yuasa, *Appl. Phys. Lett.* **50**, 673 (1987).

13. S. R. Kurtz, J. M. Olson, and A. Kibbler, *Appl. Phys. Lett.* **57**, 1922-1924 (1990).

14. M. E. Klausmeier-Brown, M. S. Lundstrom, M. R. Melloch, and S. P. Tobin, *Appl. Phys. Lett.* **52**, 2255-2257 (1988).

15. K. Meehan, F. P. Dabkowski, P. Gavrilovic, J. E. Williams, W. Stutius, K. C. Shieh, and N. Holonyak, *Appl. Phys. Lett.* **54**, 2136-2138 (1989).

16. L. W. Molenkamp and H. F. J. van't Bilk, *J. Appl. Phys.* **64**, 4253 (1988).

17. R. R. King, J. H. Ermer, D. E. Joslin, M. Haddad, J. W. Eldredge, N. H. Karam, B. Keyes, and R. K. Ahrenkiel, "Double heterostructures for characterization of bulk lifetime and interface recombination velocity in III-V multijunction solar cells," in *Proceedings of the 2nd World Conference on Photovoltaic Solar Energy Conversion*, 1998, 86.

18. R. K. Ahrenkiel, J. M. Olson, D. J. Dunlavy, B. M. Keyes, and A. E. Kibbler, *Journal of Vacuum Science and Technology B* **8**, 3002 (1990).

19. R. J. Nelson, *J. Vac. Sci. Technol.* **15**, 1475 (1978).

20. A. van Geelen, in *Solar Cells of the III-V Compounds GaAs and GaInP$_2$*, PhD thesis (1997).

21. S. R. Kurtz, J. M. Olson, K. A. Bertness, K. Sinha, B. McMahon, and S. Asher, "Hidden but important parameters in GaInP cell growth," in *Proceedings of the 25th IEEE Photovoltaic Specialists Conference*, 1996, 37-42.

22. D. G. Deppe, *Appl. Phys. Lett.* **56**, 370-372 (1990).

23. S. R. Kurtz, J. M. Olson, D. J. Friedman, and R. Reedy, "Effect of Front-Surface Doping on Back-Surface Passivation in GaInP Cells," in *Proceedings of the 26th IEEE Photovoltaic Specialists Conference*, 1997, 819-822.

24. D. J. Friedman, S. R. Kurtz, A. E. Kibbler, and J. M. Olson, "Back surface fields for GaInP$_2$ solar cells," in *Proceedings of the 22nd IEEE Photovoltaic Specialists Conference*, 1991, 358-360.

25. H. Kato, S. Adachi, H. Nakanishi, and K. Ohtsuka, *Jpn. J. Appl. Phys.* **33**, 186-192 (1994).

26. K. H. Lee, S. G. Lee, and K. J. Chang, *Phys. Rev. B.* **52**, 15862-15866 (1995).

27. S. R. Kurtz, D. J. Arent, K. A. Bertness, and J. M. Olson, "The Effect of Phosphine Pressure on the Band Gap of Ga$_{0.5}$In$_{0.5}$P," in *Proceedings of the Compound Semiconductor Epitaxy*, 1994, 117-122.

28. S. Adachi (private communication).

ELECTRICAL AND OPTICAL STUDY OF CHARGE TRAPS
AT PASSIVATED GaAs SURFACES

Yasunori Mochizuki
Fundamental Research Laboratories, NEC Corporation, 34 Miyukigaoka, Tsukuba,
Ibaraki 305-8501, Japan   mochizuk@frl.cl.nec.co.jp

ABSTRACT

Performance of GaAs-based FETs is strongly affected by the electrical properties of passivated surfaces via the surface charges induced at interface states and at the traps within the passivation films.   In order to study these two kinds of sources, a set of new characterization techniques are respectively developed for the GaAs interface system, which is in a largely different situation from that for the SiO_2/Si interfaces.   Our optical technique based on electroreflectance allows accurate evaluation of interface states and is especially useful for the material systems with strong surface pinning.   For the study of trapping centers in CVD dielectrics, use of thermally oxidized Si interfaces is found to be quite convenient. The experimental findings are further discussed in conjunction with FET characteristics and an importance of time-domain control of the surface charges is emphasized in the power FET operation.   This concept turns out to be the main motivation of a new MESFET structure with a filed-modulating plate (FPFET) which has achieved a state-of-the-art RF performance based on GaAs.

INTRODUCTION

Electrical charges on passivated GaAs surfaces play a crucial role in the performance of MESFETs.   In fact, its importance was noted, for instance, in a device simulation study: a presence of negative surface charges was predicted to yield a higher breakdown voltage for GaAs-FETs by relaxing a lateral potential gradient near the drain-side edge of gate electrode[1].   Since the breakdown voltage limits the maximum operation voltage (V_{dd}) and its product with drain current limits the available output power density, the subject to be discussed here is of particular significance in improving the performance of power-FETs. However, in order to characterize electrical property of actual GaAs surfaces, a twofold difficulty has been experienced.   First, due to the poor controllability of GaAs surfaces, a large density of interface states are inevitably present in between the passivation dielectrics and semiconductor[2].   Since they cause a strong (extrinsic) pinning of surface Fermi-level, direct application of the conventional capacitance technique gives rise to a serious ambiguity in the analysis, and thus an alternative approach for the characterization is necessary. Furthermore, since the passivation films are in most cases prepared by the CVD techniques, their chemical composition (including hydrogen contamination, *etc*) and structural stability are not as well-defined as those of thermally oxidized silicon.   Therefore, structural defects and/or fixed charges within the dielectric films might also act as a significant source of surface charges in the devices and their contribution should also be evaluated from experiment.

In this contribution, I will describe our recent effort from such a viewpoint.   For the characterization of the interface states, a new method based on electroreflectance (ER) spectroscopy is developed.   Although being an optical technique, the method can extract electrical information and allows a quantitative evaluation of interface-state density even for a strongly pinned interface systems.   The charges in the dielectrics, on the other hand, can be characterized using the conventional capacitance technique.   By properly designing the sample structure, an easy way to separate the effect of interface states is provided.   The effect of these surface charges are finally discussed in relation to the FET performance and a way to optimize the electrical behavior of the passivated device surfaces is analyzed.   In fact, the concept of surface-charge control via an additional electrode, the field-modulating plate (FP), has led to a development of state-of-the-art power FETs[3], whose maximum output power and operation voltage greatly exceed those of conventional products.

INTERFACE STATES CHARACTRIZED BY ELECTROREFLECTANCE

Principle

The surface region of GaAs-FET is passivated by dielectric films such as SiN_x or SiO_2 and a large density of trapping states are present at the passivated interface. In order to discuss the effect of the charges induced by the interface states, energy distribution of the state density, $N_{SS}(E)$, has to be characterized. The task, however, has long been known to be difficult using the conventional electrical techniques based on capacitance measurements. The main source of the uncertainty in this type of techniques is the strong pinning of surface Fermi level due to those interface states since it is not appropriate in such a situation to analyze the state density from the comparison with ideal C-V curve. As an alternative approach, we have demonstrated that measurement of the so-called Franz-Keldysh oscillation (FKO) spectra in electroreflectance provides a way to assess the interface Fermi level, and thus $N_{SS}(E)$, without relying on the external energy reference (i.e. ideal C-V curve)[4].

Energy distribution of interface-state density (N_{SS}) can be obtained by measuring the semiconductor surface potential (ψ_s) as a function of applied voltage (V_G). This is because, in a MIS capacitor, one can write

$$C_{ox} \cdot \left(V_G - \psi_s \right) = \left(C_{it} + C_D \right) \cdot \psi_s \tag{1}$$

which, then, reduces to

$$N_{SS}(\psi_s) = C_{it}(\psi_s) / q \tag{2}$$

with

$$C_{it}(\psi_s) = C_{ox}\left[\left(\frac{d\psi_s}{dV_G}\right)^{-1} - 1\right] - C_s(\psi_s) \tag{3}$$

where C_{ox}, C_D and C_{it} are the capacitance due to dielectric film, depletion layer and interface charges, respectively.

The problem of obtaining $N_{ss}(E)$ thus reduces to an experimental determination of ψ_s vs V_G. In the conventional C-V technique, ψ_s has to be estimated from the measured capacitance values after comparing them with those for an ideal MIS capacitor. Here, surface majority-carrier accumulation and minority-carrier inversion are assumed to respectively yield C_{max} and C_{min}. This assumption is equivalent to setting energy references, E_C and E_V, for the determination of ψ_s, which, however, is not necessarily justified for an interface with large $N_{ss}(E)$. This is because the response (excursion) of ψ_s to applied voltage may be pinned alternatively at some position within the band gap due to a huge density of interface-state charges. Together with a possibility of interface traps partially responding at the measurement frequency of capacitance, a serious numerical uncertainty may arise depending on the combination of doping level and dielectric film thickness.

On the contrary, ψ_s can be evaluated by measuring the surface field using the ER technique without relying on the theoretical curve for ψ_s. In modulated reflectance spectra of GaAs having finite surface band bending, an oscillatory features appear on the higher energy side of the fundamental gap transition, which are called Franz-Keldysh (FK) oscillations[5]. Since energy intervals of the oscillations is related to the magnitude of surface electric field, F_S, it is also connected with the surface band bending. The method for determining the surface field from the shape of FK oscillations has been described, for instance, by Bottka et al.[6]. From the analysis of Franz-Keldysh oscillations, we can determine the surface bending (surface potential) at a given V_G. In this way, by repeating the ER measurements at various V_G, the relationship of

108

ψ_s vs V_G is obtained to yield N_{ss} as a function of energy within the band gap.

The key consideration in our ER scheme is the employment of the so-called van Hoof (UN$^+$ and UP$^+$) structure[7], which is a double-layer structure consisting of undoped highly pure layer and underlying heavily doped one. In this case, the surface band-bending has a very simple relation with the surface filed, F_S, as

$$\psi_S = F_S \cdot d_i \tag{4}$$

with d_i being the i-layer thickness. In addition to such an advantage, this structure is known to yield a large number of oscillation peaks in the ER spectra due to a slow damping of FKO as a function of photon energy. Since the procedure of calculating $N_{ss}(E)$ requires a local differentiation of the ψ_s vs V_G curve, it is essential to enhance the numerical accuracy of surface field measurement, which is in fact possible using this structure.

Experimental

The sample structure for our ER measurement is sketched in Fig. 1. In order to achieve optical access for the reflectance measurements, transparent $In_2O_3+SnO_2$ (ITO) was used as the gate electrode of the MIS diodes. The epitaxial GaAs part has a UN$^+$ (UP$^+$) structure, an i/n (i/p)-GaAs junction structure, grown on semi-insulating GaAs substrates. Two types of layer structures were tested, in which the nominal thickness of the top undoped GaAs (d_i) was 200 and 150nm, respectively. The doping concentration of the underlying n-layer (p-layer) was kept constant at 1.7×10^{18} (6×10^{18}) cm^{-3}.

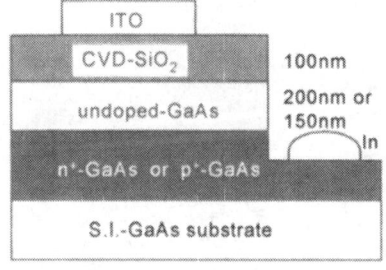

Fig. 1 Schematic structure of MIS diodes for ER measurement.

The reason for using two sets of i-layer thickness is to check if the material quality (purity of the i-layers) indeed justifies the use of eq.(4). Besides the purity, this check turns out to be useful when the deposition process of dielectrics is suspected to introduce damage in the near-surface region of GaAs. In fact, residual damage due to plasma-CVD of SiN$_x$ was detected using this method. Furthermore, we have performed experiments on both i/n and i/p structures. This was done in an attempt to check whether or not the movable ranges of surface Fermi-level are identical from the viewpoint of the surface pinning mechanism for GaAs.

To fabricate MIS diodes, SiO$_2$ films with a thickness (d_{ox}) of 100nm were deposited onto the epitaxial wafers using the conventional chemical vapor deposition (CVD) technique. The transparent ITO films with 200 nm thickness were then formed by RF sputtering. To define the diode area, which was 2mm x 3mm, these ITO films were patterned by the standard lithography and etching techniques. Ohmic contacts were obtained by first mesa-etching the outer region of the MIS diodes to expose the n$^+$-GaAs layer, and then by soldering indium. The leakage current under the largest voltage applied was 1.6×10^{-8} A/cm^2 or less for both bias polarities.

In the ER measurements, dispersed light from a halogen lamp was used. The modulated portion of reflectance was lock-in detected by applying a small-signal AC voltage (square modulation), which was superimposed on the DC bias voltage (V_G). The ER spectra were then recorded by monitoring the in-phase component of reflected light with scanning wavelength. In this way, the spectral measurements were repeated at various V_G. All the measurements were performed at room temperature. As will be discussed later, the light intensity for the ER measurement has to be weak enough to eliminate photovoltage[8]. In the experiment, therefore, the intensity was kept smaller than 5×10^{-8} Wcm^{-2}.

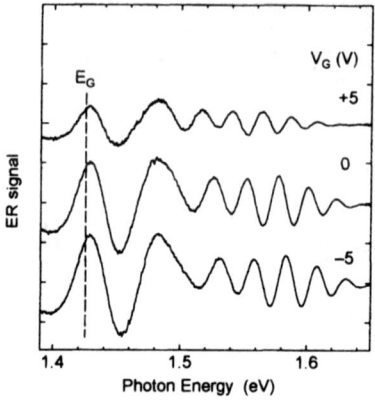

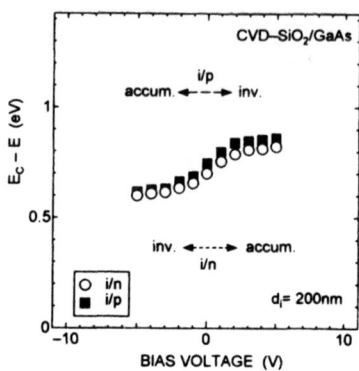

Fig.2   ER spectra of an i/n- MIS diode for various bias voltages.

Fig.3   Surface potential as a function of bias voltage for i/n and i/p diodes.

Results

Figure 2 shows a set of ER spectra for the i/n-GaAs MIS diode.   The energy intervals of the FK oscillations become larger for a more negative bias voltage.   By analyzing each spectrum, corresponding surface field is evaluated, which then is translated to the band bending at each bias. In fact, this last procedure requires a small correction to eq.(4) which takes into account the difference between the bulk Fermi-level and the conduction band edge.   Furthermore, a finite potential drop at the boundary of i- and n^+-layers takes place due to a formation of dipole by free electron and donor ions.   To make these corrections, we used a simple numerical simulation of the i/n (and i/p) structures, whose details can be found in ref. 9.

The summary of surface potential as a function of bias voltage is given in Fig. 3. (In this figure, and also others to be cited in the following, energy is measured from the conduction band minimum.)   The curves for both i/n and i/p diodes almost fall on top of each other.   Therefore, it is concluded that the movable range of surface Fermi-level is identical for these two systems. Such a conclusion turned out to be the case for all of the samples we have so far examined (up to nine different preparation conditions).   For the data shown in Fig. 3, allowed range of the surface Fermi level is strongly confined within the narrow region (~0.3eV for this sample) around midgap.

The corresponding $N_{ss}(E)$ curve is shown in Fig. 4, which is obtained using eq.(4).   It shows a U-shaped energy distribution with its minimum located

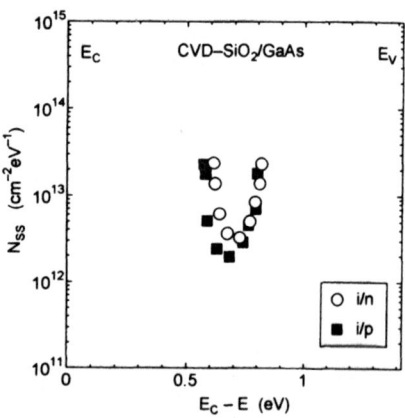

Fig.4   Interface-state density for i/n and i/p MIS structures.

nearly at the midgap. The minimum density at this point is $2\sim3\times10^{12}cm^{-2}eV^{-1}$ but rapidly increases in both sides to reach mid-10^{13} $cm^{-2}eV^{-1}$ range. Further excursion of the Fermi-level towards either band edge is strongly pinned by the charges induced at these midgap interface states.

Discussions

Since our technique is an optical one, it is necessary to check whether or not the illumination degrades the quantitative accuracy of the data. The major potential source of error is the photovoltage effect which arises from the flow of minority carriers photogenerated in the semiconductor. They recombine at the surface with majority carriers which are supplied by reducing the surface diffusion barrier. This reduction in the surface band bending is defined as photovoltage. It gives rise to an underestimation of the surface band bending, and tends to be more serious under more intense illumination conditions. Actually, Shen et al.[9] points out the importance of such a consideration through their careful examination of this effect in the photoreflectance (PR) technique applied to the characterization of GaAs surfaces. Fortunately, in general, ER suffers less from this effect than PR does since a pump light is not necessary and the probe light can be quite weak.

Actually, we have checked the absence of the effect in advance by the ER measurements by changing the illumination intensity and have found that photovoltage is negligible for the probe light intensity of less than $5\times10^{-8}W/cm^{-2}$. In Fig. 5, we have plotted our corresponding data together with analytic curves calculated from the theory of photovoltage in the PR experiments [10]. The validity of the theory is confirmed by an agreement with experimental result reported by Yin et al.[11] Also for our own data, we observe an excellent agreement between experiment and theory and the latter indicates the value of 5×10^{-8}

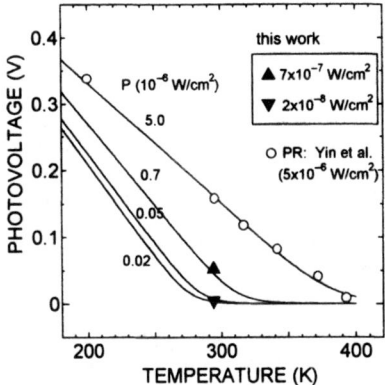

Fig.5 Temperature dependence of photovoltage as a function of probe-light intensity, P. Solid curves are calculated from theory described in ref 10. The PR data by Yin et al. [11] are also shown. For $P<5\times10^{-8}$ Wcm^{-2}, the effect is negligible even at room temperature.

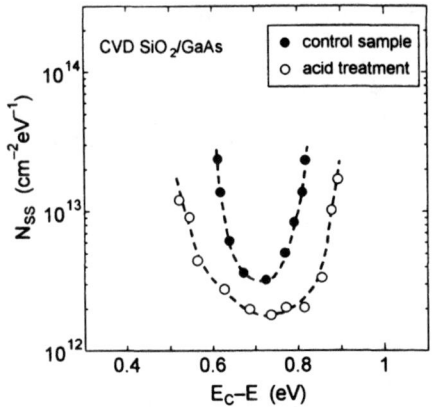

Fig. 6 Decease in N_{SS} due to pre-cleaning by acid treatment prior to the CVD process.

Wcm^{-2} for the acceptable light intensity for room temperature experiments. As further evidence, the surface potential data in Fig. 3 indicate that the they are at identical positions for both i/n and i/p samples. If the photovoltage effect were present, it tends to shift the curve for the i/n towards the conduction band whereas that for the i/p should be shifted towards the valence band. What should have resulted as a consequence is that these curves would not agree in the scanned range of surface Fermi level. The two curves, however, are almost overlapping and indicate that they are indeed free from the photovolatge effect.

Therefore, the quantitative accuracy of our ER scheme should allow detection of a change in the interface-states as a function of device process parameters. As an example, we have also characterized $N_{SS}(E)$ of CVD-SiO$_2$/GaAs interfaces as a function of fabrication process. Figure 6 shows the variation of N_{SS} due to an acid treatment of the surface prior to the SiO$_2$ film deposition. The data indicated by open circles are those for the surface of GaAs which was cleaned by the acid treatment prior to the deposition of SiO$_2$ layer. The reduction of N_{SS} in the entire energy range is observed, and the movable range of surface Fermi level increased from 0.3eV to 0.5eV. Meanwhile, the minimum in both cases are located at midgap (E_C - 0.72eV). From technological point of view, such numerical information must be useful for modeling GsAs FETs by device simulators if they can take into account the interface states with a continuous distribution in energy. Intuitively, on the other hand, the net charges induced at the interface states in a FET is not determined solely by themselves and a consideration of immobile charges in the passivation films is also necessary.

TRAPPED CHARGES IN PASSIVATION DIELECTRICS

Capacitance Measurements

The passivation films of devices are in most cases prepared by CVD techniques. As compared to silicon oxide formed by thermal oxidation, CVD dielectrics contain a large amount of structural imperfections and/or impurities such as hydrogen and hydroxyls. Depending on the charge states, such defects should also affect the device performance as fixed surface charges. It is, therefore, also important to characterize the trapping centers in the passivation films (CVD SiO$_2$ and/or SiNx) separately from the effect of the interface states. For this purpose, it is most simple and straightforward to evaluate the flat-band voltage shift of MIS diodes through capacitance measurements. However, since the interface states in the SiO$_2$/GaAs systems are too large in density and thus cause a serious distortion of the C-V characteristics, an alternative structure was found to be more convenient for the capacitance measurements. In this study, we use a Si-MOS diode with a double layer dielectrics as sketched in Fig. 7.

The samples were prepared as follows. First, a Si wafer (n~2x10^{15}cm^{-3}) was wet-oxidized after surface cleaning and an 8nm-thick thermal SiO$_2$ layer was formed. The wafer, then, was loaded into a CVD machine which is used for the passivation process of GaAs FETs, and an additional CVD-SiO$_2$ layer with a thickness of 100nm was deposited. In

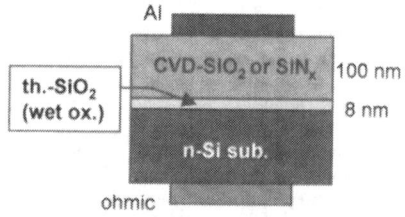

Fig. 7 Si-MOS diode structure for the C-V measurements of traps in CVD dielectric films.

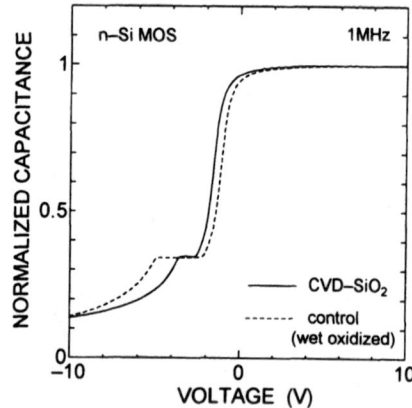

Fig. 8 C-V curve for CVD-SiO$_2$ MOS diode with a thin thermal-oxide at the interface (solid curve). The data for the control sample (100nm-thick thermal SiO$_2$) is also shown (dotted curve).

order to perform capacitance measurements, metal electrode was finally formed by evaporating Al through a metal-mask. As a control sample, a Si-MOS diode with 100nm-thick thermal oxide was also fabricated. As already pointed out, the C-V characteristics of GaAs MIS diodes suffers too much from the interface states and evaluation of oxide charges is nearly hopeless. In this diode structure using Si, however, the high-quality interface of thermal oxide/Si provides a well-defined C-V curve and an uncertainty in the interface quality for deposited dielectrics on semiconductor can be minimal. At the same time, the device yields a rather high sensitivity to the trapped charges present within the CVD dielectrics due to steep shapes of the C-V curves.

Figure 8 shows the typical high-frequency C-V curves for as-deposited samples. The one with a CVD-SiO$_2$ shows a quite similar C-V curve (*i.e.* steep rise) to that for the control sample with only a slight deviation in the flat-band voltage. As for the charges in the as-deposited dielectrics, oxide charges were certainly detected depending on the samples, whose density sometimes amounts to 10^{11}cm^{-2} range. The value, however, seems to be insufficient to cause a sizable effect on the breakdown characteristics of GsAs-FETs following the result of simulations by Barton and Landbrooke, who estimate the required surface charge density to be $10^{12} \sim 10^{13}$ cm^{-2}[1].

Effect of Carrier Injection

A drastic increase in the density of trapped charges was observed when the diode was subjected to a large negative bias-stress. The C-V curves shown in Fig. 9(a) were obtained after applying -65V to the MIS diode with a CVD-SiO$_2$ layer under room light illumination. (The illumination was simply intended to avoid the diode from working in deep depletion, in which case a significant voltage drop would appear across the semiconductor depletion region instead of the insulator.) In this case, a rapid shift of the flat-band voltage manifests generation of negative trapped charges with its density well beyond 2×10^{12}cm^{-2}. Actually, the amount of generated charges were estimated not based on the flat-band voltage but on the so-called midgap voltage because the stress slightly increases the interface-state density simultaneously. A similar behavior was also observed for CVD-SiN$_x$ films. Although a stress-induced trap generation is a well known phenomena in thermally oxidized SiO$_2$, the present result for the CVD films is likely to be due to a different mechanism because the stress

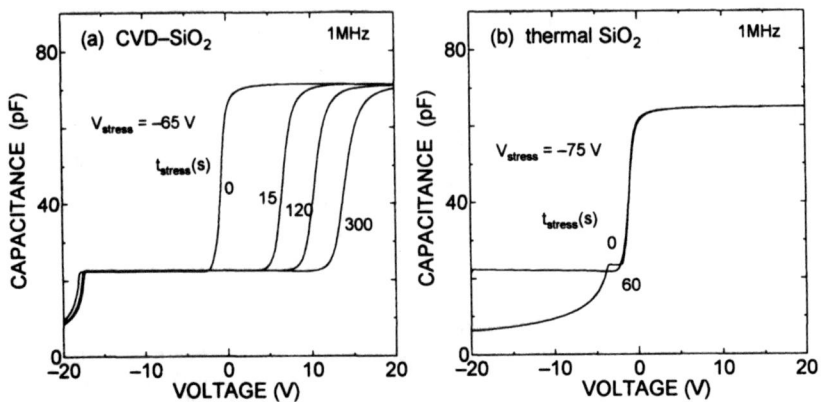

Fig. 9 (a) Shift of C-V curves for Si-MOS diode with CVD-SiO$_2$ film as a function of stress time at Vstress= -65V, indicating the generation of trapped negative charges. A shift of 1V corresponds to a sheet charge density of ~4x10^{11} cm^{-2}. (b) Such a shift is absent for thermal oxide, even with a larger voltage stress.

bias employed (~6 MV/cm) is still lower than the FN current regime. Furthermore, it was found that the cumulative injected current (total injected charges) necessary to cause the effect was 4 to 5 orders of magnitude smaller than the typical stress conditions used in the reliability study of thermal oxides. Actually, in the control sample (with wet-oxidized thermal SiO_2), such a stress-induced shift of the C-V curves was negligible under the present bias condition as shown in Fig. 9(b). Therefore, the observed effect is concluded to specifically reflect the structural instability of the CVD films.

As for the origin of these charges, we speculate that they are the electrons injected from the metal (Al) cathode which are then trapped at neutral oxide traps within the CVD-SiO_2 near the cathode region. This speculation is based on the following observations:

· Subsequent application of positive-bias stress removes the negative trapped charges, and the midgap voltage almost recovers to the original value. Therefore, defect generation during the bias-stress is rather unlikely.

· If the initial bias stress was of opposite polarity (electron injection from the thermal oxide side), the rapid increase of the trapped charges was not observed.

· Transient characteristics of the current during the bias stress shows a 1/t behavior frequently observed for tunneling electron injection in *stressed* thermal SiO_2 [12].

Taking into account the fact that the C-V measurement gives an average sheet charge density which is weighted by the distance from the insulator/semiconductor interface, the actual number of the trapped charges, and thus the density of neutral traps, should be even larger. In other words, CVD dielectric films contain a large density of neutral traps which initially are rather free from fixed charges but can act as a potential source for negative trapped charges once electrons are injected from outside.

Our findings, in fact, provides an important hint to the mechanism of metastable shift in the breakdown characteristics of FETs. When a MESFET is biased and held at the gate-drain breakdown condition for some period, the two-terminal I-V characteristics becomes shifted and the breakdown voltage of the device tends to increase[13]. Such a situation is shown by an example in Fig. 10. As evident in this figure, the device with its

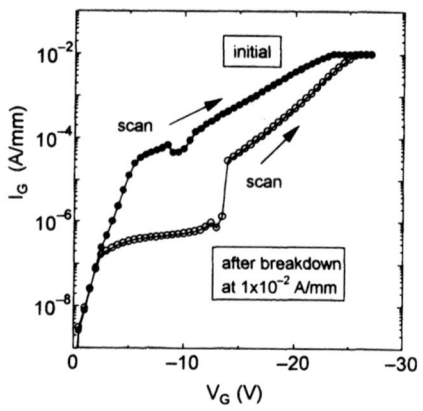

Fig. 10 An example of gate-drain I-V characteristics in FETs showing a metastable shift of BV_{gd} after holding at breakdown bias condition (10mA/mm).

first scan of the gate voltage shows a rather poor breakdown characteristics. After sustaining the breakdown condition at the end of this first scan, the two-terminal I-V characteristics was then shifted to the one indicated by open circles. This shift is metastable and the I-V curve does not recover to the original one at least under normal operation at room temperature. It is therefore speculated that, during the initial G-D breakdown, hot electrons are generated due to impact ionization, which are then most likely to surmount the potential barrier between the passivation dielectrics and GaAs to be finally captured by the neutral traps with in the passivation film. Since the generated charges are negative in polarity, the breakdown voltage should increase as a consequence, just as the device simulations predict as mentioned above.

SURFACE-CHARGE CONTROL IN GaAs-BASED MESFETs

Role of Negative Surface Charges

In fact, the speculation above was experimentally verified by Miyoshi *et al.*[14] using a FET configuration, in which an additional MIS electrode (denoted as FP) was placed adjacent to the drain-side of the gate electrode. By applying a negative voltage to this electrode, an increase in the G-D breakdown voltage was observed as shown as the shifted I-V characteristics in Fig. 11. Since the measurement was performed on the identical device, the effect arises from the surface (fixed) charges and there is no variation in the interface-state density which always used to be the major source of uncertainty. Although the large density of interface states might tend to shield the effect of surface charges, a sizable improvement of the breakdown characteristics due to the negative charges is indeed evident from the experiment. By analogy from such an effect provided by the negatively-biased MIS electrode, the negative charges trapped in the passivation dielectrics are also expected to play a similar role in the conventional device structure.

From the device point of view, however, it should be noted that a simple increase in the breakdown voltage does not usually guarantee an improvement of the overall device performance of power FETs. It has been already pointed out in an early study that the breakdown voltage of MESFETs tends to show a trade-off relation against output RF power (high-frequency response). In other words, a device with a large breakdown voltage often suffers from an increased magnitude of gate-lag[15]. However, it was rather difficult in the previous study to directly correlate the mechanisms determining the breakdown voltage and the gate-lag. Although the effect of the surface charges on the device breakdown has been considered previously as in the work of ref. 1, their further role in the dynamic response of the interface states was too complicated to discuss, especially in the passivated surface regions of FETs.

From such an interest, the effect of the negative surface charges on the gate-lag magnitude was also examined using the

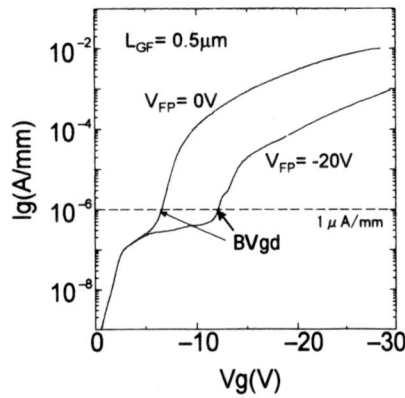

Fig. 11   Gate-drain I-V curves for a FET with additional MIS electrode (denoted as "FP") located 0.5µm away from the gate electrode.   Application of negative bias voltage (V_{FP}) increases the breakdown voltage.   After Miyoshi et al.[14]

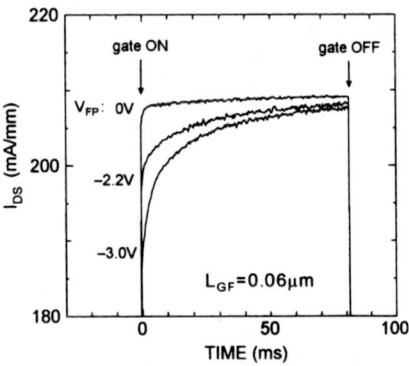

Fig. 12   Transient response of drain current when the gate is turned on (gate-lag), as a function of voltage applied to the FP (V_{FP}) during the gate-on period.

aforementioned FET structure implemented with the FP. Figure 12 compares the transient response of drain current after switching on the gate voltage. Here, it is observed that application of a negative voltage to the FP, which is equivalent to placing negatively charged oxide traps, increases the magnitude of the gate-lag, to which the emission of an increased number of electrons trapped at the interface states contribute when the FET is turned on. Thus, the results shown in Figs. 11 and 12 altogether provides a direct account for the trade-off relation between the breakdown characteristics and the gate-lag magnitude, for the first time from experiment to our knowledge. The gate-lag characteristics studied using this special FET structure will be described into further details elsewhere[16].

FPFET: A Novel GaAs-based Power FET

In order to overcome the trade-off between the breakdown voltage and the gate-lag, it is of great advantage to control the amount of surface charges in time domain. This follows from a consideration that the negative charges on the surface of channel are required only during the period of FET pinch-off, in which a large voltage drop appears between gate and drain, but they should be immediately removed as soon as the gate is subsequently turned on. In other words, with a typical load-line for power amplification, the G-D breakdown is not significant in the latter sequence. With such an idea, which is illustrated in Fig. 13, a new type of FET can be fabricated whose breakdown characteristics during the "off" cycle and rf response during the "on" cycle are both optimized even with a presently available quality for the passivated device surface.

The simplest method to realize such an operation is to connect the MIS electrode in the aforementioned MESFET structure with the gate electrode to bias the channel surface in-phase with the gate voltage. This is the FPFET (FP is for field-modulating plate) recently developed in by Asano et al.[3], whose structure is sketched in Fig. 14. The specific structure shown here has a thin inserted layer of AlGaAs, which is known as the concept of HFET, to increase the gate barrier height, but the essential improvement in the performance is due to the FP as evidenced from experiment also on the simpler MESFET version of FPFET[17].

The device characteristics are summarized in Table I. In the FP-HFET, the breakdown voltage as large as 47V was achieved as compared to the value of 28V for the same HFET without the FP. Meanwhile, the threshold voltage and maximum drain

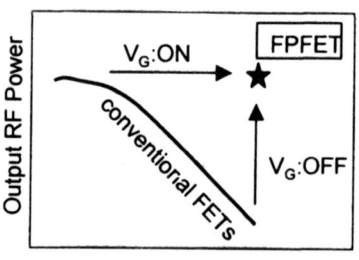

Fig. 13 Concept of time-domain control of surface negative charges achieved in FPFET.

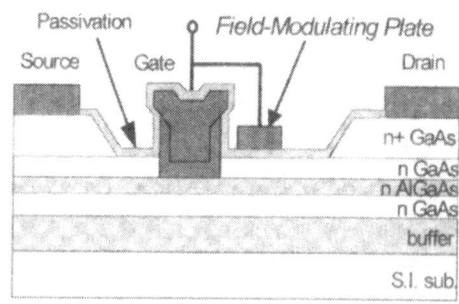

Fig. 14 Schematic device structure for the FP-HFETs. After Asano et al.[3]

116

Table I   Comparison of device parameters for FP-HFET and the conventional HFET after Asano *et al.*[3]

	DC parameters			RF (1.5 GHz)	
	BVgd (V)	Id-max (mA/mm)	Vth (V)	Pout-max (W/mm)	@Vd (V)
FP-HFET	47	326	-2.9	1.7	35
conventional HFET	27	323	-2.9	0.75	20

current were identical for these devices.   Due to its large breakdown voltage, operation drain voltage can be raised up to 35V.   In the RF (1.5GHz) operation of FP-HFET at this drain voltage, the maximum output power density per unit gate-width of 1.7W/mm was obtained which is more than 100% superior to that of the conventional HFET (0.8W/mm), whose output power already starts to saturate at ~15V due to the smaller breakdown voltage.   The output power for the FP-HFET indeed shows a linear increase with operation drain voltage up to 35V, and indicates that the degradation of drain current caused by the gate-lag is negligible for this device in spite of the largely improved breakdown characteristics.   Therefore, it was demonstrated for the first time that the time-domain control of the surface charges through FP can resolve the trade-off situation.

CONCLUSIONS

Characterization methods for charge traps at passivated GaAs surfaces are presented.   A large density of interface states cause a strong pinning of surface Fermi level and our new technique based on electroreflectance spectroscopy finds advantage over the conventional C-V method, due to its capability of directly evaluating semiconductor surface band-bending from Franz-Keldysh oscillation features.   Meanwhile, based on a bias-stress study of Si MOS diodes with double dielectric layers, the passivation films for GaAs devices are found to contain a large number of trapping sites which act as the potential source of negative charges. These findings are rationalized with FET characteristics and an importance of time-domain control of the surface charges is emphasized in the power FET operation which requires both high breakdown voltage and a lag-free current response.   For this purpose, an additional MIS electrode (a filed-modulating plate) in a FET structure (FPFET) is quite effective and its state-of-the-art performance has been verified.

ACKNOWLEDGMENTS

The author is grateful to K. Ito of Fundamental Research Laboratories, NEC, Y. Miyoshi, Y. Nashimoto, K. Asano, M. Kuzuhara and M. Mizuta of ULSI Device Development Laboratories, NEC, and H. Ishiuchi of Compound Semiconductor Device Department, NEC Kansai for their collaborations and useful discussions.

REFERENCES

1. T. M. Barton, and P. H. Ladbrooke, Solid State Electron. **29** p.807 (1986).
2. H. Hasegawa and T. Sawada, Thin Solid Films **103**, p.119 (1983).
3. K. Asano, Y. Miyoshi, K. Ishikura, Y. Nashimoto, M. Kuzuhara, and M. Mizuta, IEEE IEDM'98, Tech. Dig., p.59 (1998).
4. Y. Mochizuki and M. Mizuta, Appl. Phys. Lett. **69**, p.3051 (1996).
5. D. E. Aspnes, Phys. Rev. **B12**, p.371 (1975).
6. N. Bottka, D. K. Gaskill, R. J. M. Griffiths, R. R. Bradley, T. B. Joyce, C. Ito, and D. McIntire, J. Cryst. Growth 93, p.481 (1988).
7. C. van Hoof, K. Dennefe, J. DeBoeck, D. J. Arent, and G. Borghs, Appl. Phys. Lett. **54**, p.608 (1989).
8 H. Shen, M. Dutta, L. Foitadis, P. G. Newman, R .P .Moerkirk, W. H. Chang, and R. N. Sacks, Appl. Phys. Lett. **57**, p.2118 (1990).
9. Y. Mochizuki and M. Mizuta, Appl. Surf. Sci. **117/118**, p.614 (1997).
10. H. Shen and M. Dutta, J. Appl. Phys. **78**, p.2151 (1995).
11. X. Yin, H.-M. Chen, F. H. Pollak, Y. Chan, P. Montano, P. D. Kirchner, G. D. Pettit, and J. M. Woodall, J. Vac. Sci. Tech. **A10**, p.131 (1992).
12. D. J. Dumin, and J. R. Maddux, IEEE Trans. Electron. Dev. **40**, p.986 (1993).
13. Y. Miyoshi, Y. Mochizuki, Y. Nashimoto, and M. Mizuta, Proc. 1998 Spring Meeting of Jpn. Soc. Appl. Phys., p.III-1312 (1998), *in Japanese*.
14. Y. Miyoshi, K. Asano, Y. Nashimoto, Y. Mochizuki, K. Ishikura, M. Kuzuhara, and M. Mizuta, Int. Symp. on Compound Semiconductors, Nara, 1998.
15. J. C. Huang, G.S. Jackson, S. Shanfield, A. Platzker, P. K. Saledas, and C. Wiechert, IEEE Trans. Microwave Theory and Tech. **41**, p.752 (1993).
16. K. Ito, Y. Mochizuki, M. Mizuta, K. Asano, M. Kuzuhara, Y. Miyoshi, and Y. Nashimoto, unpublished.
17. K. Asano, Y. Miyoshi, K. Ishikura, Y. Nashimoto, M. Kuzuhara, and M. Mizuta, Extended Abs. Int. Conf. on Solid State Devices and Materials, Hiroshima, 1998, p.392.

A STUDY OF SEMICONDUCTOR QUANTUM STRUCTURES
BY MICROWAVE MODULATED PHOTOLUMENESCENCE

R.GULIAMOV*, E.LIFSHITZ*, E.COHEN**, A.RON*, H.SHTRIKMAN***
*Department of Chemistry and Solid State Institute, Technion, Haifa, 32000, Israel,
**Department of Physics and Solid State Institute, Technion, Haifa, 32000, Israel,
***Department of Condensed Matter Physics, The Weizmann Institute of Science,Rehovot,Israel
E- mail: ssefrat@tx.technion.ac.il

ABSTRACT

Mixed types I - type II multiple quantum wells (QW) structures consist of alternating narrow- and wide GaAs wells, separated by AlAs barriers. Transfer of electrons from the narrow- to the wide well results in the formation of a two-dimensional electron gas (2DEG) in the wide wells and a hole gas (2DHG) in the narrow ones. The present study investigated the effect of these gases on the various photoluminescence (PL) bands. The study utilized two modulations techniques: double beam PL and microwave modulated PL (MMPL), offering high-resolution spectroscopy, control of the 2DEG density and effective temperature. The results showed that the formation of a low density 2DEG in the wide wells cause the formation of trions. However, a large density of excess electrons makes mutual collisions with other photo-generated species, causing the dissociation of the trions and excitons. In addition, electrons transfer through the barrier gives rise to barrier-well indirect recombination emission.

INTRODUCTION

The optical properties of quantum well (QW) structures, containing a two-dimensional electron gas (2DEG) and a two-dimensional hole gas (2DHG), have a scientific and technological importance [1,2,3]. Primarily, these properties have been examined in a modulation–doped quantum wells (QW), where the electron density is controlled by applying a bias voltage [4]. Alternatively, the influence of 2DEG has been studied in mixed type I – type II QW's (MTQW) structure (vide infra), controlling the electron-density (n_e) by optical pumping [5,6,7,8].

The present study utilized MTQW's structure that is drawn schematically in figure 1. This structure consists of a sequence of alternating narrow- and wide GaAs wells, separated by AlAs barriers [1]. Furthermore, it is designed in such a way that the lowest Γ state of the GaAs narrow well (NW) is higher in energy than the lowest X state of the adjacent AlAs layer, leading to type II alignment. However, the last X state is energetically higher than the Γ state of the wide GaAs well (WW), creating type I alignment. Then, photocreated electron-hole pairs become spatially separated by rapid Γ-X-Γ electron transfer from the NW to the WW. It is already known that the initial Γ-X transfer occurs on a sub-picosecond time scale, while the subsequent X-Γ electron transfer takes about 30 picosecond [2]. Conversely, the hole tunneling through the barrier occurs in the millisecond range, thus leading to a temporary accumulation of the 2DEG in the WW. The present study describes our attempts to follow the interactions of the 2DEG and 2DHG with other photogenerated species. The results revealed the existence of the following events: (a) attachment of an excess electron to an existing exciton in the WW, to form a negatively charged specie, named a trion [5,6]; (b) direct recombination of an ensemble of free electrons with an existing hole within the WW, forming a plasma-like emission band [9]; (c) further collision of an excess electron with an existing exciton or trion within the WW, resulting

119

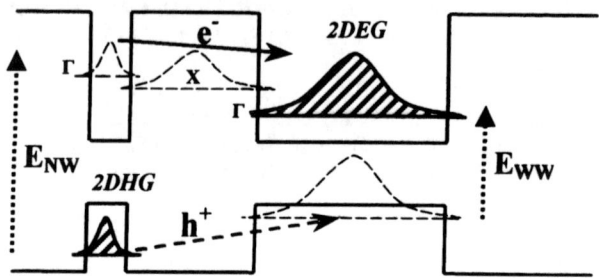

Figure 1: Schematic energy band diagram of MTQW structure

in a dissociation of the latter species [5,7,9]; (d) indirect electron-hole recombination between electron in the barrier X state and a hole either in the NW or WW. The aforementioned properties depend on n_e and on the gas effective temperature. 2DEG and 2DHG are generated by optical pumping in resonance with the NW band-gap energy (E_{NW}), while other photo-generated species were formed by optical excitation in resonance with the WW band-gap energy (E_{WW}). The effective temperature of 2DEG varied by additional irradiation of the samples with a microwave beam. Thus, the present study discuss the utilization of double beam photoluminescence (PL) and microwave modulated PL (MMPL). Detailed description of the experimental setup is given in the next section. The results and discussion are presented in the following sections.

EXPERIMENT

A MTQW structure, consisting of 30 periods of alternating, wide (198Å) and narrow (26Å) GaAs wells, separated by 102Å width AlAs barriers, have been used in the present research. The sample was placed in a microwave (mw) resonance cavity (operating at 10.755 GHz), which by itself was mounted in the liquid-He dewar with a superconductive magnet (up to 3T). The mw power dissipation within the cavity at 0-dB attenuation was about 55mW, however only 2% of it was absorbed by the sample.

The electron-hole pairs in the NW were generated by nearly resonance excitation with E_{NW}, by a He-Ne (1.96 eV) laser. The excitons in the WW were generated by nearly resonance excitation with E_{WW}, by a Ti sapphire (1.62 eV) laser. Accumulation of electrons in the barrier was formed by excitation with an Ar^+ (2.41 eV) laser. The single beam, double beam and mw modulated PL spectra, were all recorded at 1.4K. In addition, in the double beam experiment, both lasers impinged on the same spot of the sample.

RESULTS AND DISCUSSION

A representative PL spectrum of the MTQW structure, excited at E_{WW}, is shown in figure 2a. It consists of exciton (X) and trion (X^-) bands, centered at 1.524 eV and separated by 1.42 meV. The relative intensities of the indicated bands vary with a laser excitation power, and above 0.5 mW/cm^2 the trion band dominates the spectrum. A PL spectrum of the MTQW

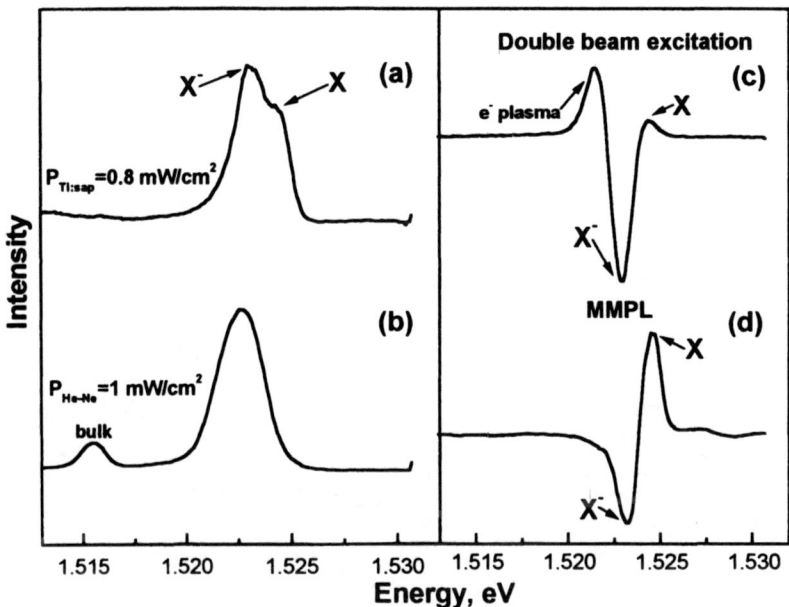

Figure 2: PL spectra under Ti:sapphire (a) and He-Ne laser (b) excitation, PL spectrum at double beam configuration and modulation of He-Ne laser (c), MMPL spectrum under Ti:sapphire laser excitation (d).

structure, excited at E_{NW}, is shown in figure 2b. It is dominated by a broad and featureless band, associated with the recombination between large density 2DEG (generated with E_{NW}) and holes in the WW. It should be noted that this plasma-like band extends to lower energy than the corresponding exciton and trion due to the re-normalization of the conduction band, occurring at large n_e [9]. The weak band at 1.515 eV corresponds to a bulk GaAs substrate and will not be discussed any further.

The double beam PL spectrum, recorded at He-Ne laser modulation (modulation of 2DEG density) is shown in figure 2c. It is seen that a modulation of n_e leads to difference spectra resolving the exciton (high energy and positive), trion (negative) and distinct plasma (low energy and positive) bands. The quenching of the trions population corresponds to their dissociation under intense flux of n_e and their conversion into excitons and plasma species [9].

Typical MMPL spectrum under Ti:sapphire laser excitation is shown in figure 2d. One can see that mw absorption causes a decrease of X^- with simultaneous increase of the X band. This is due to the induced collisions between mw-heated 2DEG with the trions that consequently lead to the dissociation of the last ones.

As indicated at the introduction, the electron transfer via the Γ-X-Γ cascade of states results in a pause of 30 psec at the barrier [2]. At this time, electrons may make indirect recombination with holes, either in the narrow or wide wells. We presume that the recombination processes observed around 1.76 eV are associated with those indirect transitions. Representative

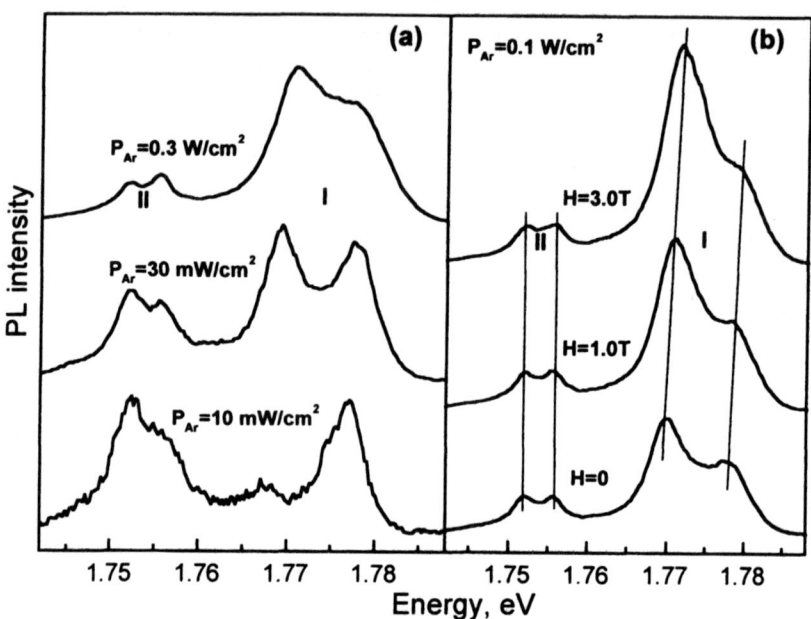

Figure 3: PL spectra at the barrier energy range at different levels of Ar⁺ laser excitation (a), and different magnetic fields (b).

PL spectra at this energy range are shown in figure 3. The spectra contain two groups, each consists of two peaks (the 1.770 and 1.778 eV correspond to group I, while 1.752 and 1.756 eV peaks correspond to group II). The spectra in figure 3a indicate that an increase in the excitation intensity causes a relative increase of group I and some variation in the relative intensities within a group. Application of static magnetic field causes a drastic increase of group I and a typical diamagnetic shift, while group II represent a negligible change (figure 3b).

The MMPL spectra at the 1.76 eV energy range, recorded at different levels of mw power are shown in figure 4a. This figure shows that group I exhibit a gradual quenching with an increase in the mw heating, while peak at 1.756 eV within group II shows an initial enhancement followed by a quenching with an increase of the mw power. The change in the PL intensity under microwave radiation versus the microwave power is plotted in figure 4b.

Overall, the PL and MMPL spectra shown in figures 3 and 4, indicate the existence of different origin for the emitting groups. Preliminary calculations (not shown) suggest a recombination mechanism between electrons in the barrier, with heavy and light holes in the narrow (group I) and wide (group II) wells, as shown schematically in figure 5. The quenching of group I in the MMPL spectra maybe associated with mw heating of electrons within the barrier and their increased transfer to the NW. Partial enhancement of the 1.756 eV peak, at certain mw powers, can be due to redistribution among the WW's hole states, while complete quenching of the indirect excitons, corresponds to an induced dissociation by the mw induced collisions.

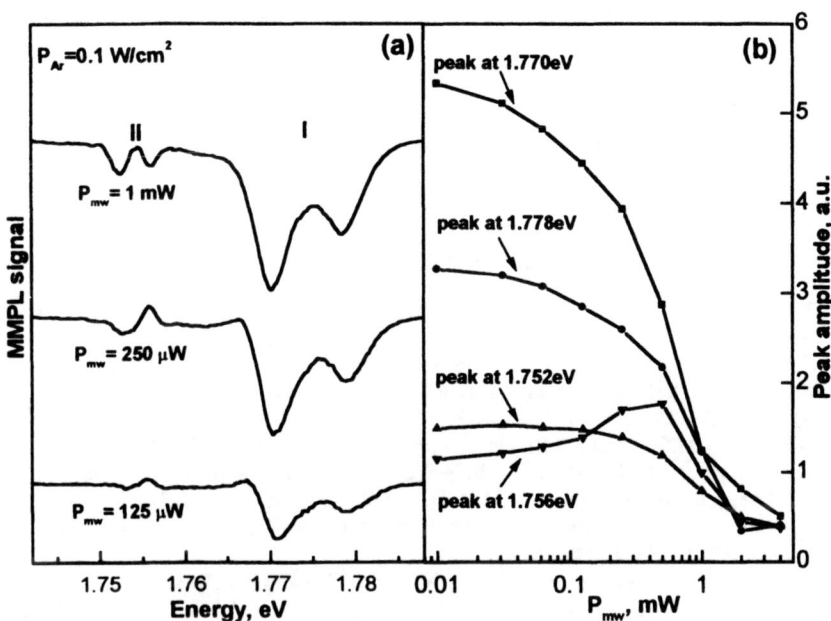

Figure 4: MMPL spectra at the barrier energy range at different mw powers (a), the behavior of the PL intensity for different peaks at different mw powers (b).

As a summary, the present study represented the occurrence of few photoluminescence events (exciton, trion and plasma) in the WW, utilizing high-resolution spectroscopic methods, such as double beam excitation and MMPL. While, the WW optical processes had been studied

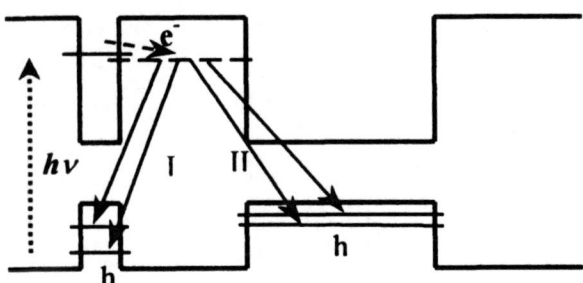

Figure 5: Recombination scheme, associated with the barrier in MTQW.

before by other means, the indicated methodologies in the present work enabled to investigate also the recombination mechanisms between the barrier and the wells. Further analysis of the indirect transitions is currently being done in our laboratory and will be published in the near future.

ACKNOWLEDMENTS

This work was done in the Barbara and Norman Seiden Center of Advanced Optoelectronics at Technion and supported by the Israel Ministry of Science.

REFERENCES

1. I.Galbraith, P.Dawson, C.T.Foxon, Phys.Rev. **B 45** (23), 13499 (1992).
2. J.Feldmann, M.Preis, E.O.Göbel, P.Dawson, C.T.Foxon, I.Galbraith, Solid State Commun. **83** (3), 245 (1992).
3. G.Finkelstein, H.Shtrikman, I.Bar-Joseph, Phys.Rev. **B 56** (16), 10326 (1997).
4. F.Hirler, R.Strenz, R.Küchler, G.Abstreiter, G.Böhm, G.Weimann, Surf.Sci. **305**, 591 (1994).
5. A.Manassen, E.Cohen, Arza Ron, E.Linder, L.N.Pfeiffer, Phys.Rev. **B 54**(15), 10609 (1996).
6. Arza Ron, H.W.Yoon, M.D.Sturge, A.Manassen, E.Cohen, L.N.Pfeiffer, Solid State Commun. **97** (9), 741 (1996).
7. R.Harel, E.Cohen, E.Linder, Arza Ron, L.N.Pfeiffer, Phys.Rev. **B 53** (12), 7868 (1996).
8. A.Manassen, E.Cohen, Arza Ron, E.Linder, L.N.Pfeiffer, Superl.& Microstr. **15** (2), 175 (1994).
9. A.Manassen, R.Harel, E.Cohen, Arza Ron, E.Linder, L.N.Pfeiffer, Surf.Sci. **361/362**, 443 (1996).

Characterization of GaS-Passivated Quantum-Well Laser Diodes

L. G. VAUGHN*, T. C. NEWELL*, L. F. LESTER*, AND A. N. MACINNES**
*Center for High Technology Materials, University of New Mexico, 1313 Goddard S. E., Albuquerque, NM 87106
**TriQuint Semiconductor, 2300 N. E. Brookwood Parkway, Hillsboro, OR 97124

ABSTRACT

The degradation of AlGaAs/GaAs diode laser performance during operation is typically due to catastrophic optical damage of the facets caused when thermal runaway occurs. These heating effects are due to the presence of non-radiative recombination sites at and near the facets. MOCVD GaS is deposited on the facets of 825-nm ridge waveguide AlGaAs/GaAs quantum-well laser diodes as an electronic passivation to reduce the number of surface states available for non-radiative recombination. For passivated devices, a peak pulsed power nearly double that of unpassivated devices was achieved. The passivated devices also exhibit a longer lifetime before degradation. The impact of the passivation process on other optical characteristics of the laser diodes will also be discussed.

INTRODUCTION

The one failure mechanism in an edge-emitting laser diode that cannot easily be mitigated by a modification in the design geometries and process controls is degradation of the cleaved facets during device operation. This permanent degradation, known as catastrophic optical damage (COD) has been studied extensively[1-3], and is due to the process of localized heating as carriers in the diode are absorbed at non-radiative recombination sites. This heating leads to bandgap narrowing allowing more carriers and photons to be absorbed at these sites, which in turn leads to more heating and more absorption. Thermal runaway occurs as this positive feedback process continues, eventually resulting in localized melting of the crystal and degradation or interruption of lasing. Non-radiative recombination sites can be crystal inhomogeneities, such as crystal defects and impurities, but moreover can be surface states at the edges of the crystal. These surface states, often termed 'dangling bonds', are sufficiently high in concentration at the cleaved surfaces of the laser diode, so that even when crystal defects and impurities are minimal, the cleaved surface itself is sufficient to nucleate damage.[1]

For GaAs, the concentration of surface states can be so large that the Fermi level is pinned. Attempts to satisfy some of these dangling bonds and hence reduce surface state density have included passivation by overlayers such as SiO_2,[4] SiN_x,[5,6] $ZnSe$,[7-9] or GaS.[10,11] The extent to which the concentration of surface states is reduced is termed electronic passivation and contrasts with physical passivation or encapsulation that merely prevents facet degradation by preventing interaction with the ambient.

The passivating effects of sulfur and sulfur containing species on GaAs and related compounds are well known.[12] In this respect, ammonium sulfide has been shown to remove the native surface oxides on GaAs and form sulfide monolayers.[13] These monolayers tend to suppress re-oxidation of the surface in air.[14] Related studies utilized P_2S_5-ammonium sulfide to treat the cleaved facets of edge-emitting diode lasers. The resulting passivation imparted a

125

significant increase in the maximum achieved output power without COD.[15] However, the significant disadvantage of this process is that the results are only temporary. In addressing the transient nature of sulfur-containing solution treatments, a more stable passivation for GaAs surfaces has been demonstrated using metalorganic chemical vapor deposition (MOCVD) grown gallium sulfide films.[10] These films demonstrated a high degree of surface passivation; so much as to unpin the Fermi level,[11] and furthermore, maintain their stability for an extended period of time.

Although many GaAs surface passivation processes have been studied, the effects of these films are more often tested on metal-insulator-semiconductor (MIS) devices and less often on laser diodes. In this presentation, we investigate the effects of GaS passivation on ridge waveguide AlGaAs/GaAs quantum-well laser diodes. Unpassivated laser diodes are compared with those passivated using different surface preparations. The passivation of the laser diode facets using an MOCVD gallium sulfide film after removing the native oxide gives results consistent with a substantial reduction in the surface states and, therefore, a measurable increase in peak output power before COD.

EXPERIMENT

The laser diodes were fabricated concurrently from the same Center for High Technology Material (CHTM)-grown AlGaAs/GaAs multiquantum-well structure wafer, which emits light at 825 nm. All samples were processed using a 9-micron ridge waveguide process. The ridges were etched in the starting material then the entire surface was covered with a plasma-enhanced chemical vapor deposition (PECVD) silicon nitride film to provide electrical isolation. P-side openings were etched through the silicon nitride on the tops of the ridges. This was followed by e-beam evaporation of two levels of Ti/Pt/Au metallization on the front side of the wafer to form the p-side contact of the devices. The backside of the wafer was then thinned and metallized with Ni/Ge/Au/Ag/Au to form the n-side contact of the device. The completed wafer of laser diodes was cleaved at 500-μm intervals to yield bars of twenty electrically isolated devices each.

For the purpose of this study, three separate facet coatings were investigated. These were designated as '374', 'TQ1' or 'TQ5' (see Table I for details) and present a matrix of facet preparations that allow the relative contributions of electronic versus physical passivation to be established for the applied GaS coatings. After cleaving, the 374 control sample and the TQ1 samples were separated and the TQ5 sample were dipped in ammonium sulfide to remove the native oxide. It must be noted that the ammonium sulfide in this case is not thought to contribute to the actual passivation but merely provides a method of surface cleaning. This is based upon an earlier comparison of GaAs surface photoluminescence for surfaces prepared by either ammonium sulfide immersion, under an inert atmosphere in a glove box or simply as received; all followed by GaS coating. The glove box and sulfide-prepared surfaces were observed to be similarly passivated whereas the as-received surface displayed only very minor improvements in surface electronic properties—this feature is exploited in the laser studies described here.

A GaS film of approximately 200 Å was then deposited on the TQ1 samples and the pre-treated TQ5 samples by MOCVD in a hot-walled chamber using [(t-Bu)GaS]$_4$ as the precursor. This deposition procedure has been described elsewhere.[16] Following the film deposition, TQ1 and TQ5 samples were annealed at 400 °C for 5 minutes in a forming gas (N$_2$/H$_2$) atmosphere using a rapid thermal anneal (RTA) process. The role of this anneal is still unclear but has been shown to enhance passivation effects on GaS/GaAs interfaces. This anneal may allow interfacial rearrangement, maximizing Ga-S bonds at the interface or removes some of the excess carbon expected in the MOCVD GaS film. In Table I, the processing matrix for the three samples is described. TQ5 may thus be viewed as the electronically passivated facet coating proposed in this investigation. TQ1 is the same coating but overlayed on the pre-existing non-passivating native oxide i.e., the GaS is not expected to provide electronic passivation, but rather a physical encapsulation. In this manner we aim to differentiate between the electronic passivating ability and simple encapsulating ability of the deposited GaS film.

Table I. Processing Matrix

Sample Set	Pre-treatment	Passivation	Post-treatment
374 (control)	None	None	None
TQ1	None	MOCVD GaS	400 °C
TQ5	Ammonium Sulfide	MOCVD GaS	400 °C

Optical testing was performed on the laser diode bars. Each bar was tested p-side up on a thermoelectric cooler maintained at 18 ± 0.1 °C and driven by an ILX Lightwave LDP-3840 pulsed current source. For every test, the pulse width was 300 ns with a 1% duty cycle. Emitted light was collected in a Labsphere integrating sphere using a silicon detector.

RESULTS

In the following discussion, the output powers are for single facet emissions. Since the passivation is a conformal coating and is applied to both sides of the laser bar, any variations in the amount of light emitted from the facets on each of the two sides is assumed to be due to minor differences in the cleaved facets. A range of optical properties is typically observed on laser structures where the failure may be attributed to thermal runaway causing COD or to laser structure defects. Lasers failing at anomalously low powers are generally due to structural faults such as poor cleaving or metallization, or localized growth problems whereas laser facet degradation is usually associated with the upper limit of performance.

Devices were monitored for power while the drive current was slowly increased. This output power versus input current behavior for the best devices from each of the samples is illustrated in Figure 1.

Lasing Threshold Current

The lasing threshold, I_{th}, for the 374 control sample of unpassivated lasers ranges from 40 to 55 mA with an average value of 47 mA. The GaS-passivated samples, both with and without the pre-treatment, yielded I_{th} values from 60 to 90 mA. The average value for TQ1, coated devices with no pre-treatment, is 77 mA, and for TQ5, coated devices with pre-treatment is 74 mA. The increase in lasing threshold for TQ1 and TQ5 is due, in part, to the decrease in facet reflectivity. The GaS film is an anti-reflective coating,[11] and thus higher bias currents are required to create the necessary cavity resonance condition.

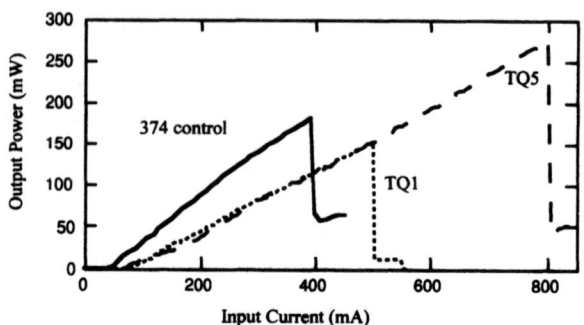

Figure 1. Best L-I curves from the 374 control samples, the GaS-passivated samples with no pre-treatment (TQ1), and the GaS-passivated samples with pre-treatment (TQ5).

External Differential Quantum Efficiency (Slope Efficiency)

The external differential quantum efficiency, η_d, is sometimes referred to as the slope efficiency, and is a measure of the change in output photon flux for a given change in injected electron flux. The average slope efficiency for the unpassivated control was 69%. The average values for the passivated with no pre-treatment, TQ1, and the passivated with ammonium sulfide pre-treatment, TQ5, were 43% and 40%, respectively. The decrease in efficiency along with the large increase in threshold current are not attributable to only the anti-reflective nature of the GaS passivation layer. These changes indicate a decrease in quantum efficiency, due to the high temperature processing.

COD Limit

The COD limit is characterized by the peak output power. The best TQ5 device, pre-treated and GaS-passivated, demonstrated a maximum output power of 280 mW, while the best 374 control sample device exhibited a peak output power of only 180 mW. This 60%+ increase in peak power is consistent with electronic passivation of the facet surface. Considering the diminished operating characteristics of TQ5 compared to the 374 control sample, this is an even more significant increase. The best TQ1 device, GaS-coated with no pre-treatment, failed after emitting a peak output power of 150 mW. This is below both the best devices from both TQ5 and 374.

It is interesting to note that the devices receiving a pre-treatment before the GaS passivation layer (TQ5) have an average lasing threshold and slope efficiency similar to those with no pre-treatment (TQ1). They differ only in the ability of the GaS to electronically passivate the laser facet due to the removal of the native oxide. Therefore, the difference between these two groups of lasers is the passivation effect of the GaS. So, the peak power of the best TQ5 device is almost double that of the best TQ1 device.

Average values for peak output power show similar results. TQ5 has an average peak power of 189 mW, nearly double the 374 average of 104 mW, and more than double the TQ1 average of only 88 mW.

High Power Life Test

The lifetime of the devices was tested by monitoring the duration a device was able to function at a constant pulsed input current before degradation. A typical device from each of the samples was tested at an input current of 275 mA using a pulse width of 300 ns, a duty cycle of 1% and a temperature of 18 °C. The devices were stressed at a constant pulsed current and operated until failure. This current is approximately 6 times the threshold current and just below the COD limit for 374, the control sample, though only about 3 1/2 times the threshold current of the two coated samples. A substantially longer lifetime was observed for the GaS-passivated TQ5 device. This TQ5 device failed after 232 minutes, while the 374 control device failed after 5 minutes and the TQ1 device, GaS-coated without pre-treatment, failed after 30 minutes. This result also allows us to differentiate the extent to which the bare facet of 374 contributes to a low observed lifetime under the applied pulsed conditions. The observation of a lifetime increase with the simple physical passivation of TQ1 indicates that even under the pulsed low duty cycle testing there is a significant facet/ambient interaction that contributes to lasing degradation. The further dramatic increase in lasing lifetime on the provision of an electronic passivation (TQ5 vs TQ1) clearly indicates the role of surface states lifetime degradation. Future tests will drive the lasers at equivalent output power instead of equivalent input current to verify this effect.

A summary of the average values for threshold current, slope efficiency, and peak power, is presented in Table II. The peak power for the best device from each sample and the time-to-failure for each sample at an input current of 275 mA are also listed.

Table II. Summary of the Test Results

	Average Threshold I_{th}	Average Efficiency η_d	Average Peak Power	Peak Power of Best Device	Lifetime at 275 mA
	mA	%	mW	mW	minutes
374 control	47	69	104	180	5
TQ1	77	43	88	150	30
TQ5	74	40	189	280	232

CONCLUSIONS

The peak output power and the lifetime of the pretreated GaS electronically passivated devices show a significant increase when compared to the unpassivated control devices and the physically passivated devices without a pre-treatment. This increase in COD limit and lifetime are due to electronic passivation of the surfaces and indicate the critical nature of surface states on laser performance and lifetime. An increase in threshold current and a decrease in slope efficiency were also observed and are consistent with device degradation unrelated to the passivation process.

The poor performance of the physically passivated devices is expected since they received an anti-reflective coating which increased the threshold current. They did not benefit from the effects of the passivation because the native oxide present prior to deposition prevented the GaS film from satisfying any of the dangling bonds on the facet surfaces, which, in turn, led to a low COD limit and a short lifetime.

Since the GaS was deposited on finished devices, metallization was present so the maximum anneal temperature had to be kept relatively low. Further studies will use etched facet laser diodes that can be passivated and annealed prior to metallization. This should allow for more latitude in process optimization.

ACKNOWLEDGMENTS

Financial support for this work was provided by the National Science Foundation, NSF project reference# DMI-9422662.

REFERENCES

1. C. H. Henry, P. M. Petroff, R. A. Logan, and F. R. Merritt, J. Appl. Phys. **50**, p. 3721-3722 (1979).
2. W. C. Tang, H. J. Rosen, P. Vettiger, and D. J. Webb, Appl. Phys. Lett. **58**, p.557-559 (1991).
3. W. C. Tang, H. J. Rosen, P. Vettiger, and D. J. Webb, Appl. Phys. Lett. **59**, p. 1005-1007 (1991).
4. G. J. Gerardi, F. C. Rong, E. H. Poindexter, M. Harmatz, H. Shen, and W. L. Warren, in *Amorphous Insulating Thin Films*, edited by J. Kanicki, W. L. Warren, R. A. B. Devine, and M. Matsumura (Mat. Res. Soc. Symp. Proc. **284**, Pittsburgh, PA 1993) pp. 601-606.
5. K. Remashan and K. N. Bhat, Electron. Lett. **32**, pp. 694-695 (1996).
6. G. W. Charache, S. Akram, E. W. Maby, and I. B. Bhat, IEEE Trans. Electron. Devices **44**, p.1837-1842 (1997).
7. C. J. Sandroff, R. N. Nottenburg, J. C. Bischoff, and R. Bhat, Appl. Phys. Lett. **51**, p. 33-35 (1987).
8. C. J. Sandroff, M. S. Hegde, L. A. Farrow, R. Bhat, J. P. Harbison, and C. C. Chang, J. Appl. Phys. **67**, p. 586-588 (1990).
9. N. Chand, W. S. Hobson, J. F. deJong, P. Parayanthal, and U. K. Chakrabarti, Electron. Lett. **32**, p. 1595-1596 (1996).
10. A. N. MacInnes, M. B. Power, A. R. Barron, P. P. Jenkins, and A. F. Hepp, Appl. Phys. Lett. **62**, p. 711-713 (1993).
11. M. Tabib-Azar, S. Kang, A. N. MacInnes, M. B. Power, A. R. Barron, P. P. Jenkins, and A. F. Hepp, Appl. Phys. Lett. **63**, p. 625-627 (1993).
12. H. Kawanishi, H. Ohno, T. Morimoto, S. Kaneiwa, N. Miyauchi, H. Hayashi, Y. Akagi, Y. Nakajima, and T. Hijikata, *21ˢᵗ Conference on Solid State Devices and Materials, Tokyo, 1989*, pp. 337-340.
13. H. Ohno, H. Kawanishi, Y. Akagi, M. Koba, and T. Hijikata, (Mat. Res. Soc. Symp. Proc. **204**, Pittsburgh, PA 1991) p. 65.
14. G. Guel, E. A. Armour, S. Z. Sun, S. T. Srinivasn, K. J. Malloy, and S. D. Hersee, J. Electron. Mater. **21**, p. 1051 (1992).
15. J. S. Yoo, H. H. Lee, and P. Zory, IEEE Photon. Technol. Lett. **3**, p. 202-203 (1991).
16. A.N. MacInnes, M.B. Power, and A.R. Barron, Chem. Mater. **4**, p. 11-14 (1992).

ENERGY BAND OFFSETS AT A Ga$_2$O$_3$(Gd$_2$O$_3$)-GaAs INTERFACE

T.S. Lay[1], M. Hong[2], J. Kwo[2], J.P. Mannaerts[2], W.H. Hung[3], D.J. Huang[3]

[1]Institute of Electro-Optical Engineering, National Sun Yat-Sen University, Taiwan
[2]Bell Laboratories, Lucent Technologies, Murray Hill, NJ07974
[3]Synchrotron Radiation Research Center, Hsinchu, Taiwan

ABSTRACT

We report the energy band offsets at a Ga$_2$O$_3$(Gd$_2$O$_3$)-GaAs interface. The valence-band offset (ΔE_V) is ~ 2.6 eV, measured by soft x-ray photoemission spectroscopy. Analysis of the current-voltage characteristics of a Pt-Ga$_2$O$_3$(Gd$_2$O$_3$)-GaAs MOS (metal-oxide-semiconductor) structure, which are dominated by Fowler-Nordheim tunneling, reveals a conduction-band offset (ΔE_C) ~ 1.4 eV at the Ga$_2$O$_3$(Gd$_2$O$_3$)-GaAs interface and an electron effective mass (m*) ~ 0.29 m$_e$ of the Ga$_2$O$_3$(Gd$_2$O$_3$) film.

INTRODUCTION

Recently, in-situ deposition of an oxide mixture of Ga$_2$O$_3$(Gd$_2$O$_3$) on GaAs substrate has drawn a great attention in compound semiconductor microelectronics [1]. The Ga$_2$O$_3$(Gd$_2$O$_3$)-GaAs interfaces are thermodynamically stable and have a low interfacial density of states. Using this novel oxide and a conventional ion implantation, both p- and n-channel enhanced-mode GaAs MOSFETs (metal-oxide-semiconductor field-effect-transistors) have been demonstrated [2,3]. However, the relevant energy scales at the Ga$_2$O$_3$(Gd$_2$O$_3$)-GaAs interface, which are the essential parameters for device modeling, have not been investigated. In this paper, we report the conduction- and valence-band offsets of an as-grown Ga$_2$O$_3$(Gd$_2$O$_3$)-GaAs interface, which are determined using current-voltage measurements and room-temperature soft x-ray photoemission spectroscopy (SXPS), respectively.

EXPERIMENT

The sample was prepared as followed. First, a Si-doped GaAs layer (n ~ 1 x 10^{18} cm^{-3}) 3000 Å-thick was grown on a 2-inch, (001)-oriented n$^+$-GaAs substrate in a solid source GaAs-based molecular beam epitaxy (MBE) chamber. The wafer is, then, transferred through an ultra-high vacuum (UHV) module (without being exposed to air) to an arsenic-free oxide deposition chamber, and a Ga$_2$O$_3$(Gd$_2$O$_3$) film of thickness (t$_{ox}$) = 8.8 nm was deposited by e-beam evaporation from a single crystal Ga$_5$Gd$_3$O$_{12}$ garnet. The detail of the oxide deposition was previously reported [4]. For electrical measurements, a metal gate 125 µm-diameter with Pt as a contact was evaporated on a piece of the as-grown wafer to form a metal-oxide-semiconductor (MOS) structure. The MOS device was characterized by current density (J) vs. gate bias (V$_G$) measured by using a HP4156A high precision semiconductor analyzer. In order to reduce the background noise, the sample was placed in a dark, nitrogen gas blown probe station.

RESULTS AND DISSCUSION

Figure 1 shows the J-V_G data at both forward (positive V_G) and reverse (negative V_G) biases. At forward bias, the leakage current density remains an almost constant level of ~ 10^{-7} A/cm^2 for the t_{ox} = 8.8 nm device at low V_G. When the V_G is increased to around at 2.5 V, an abrupt increase in current density is observed. At V_G = 5V, the leakage level J reaches 0.5 A/cm^2. Similar characteristics are also exhibited in the J- V_G trace at reverse bias, while the J abruptly increases at V_G < -3.5 V.

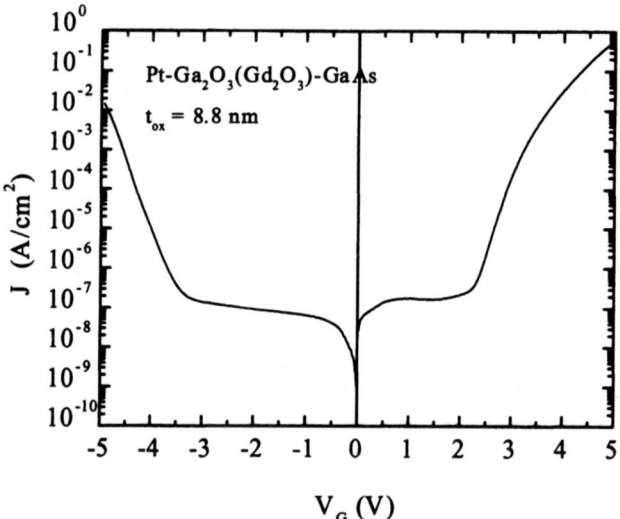

Fig. 1   Leakage current density (J) as a function of gate bias (V_G) for the Pt- Ga$_2$O$_3$(Gd$_2$O$_3$)-GaAs MOS device of t_{ox} = 8.8 nm.

For the MOS structure, the relevant energy levels are Pt work function Φ_m = 5.3 eV, oxide electron affinity χ, and GaAs electron affinity χ_s = 4.1 eV [5]. In order to determine the unknown χ, and therefore the conduction-band offset, we re-plot the electrical data in a fashion of log(J/E^2) vs. 1/E, i.e., where E {= [V_G – (Φ_m -χ_s)/e]/ t_{ox}} is the oxide electrical field [6]. As shown in Fig. 2, the curves exhibit a linear region at high E for both forward and reverse biases. The linear relation of log(J/E^2) vs. 1/E indicates a Fowler-Nordheim (FN) tunneling through the oxide film, and also indicates the high quality of the oxide film. From the FN tunneling, the slope (S) of log(J/E^2) vs. 1/E can be written as [7]:

$$S = \frac{d[\log(J/E^2)]}{d(1/E)} = -\frac{4\sqrt{2m^*}}{3q\hbar}(\Phi)^{3/2} \cdot (\log e) \quad (1),$$

where m^* is electron effective mass of the $Ga_2O_3(Gd_2O_3)$ film and Φ is the tunneling barrier height. The tunneling barrier heights can be expressed as:

$$\Phi^+ = \Delta E_C = (\chi_s - \chi) \qquad (2) \quad \text{for forward bias, and}$$

$$\Phi^- = (\Phi_m - \chi) \qquad (3) \quad \text{for reverse bias,}$$

where ΔE_C is the conduction-band offset at the $Ga_2O_3(Gd_2O_3)$-GaAs interface. Therefore, by inserting the values of S, which is extracted from Fig. 2, into Equation (1), and in conjunction with $(\Phi_m - \chi_s) = 1.2$ eV, we then obtain the conduction-band offset $\Delta E_C \sim 1.4$eV at the $Ga_2O_3(Gd_2O_3)$-GaAs interface, the oxide electron effective mass $m^* \sim 0.29$ m_e (m_e is electron rest mass), and $\chi \sim 2.7$eV. The value of $[\Delta E_C + (\Phi_m - \chi_s)] \sim 2.6$ eV is consistent with the forward bias level where the FN tunneling current arises (Fig. 1).

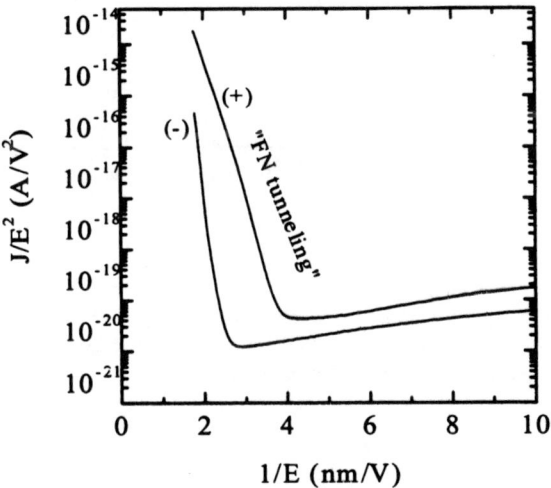

Fig. 2   Replot the J-V_G curve in the $\log(J/E^2)$ vs. $1/E$ coordinates, the linear regions indicate FN tunneling both at forward (+) and reverse (-) biases.

For the valence-band offset ΔE_V, the room-temperature SXPS measurements of the as-grown $Ga_2O_3(Gd_2O_3)$ film are performed in a UHV chamber using 70, 100, and 140 eV photons to maximize surface sensitivity and signal level. An Au foil is also measured to determine the Fermi edge (E_F). The combined resolution of the monochromator and the analyzer is about 0.2 eV at 70 eV.

Figure 3 shows the SXPS spectrum of the $Ga_2O_3(Gd_2O_3)$ film and the Fermi edge measured at 70 eV photon energy. Here we set $E_F = 0$. The spectrum shows a clear Ga(3d) and Gd(4f) core levels of binding energies at ~21.5 eV [8], and ~9.8 eV, respectively. The valence band edge (E_V) of $Ga_2O_3(Gd_2O_3)$ is extracted by linear extrapolation and has a value ($E_F - E_V$) = 4 eV. Since the GaAs substrate is heavyly doped, the Fermi level is almost lined up with the conduction-band edge of GaAs. Therefore, $\Delta E_V \sim 2.6$ eV at the $Ga_2O_3(Gd_2O_3)$-GaAs interface is obtained.

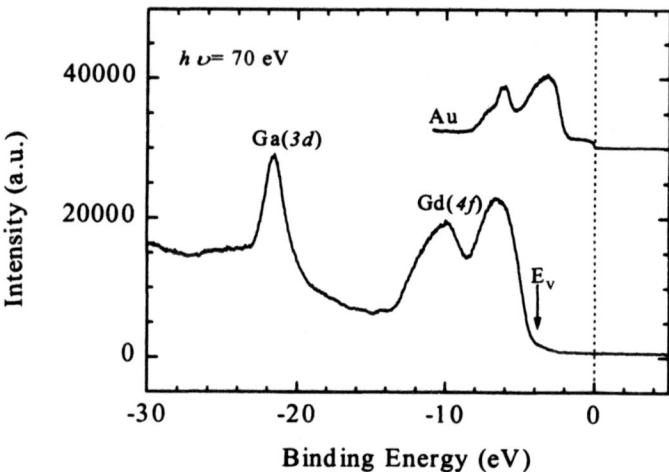

Fig. 3   The soft x-ray photoemission spectrum of the $Ga_2O_3(Gd_2O_3)$ film at photon energy $h\upsilon$ = 70 eV. The position of oxide valence-band edge E_V is linear extrapolated and the Fermi edge $E_F = 0$ is measured from an Au foil.

Table I   Energy scales of $Ga_2O_3(Gd_2O_3)$-GaAs interface

Oxide energy gap	Conduction-band offset	Valence-band offset	Oxide electron affinity	Oxide electron effective mass
$E_g = 5.4$ eV	$\Delta E_C = 1.4$ eV	$\Delta E_V = 2.6$ eV	$\chi = 2.7$ eV	$m^* = 0.29\, m_e$

SUMMARY

The relevant energy levels of the $Ga_2O_3(Gd_2O_3)$-GaAs are deduced from current-voltage characteristics and soft x-ray photoemission spectroscopy measurements. The results are listed in table I. The energy gap (E_g) of $Ga_2O_3(Gd_2O_3)$ is ~ 5.4eV, and the band offset of ΔE_C is ~1.4eV, and ΔE_V is ~ 2.6eV at the $Ga_2O_3(Gd_2O_3)$-GaAs.

ACKNOWLEDGMENTS

TSL acknowledges supports by grants of National Science Council, Taiwan (NSC87-2215-E-110-009) and Synchrotron Radiation Research Center, Taiwan.

REFERENCES

1. M. Hong, M. Passlack, J. P. Mannerts, J. Kwo, S. N. G. Chu, N. Moriya, S. Y. Hou, and V. J. Fratello, J. Vac. Sci. Technol. **B14**, 2297 (1996).
2. M. Hong, F. Ren, J. M. Kuo, W. S. Hobson, J. Kwo, J. P. Mannaerts, J. R. Lothian, and Y. K. Chen, J. Vac. Sci. Technol. **B16**, 1398 (1998).
3. F. Ren, M. Hong, W. S. Hobson, J. M. Kuo, J. R. Lothian, J. P. Mannaerts, J. Kwo, Y. K. Chen, and A. Y. Cho, *IEDM technical digest*, 943 (1996).
4. M. Hong, J. P. Mannaerts, M. A. Marcus, J. Kwo, A. M. Sergent, L. J. Chou, K. C. Hsieh, and K. Y. Cheng, J. Vac. Sci. Technol. **B16**, 1395 (1998).
5. R.S. Muller, and T.I. Kamins, *Device electronics for integrated circuits*, (John Wiley & Sons, New York, 1977), p. 32.
6. G. Lewwicki and J. Maserjian, J. Appl. Phys., **46**, 3032 (1975).
7. S.M. Sze, *Physics of semiconductor devices*, 2nd ed., (John Wiley & Sons, New York, 1981), p. 403.
8. The Ga 3d binding energy, which is associated with Ga_2O_3, is ~2eV higher than that of GaAs. *Handbook of x-ray photoelectron spectroscopy* (Perkin-Elmer Corporation, Eden Prairie, 1992).

One-Step Silicon Nitride Passivation by ECR-CVD for Heterostructure Transistors and MIS Devices

J.A.DINIZ, L.E.M. de BARROS Jr., R.T.YOSHIOKA, G.S.LUJAN, I.DANILOV, J.W. SWART
LPD/IFGW and DSIF/FEEC, State University of Campinas (UNICAMP)
P.O. BOX 6165, Campinas, SP, Brazil, 13081-970

ABSTRACT

Silicon nitride (SiN_x) films with extremely low interface charge densities have been developed by electron cyclotron resonance-chemical vapor deposition (ECR-CVD) deposition on GaAs substrates. The procedure is a one-step process and does not involve H_2 and/or N_2 pre-treatment of the sample surface. Characterization by Fourier transform infra-red (FTIR) and ellipsometry analysis indicate good properties of the film revealing N-H and Si-N bonds. Results of capacitance-voltage (C-V) measurements show surface charge densities on the order of 5×10^{10} cm^{-2}, which we believe is the lowest surface charge density achieved so far over GaAs.

INTRODUCTION

With the reduction in dimensions of high frequency devices, it is of critical importance that leakage currents be kept at a minimum. Specially, in compound semiconductor devices, such as HBTs and MODFETs, high surface charge density produces a high recombination velocity, which severely limits device performance. Although GaAs has excellent mobility characteristics, which translates into faster circuits, the poor surface quality has prevented it from a larger scale utilization as opposed to Si. Also, the development of good-quality MIS structures on GaAs has been hampered due to the lack of a suitable insulator.

Usually, this problem is dealt with by surface passivation through deposition of a thin nitride film. This process requires prior surface treatment of H_2 and/or N_2 plasmas in order to remove native oxide [1,2] and fill the surface dangling bonds forming an ultra-thin GaN layer. Moreover the nitride film is frequently deposited on a thin layer of Si over GaAs [3] in order to take advantage of the well behaved Si oxides.

In this work, we present a one-step passivation process by ECR-CVD deposition which requires no pre-treatment. The silicon-nitride film is deposited directly over GaAs. The process is described in the next section together with the results from the film characterization. Then we investigate the application of the insulator on a MIS structure where C-V as well as I-V data are discussed. Finally we describe the passivation properties of the SiN_x when applied to MIS devices.

EXPERIMENTAL PROCEDURE

The silicon nitride layers were formed on n-type GaAs (100) wafers. The substrates were cleaned with organic solvents using a Sox-let distillate. Based on previously established conditions by Diniz et al [4], the ECR depositions were carried out at a fixed substrate temperature of 20^0C, SiH$_4$/N$_2$ flow ratio of 1, Ar flow of 5sccm, pressure of 1mTorr and microwave (2.45 GHz) and RF (13.56 MHz) powers of 250W and 1W, respectively. The nitride processed without H$_2$ and N$_2$ plasma surface pre-treatment is named control nitride (CN). For comparison purposes, SiN$_x$ samples deposited on GaAs, under the same conditions, with sequential pre-treatment of H$_2$ and N$_2$ plasma surface are also considered. The plasma conditions for the pre-treatment are: temperature 20^0C, pressure of 1mTorr (H$_2$) and 3 mTorr (N$_2$), H$_2$ flow of 10 sccm and N$_2$ flow of 20 sccm. In this case, the employed RF power and microwave power are 1W and 500W, respectively. Three different samples are discussed, namely Ht1N, Ht3N and Ht5N, where the H$_2$ plasma process time for each one was 1, 3 and 5 minutes, respectively. Nitrogen plasma process time was fixed at 5 minutes. In order to characterize the control nitride composition FTIR, ellipsometric and profile measurements were performed.

Metal/nitride/GaAs capacitors were fabricated from all the samples considered with two different gate electrodes, aluminum (Al) and tungsten nitride (WN). Al/nitride/GaAs capacitors were formed by e-beam evaporation of 160 nm thick aluminum film, sintered by conventional furnace in forming gas (92%N$_2$ + 8%H$_2$) at 420^0C for 20 minutes. WN/nitride/GaAs capacitors were formed by reactive sputtering of 200 nm thick tungsten nitride film, sintered by conventional furnace in forming gas (92%N$_2$ + 8%H$_2$) at 420^0C for 5 minutes. The Al and WN electrodes were patterned with a mask composed of an array of 200 μm diameter dots. On the wafer backside a 285 nm thick AuGeNi film was evaporated to form the ohmic contact, which was sintered by conventional furnace in forming gas (92%N$_2$ + 8%H$_2$) at 420^0C for 3 minutes.

C-V measurements at 1 MHz were performed. The effective charge densities Q_0/q were calculated directly from the flatband voltage shift V_{fb} and C_{max}. All measurements were carried out at room temperature. Breakdown electric fields are determined through I-V characterization.

RESULTS AND DISCUSSIONS

FTIR spectrometry was performed in order to evaluate chemical bonds and to investigate the H incorporation in the samples without H$_2$ and N$_2$ plasma surface treatment before SiN$_x$ deposition (control nitride). The FTIR spectrum (Fig.1) revealed absorption peaks at 1100-1170 cm^{-1} and at 3340-3350 cm^{-1} which are related to the N-H and N-H$_2$ bending mode and the N-H stretching mode, respectively. No absorption at 2000-2300 cm^{-1} related to Si-H bonds was observed. This can be explained by a high degree of dissociation of silane gas molecules under the high discharge power conditions in ECR plasmas rich in nitrogen, that permit low N-H and Si-H bond incorporation in the deposited films. The N-H concentration presented was about the 1×10^{22} cm^{-3}. Furthermore, the main absorption peak occurs at 856 cm^{-1} (stretching mode) which is due to Si-N bonds [5,6], giving clear evidence of the formation of low-stress films with reduced hydrogen content.

Ellipsometric measurements performed at 632.8 nm and with incidence angle of 70^0 showed a thickness of 45 nm and refractive index (n_N) of 1.9 for the deposited control nitride. The deposition rate of 4.8 nm/min was calculated from the nitride thickness and the process time. The N/Si ratio of 1.46 was determined from the refractive index [7], which indicates our film is nitrogen rich. For the stoichiometric silicon nitride, the N/Si ratio is 1.33.

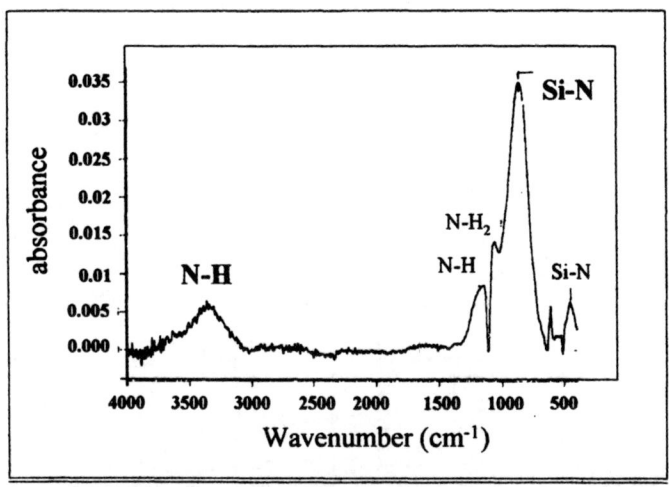

Fig.1 – FTIR spectrum of SiN$_x$ film deposited on GaAs by ECR-CVD

The etch rate determined in buffered HF solution is 26 nm/min, indicating low porosity of the film which suggests low H concentration, confirming the results obtained by FTIR analysis. This etch rate is lower than that of SiN$_x$ deposited on Si substrates by ECR-CVD with similar process conditions [8].

The C-V characteristics of Al/SiN$_x$/GaAs and WN/SiN$_x$/GaAs capacitors with CN, Ht1N, Ht3N and Ht5N nitrides were investigated. Among the pre-treated samples Ht1N presented the best results, therefore it is used for comparison. Figures 2(a) and 2(b) show C-V plots for Al/CN/GaAs and WN/CN/GaAs samples. The correspondent results with the Ht1N film are shown in Fig 2(c) and 2(d). For aluminum gates, the average effective charge density Q_{0CN}/q, Q_{0Ht1N}/q, Q_{0Ht3N}/q and Q_{0Ht5N}/q were found to be about $6\times10^{10}/cm^2$, $1\times10^{11}/cm^2$, $5\times10^{11}/cm^2$ and $2\times10^{12}/cm^2$, respectively, indicating significant decrease of the effective charge density for the structures with CN nitrides. Similar results were observed for WN gate electrodes. The average effective charge density Q_{0CN}/q and Q_{0Ht1N}/q were found to be about $5\times10^{10}/cm^2$ and $1\times10^{11}/cm^2$, respectively. Furthermore, the C-V characteristics for both types of films (CN and Ht1N) showed no hysteresis and no distortion indicating low interface state density, therefore, excellent GaAs surface passivation. Although Ht1N insulator presented good quality, the best dielectric is the control nitride due to the lowest effective charge density observed, on the order of $10^{10}/cm^2$. We believe this is the lowest effective charge density for nitride films deposited over GaAs reported so far, improving almost one order of magnitude from previous results [3,11].

The I-V characteristics obtained after successive positive gate bias high voltage ramp-up stress for samples WN/CN/GaAs and WN/Ht1N/GaAs are shown in Fig.3. Breakdown points of 13.4V and 9.2V corresponding to breakdown electric fields of 3MV/cm and 2MV/cm were determined for samples CN and Ht1N, respectively.

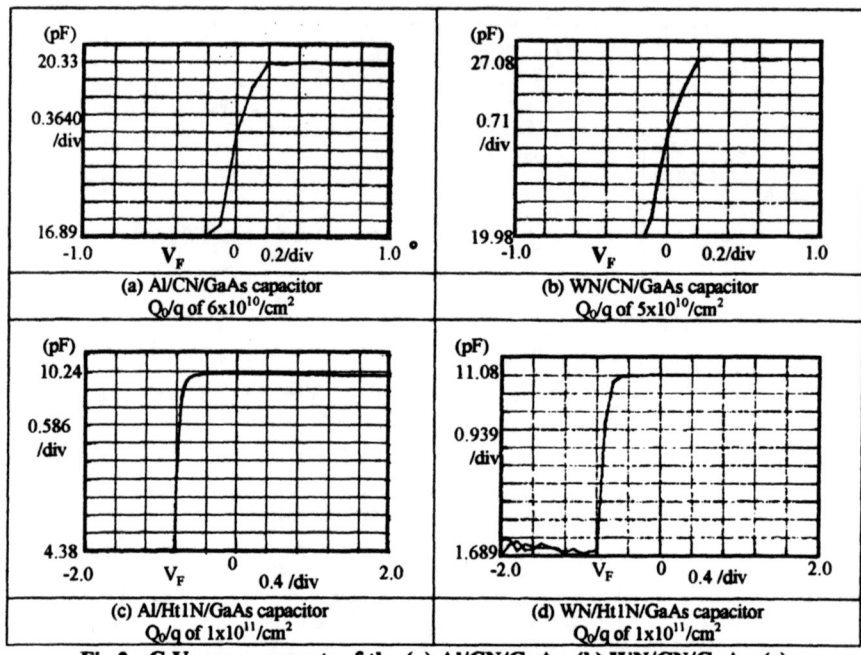

Fig.2 - C-V measurements of the (a) Al/CN/GaAs, (b) WN/CN/GaAs, (c) Al/Ht1N/GaAs and (b) WN/Ht1N/GaAs capacitors.

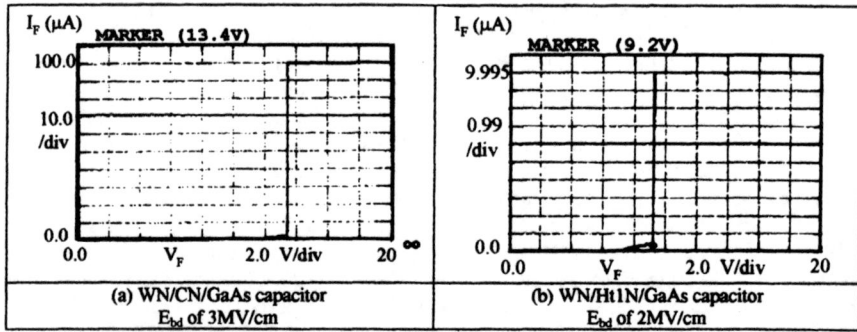

Fig. 3 - I-V measurement of the (a) WN/CN/GaAs and (b) WN/Ht1N/GaAs capacitors.

These electrical properties indicate that the CN films formed have presented high quality, very low effective charge density, comparable to nitrides deposited on silicon [5,9,10], and great GaAs surface passivation results.

The excellent GaAs surface passivation by ECR-CVD nitride deposition occurs as follow: in the deposition beginning, the high plasma density with high SiH_4 and N_2 molecule ionization and dissociation permits that hydrogen and nitrogen ions in contact with GaAs substrate surface remove the native oxide and form the ultra-thin GaN layer. The SiH_4/N_2 flow ratio must be 1, because: (i) a SiH_4 rich process, may induce higher Si-H radical concentration in the film, which reduces the correct H ion production need to remove the native oxide on the surface [4,8]; (ii) the N_2 rich plasma may induce an excess of the nitrogen radical close to the substrate . This can form a barrier against oxide removal [4,8]. Furthermore, the microwave power must be 250W, in order to generate the correct hydrogen and nitrogen ion flux to perform the surface passivation in the deposition beginning.

We have employed the silicon nitride film for AlGaAs/GaAs heterostructure bipolar transistor (HBT) passivation. Preliminary results indicate a significant performance enhancement for our small area devices due to the excellent nitride film properties. For HBTs with emitter area of $4x16\mu m^2$, current gain of 45 and cut-off frequency of 23GHz were obtained.

CONCLUSIONS

The formation of thin nitride films on GaAs substrates by ECR-CVD at low temperature, pressure and microwave power allows an excellent surface passivation, without prior plasma treatment, and the fabrication of gate insulator films for MIS structures. The FTIR analysis revealed the presence of N-H and Si-N bonds and a low H incorporation. The ellipsometric measurements presented refractive index of 1.9 and deposition rate of 4.8nm/min. The etch rate in buffered HF of 26nm/min indicates low porosity and H concentration. Electrical properties demonstrated that, due to the very low effective charge density of $5x10^{10}/cm^2$ and breakdown electric fields of 3MV/cm, these high quality films can be used as excellent passivation layer of III-V compound semiconductor transistors and gate insulator in Metal-Insulator-Semiconductor (MIS) devices on GaAs substrates.

ACKNOWLEDGEMENTS

The authors would like to thank A.C.S. Ramos (LPD/IFGW/UNICAMP) for technical assistance, Prof. Dr. N. I. Morimoto (LSI/EPUSP) for ellipsometric measurements, and Prof. Dr. Inés Pereyra (LME/EPUSP) for FTIR analysis. The work is supported by FAPESP and FINEP of Brazil.

REFERENCES

1. S. Pearton, Topics in growth and device processing of III-V semiconductors, World Scientific, 1996

2. Q.Wang, E.S.Yang, P.W.Li, Z.Lu, R.M.Osgood, Jr., W.I.Wang, IEEE Electron Devices Lett., 13 (1992), 83

3. D.M.Diatezua, Z.Wang, D.Park, Z.Chen, A.Rockett and H.Morkoc, J.Vac. Sci. Technol. B, 16 (2), Mar/Apr 1998

4. J.A. Diniz, G. S. Lujan, P.J. Tatsch and J.W. Swart, Proceedings of XIII Conference of The Brazilian Microeletronics Society, **1**, 185 (1998).

5. S.W. Hseih, C.Y. Chang, Y.S. Lee, C.W. Lin, S.C. Hsu, J.Appl.Phys. **76(6)** (1994), 3645.

6. D.V. Tsu, G. Lucovsky, J. Mantini and S.S. Chao, J. Vac. Sci. Technol., **A5** (4), 1998 (1987).

7. W.C. Mariano, Ms. Thesis/FEEC/UNICAMP, Brazil (1996).

8. S.A. Moshkalyov, J. A. Diniz, J.W. Swart, P. J. Tatsch, and M. Machida, J. Vac. Sci. Technol., **B15** (6), 2682 (1997).

9.J. Vuillod, J. Vac. Sci. Technol., **A5** (4), 1675 (1987).

10. J. A. Diniz, S.A. Moshkalyov, J.W. Swart, and P. J. Tatsch , Proceedings of XII Conference of The Brazilian Microeletronics Society (CD-ROM), artigo33.pdf (1997).

11. V. Malhotra and C.W. Wilmsen, in *Handbook of Compound Semiconductors – Growth, Processing, Characterization and Devices,* Edited by P.H. Holloway and G.E. McGuire (Noyes Publication, 1995), Chap. 7, pp. 328.

Part IV

Compound Semiconductor Surface Passivation and Novel Device Processing

OXIDATION AND CARBON CONTAMINATION IN GaAs (100) WET TREATMENTS

R. F. Elbahnasawy*, J. G. McInerney* and G. Hughes**
*Department of Physics, National University of Ireland, University College, Cork, Ireland.
**Department of Physics, Dublin City University, Ireland.

Abstract

Oxidation and carbon contamination of n-type GaAs (100) surfaces have been investigated by x-ray photoelectron spectroscopy (XPS), Auger electron spectroscopy (AES) and secondary ion mass spectroscopy (SIMS) for a variety of cleaning and etching pretreatment procedures prior to immersion in either $(NH_4)_2S$, Na_2S aqueous solutions or S_2Cl_2 solution in dichloromethane. The study has shown that sulfur passivation removes surface oxide and minimize carbon contamination for surfaces treated in $(NH_4)_2S$ and S_2Cl_2 solutions. A significant oxygen and carbon contamination in the anodic passivation of GaAs (100) in $(NH_4)_2S_x$ aqueous solution have been quantitatively measured. In addition, pretreatment in basic solutions have shown minimal oxygen and carbon level in comparison with the acidic solutions. Surface pretreatment carried out ex-situ has shown a higher risk of surface contamination prior to sulfur passivation.

1: Introduction

Surface contamination is a major problem facing semiconductor technology and its application.[1] For instance GaAs surfaces are extremely susceptible to carbon contamination after acid etching, which can induce faceting. Etching in dilute ammonium hydroxide solution has proved effective, leaving the surface relatively smooth. Sulfidation can reduce the amount of carbon contamination to a level near that of molecular beam epitaxial grown surfaces.[2] Surface treatments usually cause surface contamination and surfaces with higher oxygen contents have higher electrical non-uniformity.[3] Exposing the as-etched GaAs (100) surfaces to air for several minutes can form a thin oxide film.[4,5] Sulfur passivation in $(NH_4)_2S$ solution removes this oxide, leaving a monolayer of chemisorbed sulfur atoms which proved to be effective in protecting the clean surface from the adsorption of oxygen atoms.[6]

2: Experimental procedures

Four different procedures have been used to study the effects of pretreatment and sulfur passivation on n-type (Si) 1×10^{18} cm^{-3} GaAs (100) surface. These procedures are as follows:

Procedure I
Pretreatment:
1- As-received n-type (Si) GaAs (100)
2- 1 min ultrasonic wash in acetone
3- DI water rinse
4- 1 min ultrasonic wash in methanol
5- 1 min ultrasonic wash in DI water
Treatment:
6- Dipping in S_2Cl_2 or $(NH_4)_2S$ solution
7- N_2-blown dry
8- XPS and AES Characterization

Procedure II
Pretreatment:

1- As-received n-type (Si) GaAs (100)
2- 1 min ultrasonic wash in acetone
3- DI water rinse
4- 1 min ultrasonic wash in methanol
5- 1 min ultrasonic wash in DI water
6- 1 min etch in 1M $H_2SO_4:H_2O_2$ solution
7- 1 min ultrasonic wash DI water
8- N_2-blown dry
Treatment:
9- Dipping in S_2Cl_2 or $(NH_4)_2S$ solution
10- N_2-blown dry
11- XPS and AES Characterization

145

Procedure III

Pretreatment:
1- As-received n-type (Si) GaAs (100)
2- 1 min ultrasonic wash in acetone
3- DI water rinse
4- 1 min ultrasonic wash in methanol
5- 1 min ultrasonic wash in DI water
6- 1 min etch in 1M $NH_4OH:H_2O_2$ solution
7- 10 min ultrasonic wash in 1M NH_4OH solution
8- 1 min ultrasonic wash in DI water
9- N_2-blown dry

Treatment:
10- Dipping in S_2Cl_2 or $(NH_4)_2S$ or Na_2S solution

11- N_2-blown dry
12- XPS and AES Characterization

Procedure IV

Pretreatment:
1- As-received n-type (Si) GaAs (100)
2- 1 min ultrasonic wash in acetone
3- DI water rinse
4- 1 min ultrasonic wash in methanol
5- 1 min ultrasonic wash in DI water

Treatment:
6- Anodic treatment in $(NH_4)_2S_x$ solution
7- DI water wash
8- N_2-blown dry
9- XPS, AES, SIMS etc. Characterization

The wafers were cut into 4x4 and 9x9 mm^2 for dipping and anodic treatments respectively. X-ray photoelectron spectroscopy (XPS) was performed using a VG Microtech x-ray source (Al K_α) Depth profiling was performed with Ar$^+$ bombardment at 2 nm/min milling rate. The secondary ion mass spectroscopy (SIMS) has been performed on a SIMS analyzer (cameca IMS-3f) with primary ion beam current 50 nA, energy 14.5 keV and heavy bombardment cesium ions.

3: Oxidation and carbon contamination

XPS data has shown a marked reduction of carbon contamination and oxygen content for GaAs (100) surfaces pretreated in basic solution (procedure III) rather than in acidic (procedure II). The surfaces were found stable against oxidation and smoother in the alkaline solutions because of the threefold coordination of surface Ga atoms by chemisorption of hydroxyl groups. Dipping GaAs surfaces in 1M S_2Cl_2 solution in dichloromethane after pretreatment in procedures I and III has shown an effective etch[7] and minimised carbon and oxygen level at the interface in comparison to just pretreatment in procedure III. On the other hand, a very mild etch has been shown in dipping GaAs in 3M $(NH_4)_2S$ aqueous solution. Better surface quality is obtained with minimum carbon contamination similar to the basic pretreatment in procedure III, which is more consistent with Wang's results.[2] In contrast, surfaces treated in 1M Na_2S aqueous solution displayed a huge carbon peak and severe oxygen contamination for both procedures I and III. Rinsing the sample in DI water after treatment has completely removed the whole passivation and sulfur crust. In the anodic treatment of GaAs (100) in 3M $(NH_4)_2S_x$ solution (procedure VI), SIMS depth profiling has shown a significant continued evidence of both C and O approaching the substrate. Similarly, XPS (Figure 1) has identified both carbon and oxygen as a part of the deposited overlayer. The deposited contaminant overlayer showed no significant effect on the improved surface electronic properties of GaAs surfaces. The anodic passivation was stable chemically for at least four months in ambient air. Presumably the enhancement in the surface electronic properties after anodic sulfidation is due to the stable Ga-S and As-S bonding at the interface. This stable interface is protected by a thick layer of mixed compound of mostly Ga, As and S as well as defective O and C. Over 30% reduction in the surface barrier height has been measured after anodic sulfidation. In oxidative reactions carbon could be neutral, while oxygen provides a major threat to GaAs technology because of its high electronegativity. It seems practical that the in-situ treatment could minimise carbon contamination in the deposited layer, while oxygen should be protected against.

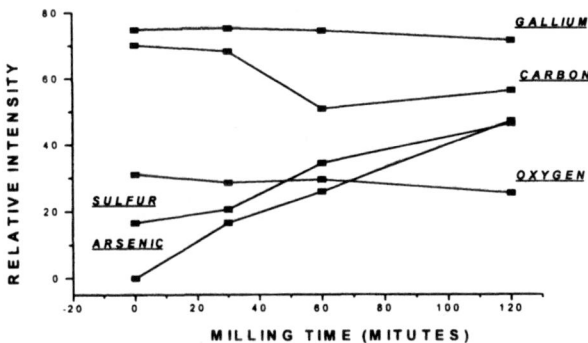

Figure 1: XPS depth profiling for n-type (Si) 1×10^{18} cm^{-3} GaAs (100) anodically treated in 3M $(NH_4)_2S_x$ (x=5g/100 ml) solution (procedure IV).

4: Gallium and arsenic oxides

As3d, As2p, Ga3d and Ga2p core level spectra in oxide-free GaAs surface (after Ar$^+$-milling for 3 hours at 2 nm/min) have been used for reference in the fitting analysis. The oxidized As3d and Ga3d states (Figure 2) have been used for chemical shift calculations (Table I). In this investigation, dipping GaAs in 1M S_2Cl_2 solution in dichloromethane has replaced both the As-O bond (3.30 eV) and the Ga-O bond (1.60 eV) by significant As-S bonds (1.33 eV and 2.47 eV) and strong Ga-S bonds (1.47 eV) as shown in Figure 3. Comparative chemical shifts have been displayed in Table II and shown consistent with previously recorded values. Similarly, dipping GaAs (100) in $(NH_4)_2S$ solution has removed the arsenic and gallium oxides and replaced them with a strong As-S bond at 1.44 eV and an insignificant Ga-S bond at 1.50 eV. Table III shows the chemical shifts of GaAs in $(NH_4)_2S$ solution in broad agreement with previous work.

Table I: XPS chemical shifts for GaAs (100) treatment in H_2O_2 solution. The present work is n-type (Si) 1×10^{18} cm^{-3} GaAs pretreated as in procedure III and dipped in 1M H_2O_2 solution for 12 hours at room temperature.

Ga-O bond	As-O bond	
1.6 ± 0.2 eV (Ga3d)	3.3 ± 0.15 eV (As3d)	Present work
1.5 ± 0.2 eV (Ga2p)	3.14 ± 0.25 eV (As2p)	Present work
1.30 eV (Ga2p)	3.2 eV (As3d)	Massies et al. [11]

Table II: XPS chemical shifts for GaAs (100) treatment in S_2Cl_2 solution. The present work is n-type (Si) 1×10^{18} cm^{-3} GaAs pretreated as in procedure III and dipped in 1M S_2Cl_2 solution for 30 seconds at room temperature.

Ga-S bond	As-S bond	
1.47 ± 0.2 eV (Ga3d)	1.33 ± 0.2 & 2.47 ± 0.3 eV (As3d)	Present work
1.53 ± 0.2 eV (Ga2p)	1.67 ± 0.1 & 2.92 ± 0.3 eV (As2p)	Present work
1.3 eV	1.54 eV	Cai et al. [7]

Table III: XPS chemical shifts for GaAs (100) treatment in $(NH_4)_2S$ solution. The present work is n-type (Si) $1x10^{18}$ cm^{-3} GaAs pretreated as in procedure III and dipped in 3M $(NH_4)_2S$ solution for 30 min. at 50°C temperature.

Ga-S bond	As-S bond	
1.5 ± 0.1 eV (Ga2p)	1.44 ± 0.15 & 2.84 ± 0.15 eV (As2p)	Present work
1.5 eV (Ga2p)		Kang et al.[8]
	1.5 eV (As2p)	Jianhong et al.[6]
	1.7 eV (As2p)	Carpenter et al.[9]
	2.7 eV (As2p)	Sandroff et al.[10]

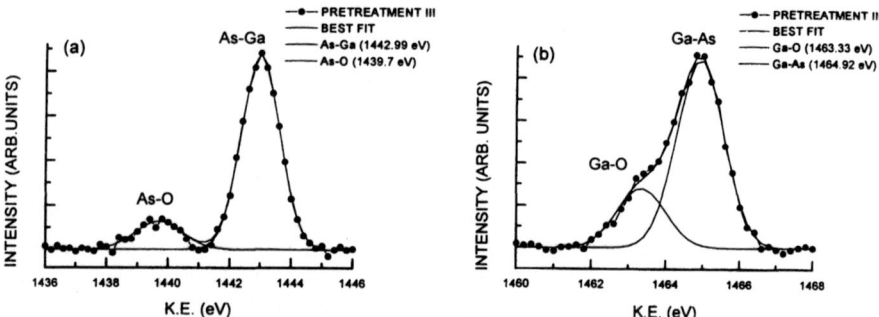

Figure 2: Core level spectra of n-type (Si) $1x10^{18}$ cm^{-3} GaAs (100), pretreated (procedure III) and dipped for 12 hours in 1M H_2O_2 solution. (a) As3d, and (b) Ga3d

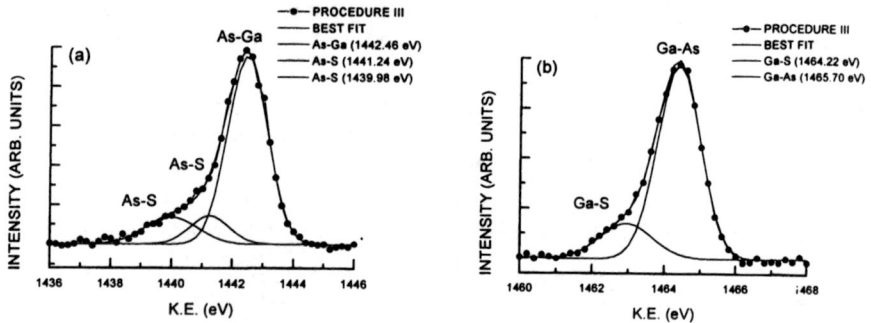

Figure 3: Core level spectra of n-type (Si) $1x10^{18}$ cm^{-3} GaAs (100) dipped at room temperature in 1M S_2Cl_2 solution for 30 second (procedure III). (a) As3d, and (b) Ga3d

5: Structural model for dipping treatment

The experiments were performed in a moderate room light and procedure III was considered the best procedure suitable for most of the sulfide treatment. In the first experiment, n-type GaAs (100) was dipped in 3M $(NH_4)_2S$ solution for 30 minutes at 50°C. In the second experiment, the sample was dipped in 1M S_2Cl_2 non-aqueous solution for 30 seconds at room temperature. A thin yellowish film of amorphous sulfur after blow-dry using nitrogen, has been observed in both experiments. Combining the surface analysis data obtained by AES and XPS a model has been proposed to explain the effectiveness of $(NH_4)_2S$ and S_2Cl_2 dipping treatments. According to this model, the pretreated n-type GaAs (100) surface has been covered with a thin layer of native oxides (Figure 4(a)). When the pretreated GaAs sample was dipped in $(NH_4)_2S$ or S_2Cl_2 solution, the oxide film is quickly etched to reveal an oxygen free surface and then the surface is instantly covered by reactive sulfur as shown in Figure 4(b). In vacuum most of the amorphous sulfur were sublimed leaving a shiny crystalline surface with a detectable sulfur layer as shown in Figure 4(c). In this study S_2Cl_2 solution in dichloromethane has shown to have a high etching and oxidizing rates as well as having the unique advantage of being an oxygen-free solution. The S_2Cl_2 treatment has also shown to have an effective passivation in comparison with the other dipping treatment. Dipping GaAs (100) in $(NH_4)_2S$ aqueous solution creates two competitive chemical reaction mechanisms. The reaction of the substrate with the sulfur ions in the aqueous solution to form gallium and arsenic sulfide as in equation (1), and also a reaction of the formed sulfides with H_2O as in equation (2) and equation (3). The second reaction is quite strong, so that the sulfides are mostly converted into oxides that are soluble in $(NH_4)_2S$ solution. Always a monolayer of S atoms remains on the surface as long as the substrate in the $(NH_4)_2S$ solution.

$$Ga_2O_3 + As_2O_3 + (NH_4)_2S \longrightarrow Ga_2S_3 + As_2S_3 + (NH_4)_2O \quad (1)$$

$$Ga_2S_3 + 6H_2O \longrightarrow 2Ga(OH)_3 + 3 H_2S \uparrow \qquad (2)$$

$$As_2S_3 + 6H_2O \longrightarrow 2H_3AsO_3 + 3H_2S \uparrow \qquad (3)$$

As gallium and arsenic sulfides dissolves in $(NH_4)_2S$ solution, the reactions continue during the soaking of the sample and thus the etching of GaAs (100) resumes. This is useful for flattening the surface, etching defects and restricting carbon contamination to the minimum. After nitrogen blown dry process, the amorphous layer of sulfur isolates the GaAs (100) surface from the active atmosphere. In vacuum the volatile amorphous layer sublimates and gradually disappears, leaving a monolayer of sulfur on the top that confirms the chemical bonding between sulfur and GaAs substrate. The etching effect of $(NH_4)_2S$ or S_2Cl_2 solution could provide a defect-free surface at which the sulfur-terminated surface has no dangling bonds so that the surface does not accept chemically adsorbing atoms.[12] In the advanced unified defect model by Spicer et al[13], this passivity of the sulfur-treated GaAs (100) seems essential to the reduction in the interface state density and significantly protect and inhibit the surface oxidation.[14]

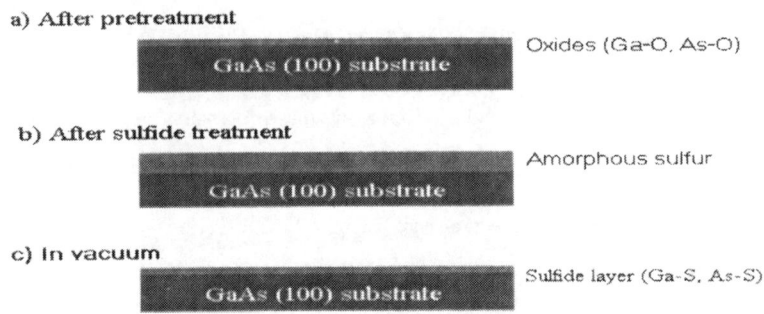

a) After pretreatment
GaAs (100) substrate
Oxides (Ga-O, As-O)

b) After sulfide treatment
GaAs (100) substrate
Amorphous sulfur

c) In vacuum
GaAs (100) substrate
Sulfide layer (Ga-S, As-S)

Figure 4: Illustrated model of GaAs (100) surface. (a) Native oxides cover the GaAs surface, (b) amorphous layer of sulfur is covering the surface after dipping in 1M S_2Cl_2 or 3M $(NH_4)_2S$ and nitrogen-blown dry, and (c) amorphous sulfur sublimates under vacuum, leaving behind a monolayer of gallium and arsenic sulfide.

6: Conclusions

Dipping GaAs in S_2Cl_2 solution was proved to be very effective in removing gallium and arsenic surface oxides and minimize carbon contamination at the interface. The etching rate was too fast in oxygen-free solution, which would be recommended for etching applications. GaAs treatment in $(NH_4)_2S$ solution displayed insignificant Ga-S and As-S bonds in comparison with S_2Cl_2 treatment. The treatment exhibited milder reaction as well as minimum surface contamination, which is more suitable for cleaning GaAs surfaces prior to further processing. The experiment has also shown that although high carbon and oxygen atomic concentrations were recorded in the depth analysis of anodically passivated GaAs in $(NH_4)_2S_x$ solution, the surface was chemically more stable and electronically enhanced. The chemical stability is due to the formation of bonds such as Ga-S, As-S and probably As/Ga-O bonds. Over 30% reduction in the surface barrier height has been recorded. An hour exposure to Ar^+ laser beam at power density $5mW/\mu m^2$ has also been applied. No chemical or electrical changes have been observed, which seems promising for GaAs device applications.

Acknowledgment
The authors would like to thank Dr. Tony Deeney and Robin Gillen (Physics Department, UCC) for their helpful discussion and Forbairt Ireland for their financial support.

References

[1] Y. Wang, Y. Darici, and P. H. Holloway, J. Appl. Phys. **71**(6), 2746 (1992).

[2] X. S. Wang, K. W. Self, R. Maboudian, C. Huang, V. Bressler-Hill, and W. H. Weinberg, J. Vac. Sci. Technol. **A11**(4), 1089 (1993).

[3] R. Rochter, H. L. Hartnagel, J. Electrochem. Soc. **137**(9), 2879 (1990).

[4] J. Lee, L. Wei, S. Tanigawa, H. Oigawa and Y. Nannichi, Appl. Phys. Lett. **58**(11), 1167 (1991).

5 R. Richter and H. L. Hartnagel, J. Electrochem. Soc. **137**(9), 2879 (1990).

[6] Z. Jianhong, H. Xiaoyuan, D. Xunmin, J. Xiaofeng, C. Ping, Chinese Physics **12**(3), 753 (1992).

[7] W. Z. Cai, Z. S. Li, R.Z. Su, G. S. Dong, D. M. Huang, X. M. Ding, X. Y. Hou and X. Wang, Appl. Phys. Lett. **64**(25), 3425 (1994).

[8] M. Kang, S. Sa, H. Park, K. Suh and J. Lee, Materials Science and Engineering **B46**, 65 (1997).

[9] M. S. Carpenter, M. R. Melloch, M. S. Lundstrom and S. P. Jobin, Appl. Phys. Lett. **52**(25), 2157 (1988).

[10] C. J. Sandroff, M. S. Hegde, L. A. Farrow, C. C. Chang, and J. P. Harbison, Appl. Phys. Lett. **54**(4), 362 (1989).

[11] J. Massies and J. P. Contour, J. Appl. Phys. **58**(2), 806 (1985).

[12] Y. Nannichi, Jia-Fa Fan, H. Oigawa and A. Koma, Japanese Journal of Applied Physics **27**(12), L2367 (1988).

[13] W. E. Spicer, Z. Liliental-Weber, N. Newman, T. Kendelewicz, R. Cao, C. McCants, P. Mahowald, K. Miyano, I. Lindau, J. Vac. Sci. & Technol. **B6**, 1245 (1988).

[14] H. Oigawa, H. Shigekawa, Y. Nannichi, Material Science Forum **185-188**, 191 (1995).

DX CENTER ENERGY LEVEL IN In$_x$Al$_{1-x}$As COMPOUNDS

HÜSEYIN SARI, HARRY H. WIEDER
ECE Department University of California, San Diego, CA 92093

ABSTRACT

The presence of DX centers in In$_x$Al$_{1-x}$As, primarily in the indirect portion of the In$_x$Al$_{1-x}$As bandgap, has been determined using modulation doped In$_x$Al$_{1-x}$As/In$_y$Ga$_{1-y}$As heterostructures by means of persistent photoconductivity (PPC) and galvanomagnetic measurements. From the cooling bias experiment, the PPC, and self consistent Poisson and Schrödinger simulations the ratio of the ionized shallow donors to the DX centers is obtained. Using this ratio in the grand canonical ensemble (GCE) the energy level of DX centers is determined. It is found that the DX energy level merges with the conduction band at x≅0.42 and is resonant with the conduction band in higher indium concentration.

INTRODUCTION

DX centers are deep energy levels with unique properties [1] which include a large difference between their optical and thermal ionization energies, and a temperature dependence of the capture cross-section. As a result of these properties DX centers exhibit a reduction in free carrier concentration, a large persistent photoconductivity (PPC) effect, and a shift in the threshold voltage of modulation doped field effect transistors (MODFET) structures, at low temperatures [2]. Although early experimental results supported the view that DX centers are effective mass-type deep levels associated with the L minimum of the conduction band the presently accepted model of Chadi and Chang predicts that DX energy level follows the weighted average of the conduction band [3]. Now it is believed that the DX centers is the ground state of donor atoms [4] and occupation of these state is accompanied with a large lattice relaxation by capturing two electrons from the conduction band [3]. Since the electron-electron correlation energy or Hubbard energy is negative due to a large lattice relaxation and to capture of the two electrons by the DX centers this model is also known as a negative-U DX center model.

DX centers have been observed in many n-type doped III-V compounds. Most of the studies on this defect have been carried out on the Ga$_x$Al$_{1-x}$As material system [5]. However, to date there is significantly less work on DX centers in In$_x$Al$_{1-x}$As compounds. This is partly due to difficulties associated with the growth of defect free materials other than lattice matched In$_{0.52}$Al$_{0.48}$As on InP and partly because the energy level of the DX center is in resonance with the conduction band in In$_{0.52}$Al$_{0.48}$As. The work of Calleja et. al. showed no sign of the DX centers in In$_{0.52}$Al$_{0.48}$As lattice matched to InP under the 12 kbar pressure and doped to 1x10^{18} cm^{-3} [6]. On the other hand, Früh et al. observed DX center-like behavior in In$_{0.52}$Al$_{0.48}$As lattice matched to InP under 13 kbar hydrostatic pressure and doped to 8x10^{18} cm^{-3}. They measured the energy level to be 180 meV above the conduction band edge and the critical temperature to be T$_C$=180 °K [7]. For the lattice mismatched case there are only two references. Hong et al. carried out DLTS measurements on lattice mismatched In$_{0.43}$Al$_{0.57}$As grown on InP [8]. They observed a trap labeled ES1, which shows DX-like deep level character. They found the thermal ionization energy to be 350 meV above to the conduction band edge. Young and Wieder [9] measured the DX center energy level in $In_{0.29}Al_{0.71}As$ by using the properties of the heterostructure. They found the thermal ionization energy of DX centers to be 230 meV below the minimum of the conduction band. In order to fully understand the dependence of DX centers energy level in InAlAs on indium composition more experimental data is needed.

EXPERIMENT

Several samples with indium compositions x=0.10, 0.15, and 0.20 in a MODFET-like heterostructure were grown by Varian Gen 1.5 solid source Molecular Beam Epitaxy (MBE) on semi-insulating (001) GaAs wafers. One of the heterostructures and corresponding band diagram are shown schematically in Figure 1. It consists of: a 200 nm undoped GaAs layer, several 200 nm undoped step graded buffer layers of $In_yGa_{1-y}As$ with y=0.10 indium increment to compensate for the lattice mismatch between the substrate and the active layer, a 50 nm nominally undoped $In_xGa_{1-x}As$ channel layer lattice matched to the barrier layer, a 40 nm $In_xAl_{1-x}As$ barrier layer with a Si doped ($8.0x10^{12}$ cm^{-2}) δ-doping layer 3 nm away from the channel, and a 10 nm $In_xGa_{1-x}As$ cap layer. For electrical and optical measurements eight-armed Hall bars were photolitographically defined by using standard wet chemical etching techniques. Ohmic contacts to the two dimensional electron gas (2DEG) were made by thermal evaporation of 5 nm Ni/120 nm AuGe/15 nm Ni. To create a MODFET device the cap layer on top of the structure is selectively etched away and a gate of 15 nm Ti and 150 nm Au was deposited by e-beam evaporation.

The electrical characteristics of the samples were investigated using Hall measurements with magnetic field 1.22 kG. PPC measurements were conducted by cooling the samples from room temperature to 25 °K in the dark. At 25 °K the sample was illuminated with a GaAs LED for 20 minutes. After the light was turned off we waited for 10 minutes to make sure that the carriers reach to a new steady state. The sample was then heated up to room temperature in

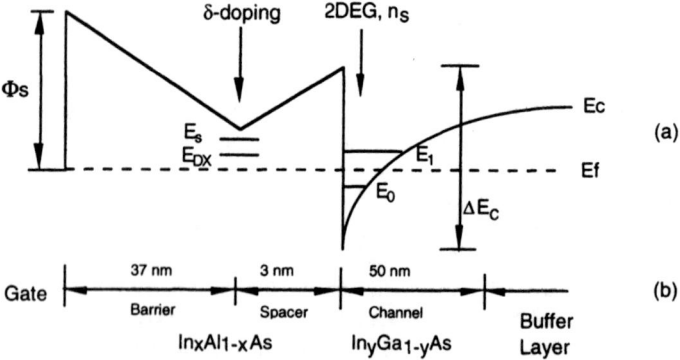

Figure 1: The energy band diagram (a) and the structure of the samples (b) used in this study. All samples have a similar dimensions except for their buffer layer.

152

the dark by remeasuring the carrier concentration. The carrier concentration measurement during both cooling and heating was done in 10 °K steps from room temperature to 25 °K; we waited approximately 10 minutes at each data point to make sure that the sample is in thermal equilibrium. The gate bias dependence of the carrier concentration was measured by applying a gate voltage to the MODFET structures at room temperature and maintaining the bias during cooling of the sample. After the sample was cooled to 25 °K, the cooling bias voltage was removed and the carrier concentration was measured by sweeping the gate voltage. The details of the experimental technique are discussed elsewhere [10].

RESULTS

The first indication of the existence of DX centers in the $In_xAl_{1-x}As$ material system investigated in this study comes from the PPC experiment. Figure 2 shows the PPC effect in $In_{0.10}Al_{0.90}As$. All samples, regardless of the indium composition, show a similar PPC effect. It is clearly seen that above T_C there is freeze out of the carriers into the DX centers. Below T_C, further occupation of the DX centers is prevented by a thermal barrier and the electron concentration in the 2DEG becomes constant. After the illumination, all the electrons in the DX centers are excited to the conduction band and stay there as long as the temperature is kept below T_C. From the PPC curve the critical temperature is found as 200±10 °K which is higher than that the GaAlAs compounds.

Figure 3 shows the sheet electron concentration in the 2DEG vs the gate voltage, V_g, applied to a MODFET structure at 25 °K for different bias voltages V_C, for the $In_{0.10}Al_{0.90}As$. The shift in threshold voltage depends on the magnitude of the cooling bias voltage and it is observed in all InAlAs samples above certain doping concentrations (8.0×10^{12} cm^{-2}). The necessity of such a high doping concentration is attributed to filling the other deep states besides DX centers [11]. The shift is a strong indication of the occupation of the DX centers especially when combined with the PPC effect observed in the same sample. Applying a gate

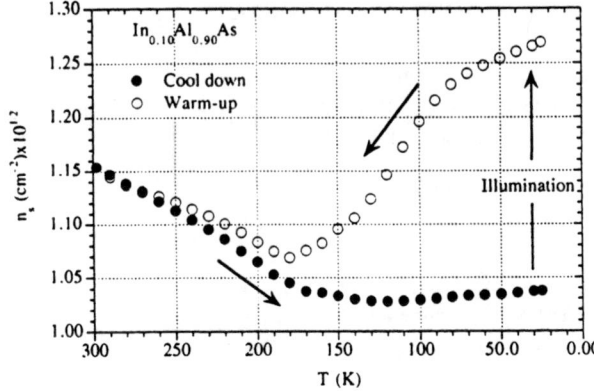

Figure 2: Persistent Photoconductivity (PPC) observed in $In_{0.10}Al_{0.90}As$. Filled points correspond to the carrier concentration during cool down in the dark and open points are after illumination with GaAs LED for 20 minutes. The percent change is significant and indication of excitation of the DX centers.

bias at room temperature causes a change in the ratio of the DX centers to the shallow donor density. The thermal capture barrier of the DX centers prevents further capturing of the electrons from the conduction band when the sample is cooled down past the critical temperature T_C [12]. As a result of this thermal capture barrier a low electron concentration results in the 2DEG when the sample is cooled under a large applied forward bias. The opposite is true when a negative bias is applied. The increase in the carrier concentration in the 2DEG under the negative bias voltage is an indication of the increase of the ionized shallow donors. The n_S vs gate voltage curve has two distinct regions: a linear region at low gate voltages and a saturation region at higher positive gate voltages. The linear region is an indication of the usual charge control law, a result of having a constant capacitance between the gate and the 2DEG. The saturation region, on the other hand, is an indication of the pinning of the Fermi level to the lowest deep level. Buks et al. have used the pinning of the Fermi level in GaAlAs/GaAs heterostructures to estimate the DX energy level to study the mobility enhancement due to the correlation between DX centers and shallow donors [13].

The effective ionized shallow donor density, N_T, as a function of cooling bias voltage V_C can be obtained from the relation [14].

$$\left(\frac{dN_T}{dV_c}\right)_{V_g} = \left(\frac{\partial N_T}{\partial n_s}\right)\left(\frac{\partial n_s}{\partial V_c}\right)_{V_g} \tag{1}$$

In this equation Δn_s is the change in carrier concentration in the 2DEG after illumination, and terms $(\partial N_T/\partial V_C)$ and $(\partial n_s/\partial V_C)$ are the change in ionized shallow donor density and the carrier concentration in the 2DEG, respectively, both measured with cooling bias V_C for a constant gate voltage V_g. The ionized donor density for each cooling bias cycle can be determined by solving the Poisson and Schrödinger equations [15]. The simulation results for different cooling biases in the linear region of the $n_s(V_g)$ curve are shown in Figure 3 with dashed lines. Since light converts the DX centers into ionized shallow donors the following relations for the N_T, shallow ionized donor density, N^+, and N^{DX} can be expressed as $N_T(PPC)=N^+-N^{DX}$ and $\Delta N_T=N_T(PPC)-N_T=2N^{DX}$.

Using the ratio of ionized donor concentration to the total donor concentration in the grand canonical ensemble of Lazzouni and Sham [16], Eqs. 2, the energy level of the DX centers relative to the conduction band, E_{DX}, can be obtained.

$$\frac{N^+ - N^{DX}}{N^+ + N^{DX}} = \frac{1 - 4\exp[\beta(2E_F - E_S - E_{DX})]}{1 + 4\exp[\beta(2E_F - E_S - E_{DX})]} \tag{2}$$

Here $\beta=1/kT$, E_f, E_s, and E_{DX} are the Fermi energy, shallow energy of donor, and the DX center energy, respectively. The Fermi energy, E_f, can be obtained from self-consistent Poisson and Schrödinger equations for each cooling cycle at $V_g=0$ V. For the composition range under investigation the shallow energy level is assumed to be 50 meV.

Figure 4 shows the estimated energy level of DX centers obtained from the GCE in $In_xAl_{1-x}As$ as a function of indium composition. This figure also shows the composition dependence of the Γ, L, and X conduction band edges relative to the valance band edge and the

154

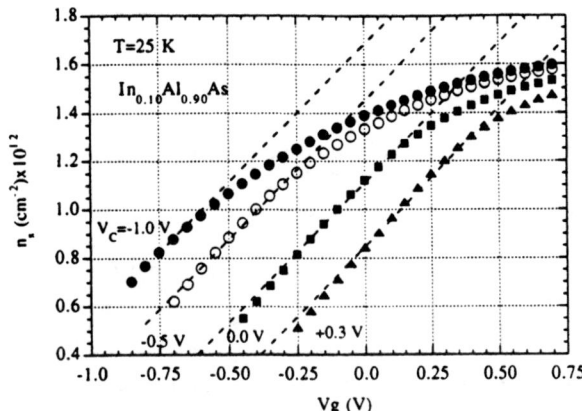

Figure 3: Threshold voltage shift in $In_{0.10}Al_{0.90}As$ as a function of the cooling bias (V_C). The sample was cooled down from room temperature by applying a bias voltage on the gate. At 25 °K the cooling bias voltage was removed and the carrier concentration in the 2DEG was measured by sweeping the gate voltage V_g.

data of Chand et al. for AlAs [5], Wieder et al. for $In_{0.34}Al_{0.66}As$ [14], and the data of Fruh et al. for $In_{0.52}Al_{0.48}As$ lattice matched to InP [7]. This dependence of E_{DX} on indium composition suggests that the energy level of the DX centers becomes resonant with the conduction band edge at $x \cong 0.42$. Similar dependence of the DX energy level on indium concentration in this composition range is also indicated by Fermi level pinning [17]. For comparision, it is well known that in $Ga_{1-x}Al_xAs$ when the Al concentration is 22 percent, the DX energy level becomes resonant with the conduction band [2].

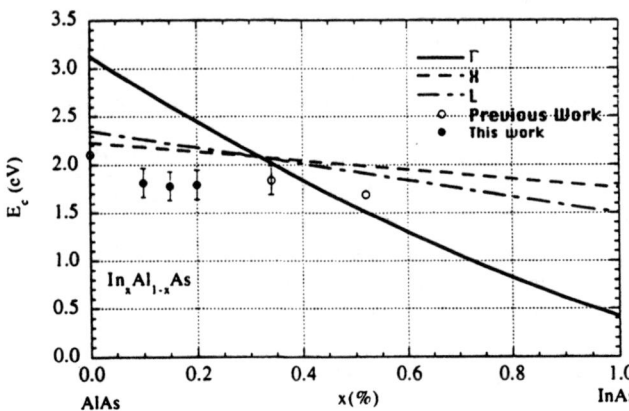

Figure 4: The estimated energy level of DX centers relative to the valence band obtained in this study for $0.10 < x < 0.34$ (solid circles). The graphic also contains the data from Ref. 5 for AlAs, from Ref. 14 for $In_{0.34}Al_{0.66}As$, and from Ref. 7 for $In_{0.52}Al_{0.48}As$ (open circles). Linear fitting suggests that the DX energy level merges into the conduction band around x=0.42 and becomes resonant at higher indium composition.

CONCLUSION

The DX center energy level in $In_xAl_{1-x}As$ as a function of indium composition was investigated by using the MODFET test structures. We clearly observed a PPC effect and a shift in the threshold voltage, which are indications of the presence of DX centers. Using the experimentally observed shift in threshold voltage and PPC effect as well as the Poisson-Schrödinger simulation and the grand canonical ensemble we determined the DX energy level in this composition range. The DX center energy level merges with the conduction band around x=0.42 and becomes resonant with the conduction band at higher indium compositions.

ACKNOWLEDGEMENTS

Author H. S. acknowledges with gratitude the fellowship support from the Turkish Education Ministry.

REFERENCES

1- D. V. Lang and R. A. Logan, Phys. Rev. Lett. 39, 635(1977).

2- P. M. Mooney, J. Appl. Phys. 67(3) 1990.

3- D. J. Chadi & K. J. Chang, Phys. Rev. B, 39(10063) 1987.

4- M. Mizuta, M. Tachikawa, H. Kukimoto, and Minomura, Japn. J. Appl. Phys. 24, L143(1985).

5- N. Chand, Henderson T., Klem J., Masselink W. T., Fischer R., Chang Y-C., H. Morkoc, Phys. Rev. B 30, 4481 (1984).

6- Calleja, A. L. Romero, S. Fernandez de Avila, E. Monuz, and J. Castagne, Semicond. Sci. Technol. 206 (1984).

7- F. E. Früh, J. M. Sallese, M. Beck, D. K. Maude, U. Willke, M. Rabary, J. C. Portal, M. Ilegems, Solid State Comm. 89, 323(1994).

8- W-P Hong, S. Dhar, P. K. Bhattacharya, and A. Chin, J. Electronics Material vol.16, No.4, 271(1987).

9- A. P. Young, and H. H. Wieder, J. Vac. Sci. Technol. B 14(4), 2944 (1996).

10- A. P. Young, DX Centers in InAlAs/InGaAs Based Heterostructures. University of California San Diego 1997.

11- L. Burstein, Y. Shapira, B. R. Bennett, J. A. del Alamo, Appl. Phys. 78 (12), 7163(1995).

12- P. M. Mooney, N. S. Caswell, and S. L. Wright, J. Appl. Phys. 62(12), 4786(1987).

13- E. Buks, M. Heiblum, Y. Levinson, and H. Shtrikman, Semi. Sci. and Technol. 9, 2031(1994).

14- H. H. Wieder and H. Sari, 26th Conference on the Physics and Chemistry of Semiconductor Interfaces, San Diego, CA 1999.

15- Software courtesy of G. Snider, Cornell University, Ithaca, NY 14853.

16- M. E. Lazzouni and L. J. Sham, Phys. Rev. B48, 8948(1993).

17- H. Sari, H. H. Wieder, J. Appl. Phys. 85, 3380(1999).

$Si_{1-x}Ge_x$ OXIDATION BY PLASMA ASSISTED PROCESSING : OXIDE UNIFORMITY AND ELECTRICAL PROPERTIES

T. BUSANI, H. PLANTIER*, R. A. B. DEVINE, C. HERNANDEZ, Y. CAMPIDELLI,
France Telecom-CNET, BP 98, 38243 Meylan, France;
*LEMD, CNRS-UJF, UMR C5517, BP 166, 38042 Grenoble Cedex

ABSTRACT

Anodic oxides of Si_xGe_{1-x} ($0 \leq x \leq 1$) alloys have been made by plasma assisted oxidation in a microwave frequency (2.45 GHz) reactor working in the constant current bias mode. Oxide films ~ 15 - 40 nm (depending upon the Ge concentration) were obtained in 10 minutes without a temperature rise of the substrate of more than 100 °C. Detailed infrared absorption studies of the oxides enabled the Si-O-Si, Ge-O-Ge and Si-O-Ge vibrational modes to be identified, the strongest being at 1056, 858 and 1000 cm^{-1} respectively. These modes are associated with the O asymmetric stretch, their values are at lower wavenumbers than in bulk oxides due partly to ultraviolet radiation induced structural modification and partly to thin film optic effects. A statistical model for the different bonds present in $Si_xGe_{1-x}O_2$, when used to simulate the infrared spectrum does not predict the experimentally observed form, the Ge-O-Ge peak is in general too intense in the experimental spectrum. Auger electron spectroscopy profiling of the Si_xGe_{1-x} oxides suggests that there is a build-up of Ge close to the surface/oxide interface so that when combined with the infrared data, we conclude that there is a GeO_2 rich region at the surface/oxide interface. The oxide is, however, globally stoichiometric. Electrical measurements (C(V) and interface state density) were begun on metal-oxide-semiconductor (MOS) capacitors for $Si_{1-x}Ge_x$ oxides over the range of concentrations $0 \leq x \leq 1$. Only $Si_{1-x}Ge_x$ oxides with $x \leq 0.15$ appear to yield satisfactory MOS capacitor curves.

INTRODUCTION

Strained $Si_{1-x}Ge_x$ layers are potentially attractive for advanced microelectronics devices since they have a variable bandgap which spans from that of Si (1.12 eV) to Ge(0.67 eV)[1] and higher hole and electron mobility [2] as compared to pure silicon. This offers the advantage of hetero-transistor structures for Si technology such as hetero-junction bipolar transistors (HBTs) with higher current gains and cut-off frequencies (f_t < 40 GHz)[3], n- and p-channel modulation doped field effect transistors (MODFETs)[4] and applications in optical devices such as photodetectors[5]. However, whereas in optical engineering $Si_{1-x}Ge_x$ has demonstrated its superiority[6], for hetero-transistor devices many problems remain. For example it is essential to preserve the built-in strain and to avoid dopant outdiffusion and Ge segregation which would result in mobility degradation. Furthermore, when considering oxide used as masks, as passivation layers and as a gate insulator in (MOS) structures many factors such as the growth technique and quality of the oxide interface with the $Si_{1-x}Ge_x$ become important. It is crucial to obtain a high quality $Si_{1-x}Ge_x$ oxide film for successful device application. Considerable efforts have been made to study the thermal oxidation of $Si_{1-x}Ge_x$ [7][8]. To understand the oxidation it must be underlined that the nature of the oxide formed depends on both thermodynamic (temperature) and kinetic (oxygen diffusion) factors. In the case of dry oxidation a pure SiO_2 layer is formed and a pile up of nanocrystalline Ge occurs at

157

the oxide/$Si_{1-x}Ge_x$ interface due to precipitation of Ge atoms from the oxide[8], no increase of oxidation rate of $Si_{1-x}Ge_x$ alloys with respect to Si has been noted [9]. In the case of wet oxidation, the formation of a SiO_2 top layer with a Ge pile up at the interface has also been reported[8] the initial oxidation rate of $Si_{1-x}Ge_x$ has been found to be about three times larger than that of pure Si[9]. Apart from the Ge precipitation problem, thermal oxidation implies high temperature (700-1000 °C) so that substrate lattice relaxation can be expected to occur introducing additional dislocations at the oxide/semiconductor interface and degrading electrical proprieties. The natural way to solve at least some of these problems is to decrease the processing temperature. Low temperature oxidation of $Si_{1-x}Ge_x$ using ultra-violet (UV) exposure[10] or radio frequency [11] and microwave electron cyclotron resonance(ECR) excited plasmas [12-[15] has shown that oxides can be obtained with no Ge pile up at the interface since diffusion of the Ge during oxidation is avoided. Floating potential plasma processing in the temperature range 80-150 °C results in an an oxide no thicker than 6 nm [12] which is of poor electrical quality and not generally useful technologically. An alternative solution to obtain better and thicker oxides may be plasma assisted, anodic oxidation [14]. In the present work we have extended these earlier studies to improve our understanding of the physical nature of anodic oxides of $Si_{1-x}Ge_x$ and to examine their electrical qualities.

EXPERIMENTS

$Si_{1-x}Ge_x$ layers (0<x<1) were grown epitaxially on p-type Si(100) substrates using an industrial single wafer (200 mm diameter) chemical vapor deposition (CVD) reactor. CVD was carried out at 650 °C using SiH_4 and GeH_4 (10 % diluted in H_2). The thickness of the $Si_{1-x}Ge_x$ layer was typically 500 nm. A 300 nm Al film was evaporated onto the rear face of the wafer and heated for 90 s at 450 °C in H_2N_2 to ensure good electrical contact to the substrate during the plasma assisted anodization process.

$Si_{1-x}Ge_xO_2$ (x = 0, 0.08, 0.15, 0.25, 0.5) samples were grown using an ECR microwave plasma source with a typical power of 500 W at 2.45 GHz. The substrates were 4 cm by 4 cm squares cleaved from the 200 mm wafers and mounted on a substrate holder which enabled them to be biased electrically positive with respect to the plasma potential (~12 V respect to the earth). Anodic oxidations were carried out with a typical current flowing through the sample of 0.3-0.5 A, the substrate bias potential was adjusted continually during oxide growth to ensure a constant total current. The oxygen pressure in the chamber (37 l) during the oxidation was 1 mTorr and the temperature (due only to the plasma discharge itself) in the range of 60-100 °C depending on the oxidation time (10-28 minutes) . The thickness of the growing oxide was monitored in-situ using a SOPRA spectroscopic ellipsometer working in the photon energy range 1.5-4.5 eV. To ensure rapid measurements a single energy was used during the growth cycle. Post-oxidation, full energy range spectroscopy studies were carried out to confirm the validity of the single photon energy measurement. A second set of samples was made using amorphous Ta_2O_5 deposited at room temperature by plasma assisted CVD on the $Si_{1-x}Ge_x$ surface. The reactor used for this was also ECR excited and used TaF_5 as the Ta based gas precursor and an H_2/O_2 plasma. To make MOS capacitors, 1.42, 1.08, 0.84 mm diameter Al dots were deposited onto the $Si_{1-x}Ge_xO_2$ and Ta_2O_5 surfaces by evaporating Al through a shadow mask.

The oxide films were characterized physically using Fourier transform infrared spectroscopy and Auger profiling. Infrared measurements were carried out using a Bruker IFS 66 with a resolution of 4 cm^{-1} in the range 400-1300 cm^{-1}. Auger electron spectroscopy was performed using a Phi model 670 spectrometer to deduce the Si, Ge and O profiles in the oxide. An Ar beam was used to erode the oxide and perform depth profiling with an estimated resolution of 2 nm. Standard high-frequency (1 MHz) capacitance-voltage (C-V) measurements were carried out at room temperature to characterize the MOS capacitors.

RESULTS AND DISCUSSION

In fig 1 and 2 we show typical plasma anodization curves for of $Si_{1-x}Ge_x$ films in the constant current mode, the current used was 0.3 A.

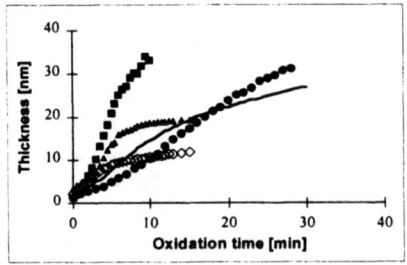

Fig. 1 : Anodic oxide growth curves for the same substrate current (0.3 A) but different concentration of Ge: SiO_2(-), 8 % of Ge($\diamond$), 15 % of Ge($\blacktriangle$), 50 % of Ge($\blacksquare$), GeO_2($\bullet$).

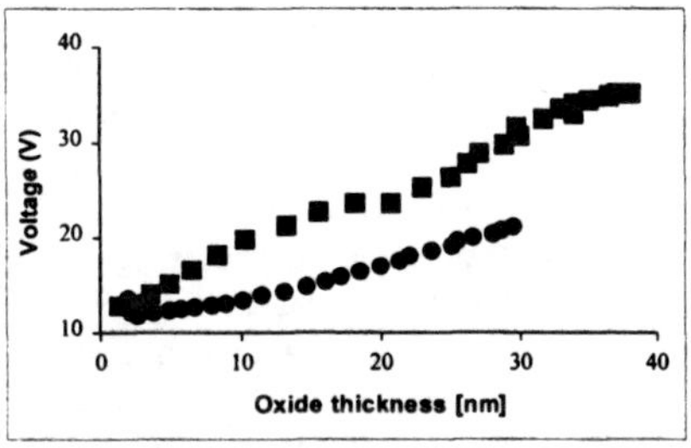

Fig.2: Potential versus oxide thickness for a sample of $Si_{0.5}Ge_{0.5}$ ($\blacksquare$) and pure Ge($\bullet$) grown with total substrate current of 0.3 A

The principle of this mode of oxidation has been described by Peeters and Li[16] : if the substrate is positively biased, a layer of electrons develops on the top of the growing oxide. Neutral oxygen atoms present in the plasma collide with the growing dielectric surface and capture these electrons and become O⁻ ions. The electric field, across the growing oxide, accelerates the O⁻ ions through it to the oxide/semiconductor interface and oxidation occurs. In our experiments we assume is that the total current density, both electron and O⁻ ion contributions, through the oxide is constant as is the electric field in the oxide. Experimental studies of Si and Ge oxidation confirm that the voltage follows the change in thickness of the oxide layer, resulting an essentially constant electric field. Using the steady state solution[16] one derives the variation of oxide thickness with time:

$$d_{ox} = A \ln (1+Bt) \qquad (1)$$

where $A = \mu E/k$, μ is the ionic mobility in the oxide, E the electric field, k is the loss factor which allows for the fact that all the O⁻ ions starting at the plasma/oxide interface may not reach the oxide/semiconductor interface. $B = \alpha N_0 k / n_b$, n_b is the O atom density in the grown oxide , N_0 is the O⁻ ion concentration at the plasma/oxide interface and α is a parameter which allows for the fact that all O⁻ ions arriving at the oxide /semiconductor interface do not necessarily result in the formation of an Si-O bond. Derivation of the equation 1 in the limit t →0 and α→1 gives the initial slope of the oxide growth as a function of time:

$$d\, d_{ox} / dt = A\, b = \mu E\, N_0 / n_b \qquad (2)$$

so that we see that the ionic mobility $\mu \propto (dd_{ox} / dt)/E$, if the others terms are constant. This model has been demonstrated to apply for pure semiconductors [15][16]. In fig. 1 we compare the oxidation curves for different substrates. We clearly see that the oxidation behaviour of $Si_{1-x}Ge_x$ is very different from pure Si or Ge. A quick examination of the curves indicates that the oxidation occurs more rapidly in pure Ge than in pure Si but slower than in the alloys. With increasing Ge concentration in the $Si_{1-x}Ge_x$ films, a thicker oxide is grown in a shorter time. The initial growth rate ($\Delta d_{ox}/\Delta t$ when t→0) was estimated taking the shortntime behaviour of the curves from fig. 1. We obtain, respectively for x = 0, 0.08, 0.15, 0.25, 0.5 and 1, the following values: 1.32, 2.74, 3.33, 4.2 and 1.28 nm per minute. The error was estimated to be ~ ±0.3 [nm/min.]. It is clear that the alloys behave differently as compared to a pure semiconductor suggesting that the Peeters and Li model will only give an approximate idea of what happens for the alloy case. According to the model, the voltage versus thickness should be constant since the electric field is constant. For the pure Ge sample this is true over a wide range of oxide thickness (fig. 2 for $d_{ox}>5$ nm), while for the sample with x = 0.5 one may distinguish two regions. The first part of the curve ($d_{ox}<15$ nm), in fact, has a different slope as compared to the second ($d_{ox}>21$ nm). Two different electric fields may be involved suggesting that oxidation occurs in two steps. Taking the experimental data from fig. 2 we have estimated the values of E [MV/cm] and substitution in eq. 2 yields the mobility μ_{Ge} [cm^2/Vs] of O$^-$ ions in GeO$_2$ and μ_1 and μ_2 for $Si_{0.5}Ge_{0.5}O_2$ in the first and in the second parts of the curve. We find $E_{Ge} = 4.7$ and $\mu_{Ge} = 0.27$ whereas for the alloy, $E_1 = 7.1$ and $\mu_1 = 0.54$ and $E_2 = 7.9$ and $\mu_2 = 0.28$. As demonstrated previously [15], the ionic mobility in the $Si_{1-x}Ge_xO_2$ alloys increases with Ge concentration. However, $\mu_1 > \mu_2$ but is not close to Ge or Si values(0.17 [16]). One might hypothesise that at the beginning of the oxidation the oxide is more homogeneous in the sense that the Si/Ge ratio remains the same in the oxide and in the substrate. As the oxidation continues the mobility decreases, becoming close to μ_{ge}. For the same reason we conclude that at the end the oxide is inhomogeneous and probably two layered, one rich in Ge oxide and the other in Si. Finally we have to underline that this behaviour is not obvious in $Si_{1-x}Ge_x$ samples with x ≤ 0.15. To be sure of this hypothesis I. R. spectra and Auger profiles were measured. A typical IR absorbance spectrum for normal incidence beams is shown in fig. 3 for 0 %15 %, 50% and 100% of Ge concentration substrates, fig 4 shows a typical Auger profile for the same $Si_{0.5}Ge_{0.5}$ oxide samples

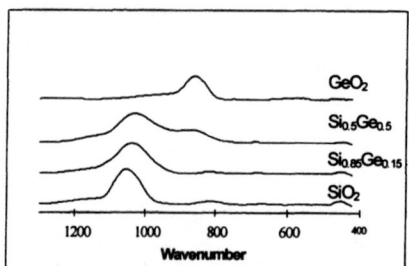

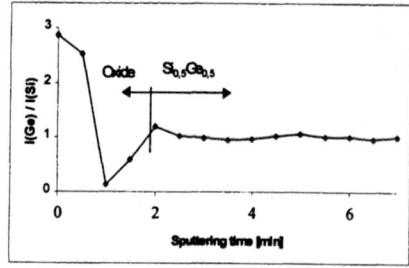

Fig 3: Infrared absorption using normal incidence for different plasma oxidised $Si_{1-x}Ge_xO_2$ samples compared with plasma oxidised Si and Ge

Fig 4: Auger electron spectroscopy profile for $Si_{0.5}Ge_{0.5}O_2$ on $Si_{0.5}Ge_{0.5}$

The main peaks observed at 858 cm^{-1} and 1000 cm^{-1} correspond to the asymmetric O stretching modes (transverse optical, TO) respectively of O atoms in Ge-O-Ge and Si-O-Ge bonds[15]. One expects that, starting from pure Si, with increasing concentration of Ge, a larger number of Ge-O-Si and Ge-O-Ge bonds should be formed. On a statistical basis[16] the normalized intensity of the Ge-O-Ge TO peak at 858 cm^{-1} (which corresponds essentially to the number of bonds) for 100%, 50%, 15%, and 0 % of Ge should be 1, 0.25, 0.025, and 0 ,while from the spectra we have found respectively 1, 0.35, 0.03, and 0 (error ±0.02). This suggests that too many Ge-O-Ge bonds are present in the samples with high Ge concentration ($x \geq 0.25$) and that probably a germanium dioxide rich layer has formed. The Auger profile shown in fig. 4, provides clear evidence that accumulation of Ge at the oxide surface occurs and follows Ge depletion near the oxide/semiconductor interface. The migration of the Ge atoms to the oxide surface is the opposite of what occurs during thermal oxidation of $Si_{1-x}Ge_x$ films and is presumably related to the presence of the electric field during anodic oxidation.

Fig. 5 shows the high frequency C-V curves for $Si_{1-x}Ge_x$ plasma oxide with x = 0.08 and 0.15. Typically, unannealed $Si_{0.92}Ge_{0.08}O_2$ samples exhibit a standard curve with corresponding fixed oxide charge density $Q_{ox} = -1.3*10^{12}$ cm^{-2}. The capacitor curve for $Si_{0.85}Ge_{0.15}O_2$ appears more like a "metal-oxide-metal" or a strongly doped semiconductor - oxide - metal structure. No differences were observed in these curves following annealing in N_2 (95%) H_2 (5%) at 425 °C for 40 minutes. The data strongly suggests that the plasma oxidation process has modified the dopant character of the $Si_{1-x}Ge_x$ substrate in the high Ge concentration samples. To test this hypothesis we formed capacitors by depositing Ta_2O_5 as a dielectric on $Si_{1-x}Ge_x$ samples cut from the same wafers as used for oxidation, the Ta_2O_5 capacitor curves are shown in Fig. 6.

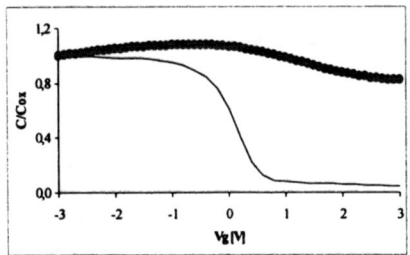

Fig 5: High-frequencies C-V curves for $Si_{1-x}Ge_x$ oxide grown in ECR plasma The Ge concentration was 8% (-) and 15 % (•).

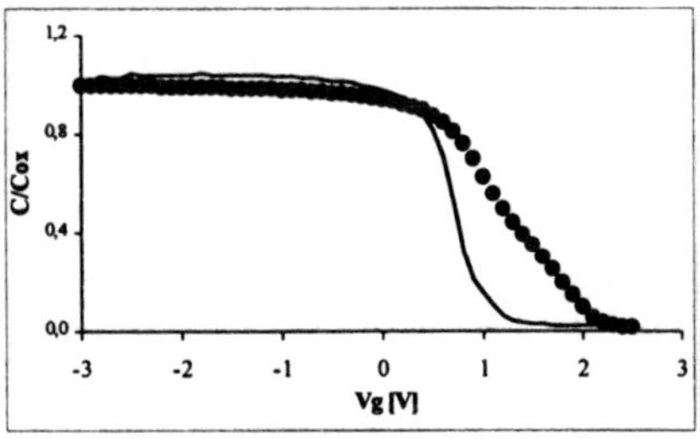

Fig 6: High-frequencies C-V curves for $Si_{1-x}Ge_x$ oxide deposited with ECR plasma reactor. The Ge concentration was 8% (-) and 15 % (•).

We see that now, even for the $Si_{0.85}Ge_{0.15}$, we have a C(V) curve reminiscent of an MOS capacitor though significant interface state densities are present. These results suggest that the plasma oxidation indeed modifies the nature of the substrate, at least superficially. (Note that during plasma assisted Ta_2O_5 deposition, some minor oxidation occurs at the substrate surface before the deposited film becomes thick enough to prevent it).

CONCLUSIONS

ECR microwave plasma oxidation of $Si_{1-x}Ge_x$ has been investigated as a potential alternative for thermal oxidation. The oxide layer grown on the $Si_{1-x}Ge_x$ semiconductor shows the oxide is "mixed". I.R. spectra were measured and studied and the main peaks and their shifts confirm[17][18] that the mixed oxide is chemically bonded but inhomogeneous. Auger profiling confirms this hypothesis and reinforces the idea that I.R. measurements can be used, in certain circumstances, to provide evidence for sample inhomogeneity. The electrical measurements suggest that plasma damage to the substrate surface occurs during oxidation resulting in effective enhanced doping. Capacitors formed by deposition of Ta_2O_5 confirm this hypothesis. The present results suggest that plasma oxidation could only be used technologically for $Si_{1-x}Ge_x$ samples with x<0.15.

REFERENCES

[1] J. I. Pankove, **Optical processes in semiconductors**, (1971)

[2] D. K. Nayak, J. C. S. Woo, K. L. Wang and K. P. MacWilliam Appl. Phys. Lett, **62**, 2853, (1993)

[3] W. Geppert and H.-U. Schreiber, Electron. Lett. **32**, 2228 (1996)

[4] V. I. Kuznetsov, R. V. Veen, E. van der Drift, K. Werner, A. H. Verbruggen, and S. Radelaar J. Vac. Sci. Technol. B **13**, 2892 (1995)

[5] T. P. Pearsall , H. Temkim , J. C.Bean, and S.Luryi, IEEE Electron Dev. Lett. **7**, 330 (1986)

[6] H. Hosono, M. Mizuzuguchi, H. Kawazoe, and J. Nishii, Jpn. J. Appl. Phy. **35**, L 236 (1996)

[7] P.-E Hellberg, S.-L. Zhang, F. M. d'Heurle, and C. S. Petersson, J. Appl. Phys. **82**, 5773 (1997)

[8] W. S. Liu, J. S. Chen, and M. A. Nicolet J. Appl. Phys. **72**, 4444, 1992

[9] F. K. LeGoues, R. Rosemberg, T.N. Nguyen, F. Himpsel and B. S. Meyerson, J. Appl. Phys. **65**, 1724 (1989)

[10] A. Agarwal, J. K. Patterson, J. E. Greene and A. Rockett Appl. Phys. Lett. **63**, 518 (1993)

[11] I. S. Goh, S. Hall, W. Eccleston, J. F. Zhang and K. Warner, Electron. Lett. **30**, 1988 (1998)

[12] M. Mukhopadhyay, S. K. Ray, T. B. Ghosh, M. Sreemany and C. K. Maiti, Semicon. Sci. Techol. **11**, 360 (1996)

[13] I. S. Goh, S. Hall, J. F. Zhang, S. Hall, W. Eccleston and K. Warner, Microelec. Eng. **28**, 221 (1995)

[14] M. Seck, R.A.B. Devine, C. Hernandez, Y. Campidelli and J-C Dupuy, Appl. Phys. Lett. **72**, 2748 (1998)

[15] T. Busani, H. Plantier, R.A.B. Devine, C. Hernandez, Y. Campidelli, J. Non-Cryst. Solids (in press, 1999)

[16] J. Peeters and L. Li, J. Appl. Phys. **72**, 719 (1992)

[17] C. Martinet and R.A.B. Devine and M. Brunel, J. Appl. Phys. **81**, 6996 (1997).

[18] T. Busani, H. Plantier, , R.A.B. Devine, C. Hernandez, Y. Campidelli, J. Appl. Phys. (in press 1999)

MODELING OF SURFACE MANIPULATION BY FEMTOSECOND LASER PULSES

A. S. GRUZDEVA, V. E. GRUZDEV
State Research Center "S.I.Vavilov State Optical Institute"
Birzhevaya Liniya 12, St.Petersburg, 199034, Russia, gru@mailbox.alkor.ru

ABSTRACT

There are discussed some fundamental physical aspects of surface manipulation by femtosecond laser pulses. Among touched on problems are formation processes of shock electromagnetic waves and surface ripple structures. Proposed theoretical model of femtosecond laser-pulse interaction with matter is illustrated by results of FDTD modeling of linear and nonlinear light scattering by rough surface. Possibility of surface roughness modification and required for that optimal laser-pulse parameters are discussed on the bases of obtained results.

INTRODUCTION

Laser-based technologies are getting a wide rage of applications in various branches of semiconductor manufacturing. Among them femtosecond laser technologies are the fastest developing and the most attractive. Their great potential applications for material processing have been shown by a lot of experiments [1-4]. Among advantages of femtosecond lasers the following should be mentioned first of all: possibility of surface manipulation with super-high resolution [1], possibility to perform selected chemical reactions at the surface [2, 3]. They make femtosecond laser-surface interaction the most attractive method for surface manipulation and conditioning [4].

One of general and the most frequent difficulties in laser processing of surface is choosing of suitable parameters of laser radiation. In many cases it slows down developing of laser technologies. Experimental investigation of this problem is often very difficult due to many uncontrolled parameters and factors influencing experimental results. Theoretical approaches to the problem are still not effective because till now there is no clear understanding of nature and regularities of femtosecond laser-matter interactions.

Main characteristic features of femtosecond interactions with transparent materials are
1) high intensity of radiation within focal spot which can exceed 10^{13} W/cm^2 [2, 3];
2) little pulse duration which is below relaxation time for electronic and phonon sub-systems. The first of them points at critical role of nonlinear electrodynamic effects in femtosecond interactions because laser-induced variations of refraction in focal area are about $0.01 - 0.001$ and nonlinear distortions of laser beam are enough to change its space structure sufficiently. Thus, the first question is as follows: what processes can result such distortions in?

The second feature of femtosecond pulses points at need to look for fast mechanisms of laser-matter interactions because traditional models of laser-induced heating, cannot explain rather large ablation rates [2], formation of surface ripple patterns and other peculiarities of laser-surface interaction for femtosecond pulses. Recent attempt of the authors to generalize two-temperature model for the case of dielectrics and semiconductors was not correct as it was pointed out by Prof. M.N.Libenson [5]. In this connection general problem arises – how is radiation energy transported to crystal lattice to damage it? The mechanism of energy transport must be fast and can be connected with nonlinear wave propagation.

Here we present a new model for description of nonthermal femtosecond laser interaction with surface of low-absorbing materials (e.g., dielectric or wide-bandgap semiconductor). It is based on results of correct calculations of nonlinear wave propagation in transparent media.

MODELING OF FEMTOSECOND LASER-MATTER INTERACTION

We start with modeling of nonlinear electrodynamical processes because they are the most suitable candidates for being a mechanism of femtosecond ablation and surface manipulation. Nonlinear radiation propagation is highly complicated process described by nonlinear wave equation that it difficult for analytical solving. Numerical methods, for example, finite-difference time domain technique seems to be the best tool to attack this problem. In this work field evolution was simulated according to developed by the authors technique [6, 7] only for the case of electric-field vector of incident wave parallel to surface ripples (TE polarization in Fig. 1). This allows reducing general nonlinear wave equation to simpler one for z-projection of electric field vector

$$\frac{\partial^2 E}{\partial x^2} + \frac{\partial^2 E}{\partial y^2} - \frac{\varepsilon_0}{c_0^2} \cdot \frac{\partial^2 E}{\partial t^2} = \frac{4\pi}{c_0^2} \cdot \frac{\partial^2 P_{NL}}{\partial t^2}, \tag{1}$$

where c_0 is light speed in vacuum, ε_0 is constant part of dielectric function.

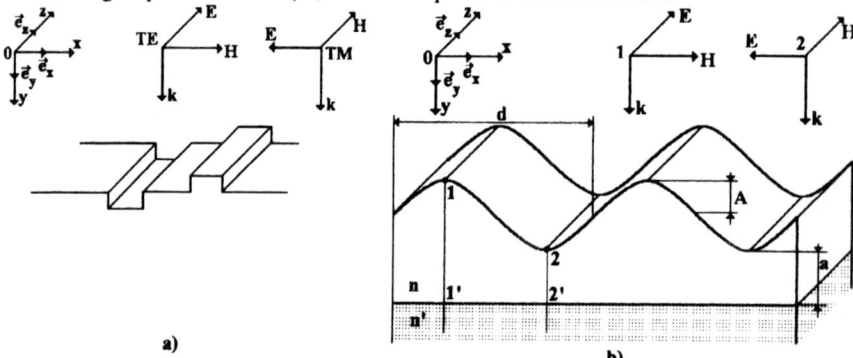

Fig. 1. Geometry of the problems for modeling: a) scattering by single scratch or surface defect; b) light scattering by dielectric or semiconductor layer of thickness a and refractive index n on dielectric substrate (refractive index n') with sine roughness of period d and amplitude A. Two linear polarizations (1 - TE; 2 - TM) of incident monochromatic plane wave were considered throughout the paper. The scratch and surface ripples are assumed to be infinitely long (so called 1D surface). It was considered normal incidence of light.

Modeling of nonlinear processes for the other polarization requires more computer resources than the authors have and was not fulfilled. Nonlinear response of the medium P_{NL} depends on electric field in the following way [8] $P_{NL} = (6\chi_{1122} + 3\chi_{1221})E^2 E$. According to this laser-induced variation of refractive index is given by [8]

$$\Delta n = n_2 E^2 = (2\pi/\varepsilon_0)(6\chi_{1122} + 3\chi_{1221})E^2. \tag{2}$$

where $\chi = (6\chi_{1122} + 3\chi_{1221})$ is an element of nonlinear response tensor. Laser-induced surface evolution under action of light pressure and gradient forces can be modeled within developed approach too: scattered field can induce inhomogeneous distribution of refraction what results in appearing of light-induced forces in bulk leading to medium deformation. The force is calculated as [9]

$$\vec{f} = -\frac{E^2}{8\pi} \cdot \nabla \varepsilon + \nabla \left(\rho \frac{E^2}{8\pi} \left[\frac{\partial \varepsilon}{\partial \rho} \right]_{T=const} \right) + \frac{\varepsilon\mu - 1}{4\pi c} \frac{\partial}{\partial t} \left[\vec{E} \times \vec{H} \right]$$ (3)

where $\varepsilon = n^2 \cdot n_0^2 + 2n_0 \cdot \Delta n$ and Δn is given by (2), ρ is medium density and μ=const. Thus, for correct modeling one should calculate space distribution of magnetic according to one of Maxwell's equations

$$\frac{\partial \vec{H}}{\partial t} = -\frac{c_0}{\mu_0} \cdot \vec{\nabla} \times \vec{E}.$$ (4)

RESULTS

Formation of Shock Electromagnetic Waves

One of the most outstanding results of modeling is showing possibility of formation of shock electromagnetic wave (SEW) [10, 11]. Threshold of that process is given by [11]

$$E_{th} \approx \sqrt{\frac{n_0}{n_2} \left(\exp\left\{ \frac{\Delta x}{s} \right\} - 1 \right)},$$ (5)

where s – is distance in nonlinear medium passed by the wave till field disruption appears, n_0 – is constant linear part of refraction, n_2 – is nonlinear coefficient of refraction, $\Delta x > 0$ – initial distance between two points of electromagnetic wave profile where electric field strengths are correspondingly $E=0$ and $E=E_{th}$. In case of homogeneous medium $\Delta x = \lambda/4$, where λ is radiation wavelength in the medium. According to (5) threshold of SEW formation in homogeneous medium is $E_{th} = 16.2 \cdot 10^6$ V/cm ($I_{th} = 6.95 \cdot 10^{11}$ W/cm^2) for $s = 2\Delta x$, $n_0 = 1.41$ and $n_2 \sim 3.14 \cdot 10^{-10}$ esu. Several consequent stages of SEW formation are depicted in Fig. 2.

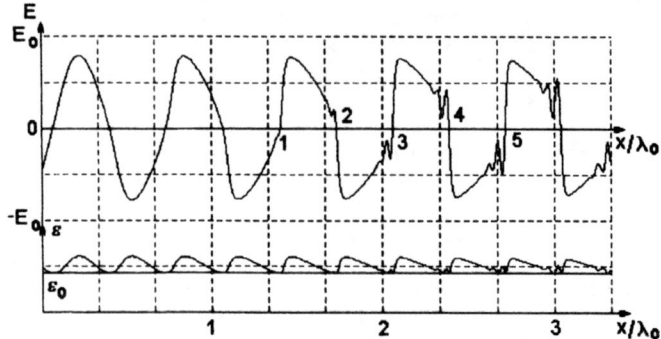

Fig. 2. Consequent stages of formation and evolution of field disruptions and SEW at instant profile of high-power wave propagating in homogeneous nonabsorbing medium with positive nonlinear response. 1 – falling of wave profile top on area of zero field strength, 2 – appearing of field disruption, 3, 4, 5 – appearing of higher harmonics in the area near SEW front and smoothening of the front. Laser-induced variations of dielectric function are depicted in the lower part of the figure.

The most important property of SEW is formation of field disruptions during wave propagation. On the one hand, such disruption is accompanied by generation of high harmonics what results in ionization of atoms. This can result in formation of localized hot-electron plasma inside transparent materials. On the other hand, space electric-field variations near the front of field disruption are so large that electric field can move atoms and ions away from crystal lattice. This process can result in phase transition from crystal to amorphous matter [3] and even in total removal of material from focal area [3]. Thus, SEW can break crystal lattice without heating it.

Mechanisms of Surface Roughness Manipulation by Femtosecond Pulses

The second step is investigation of field structure at rough surface. Roughness was modeled by sine variations of surface relief. Obtained for linear scattering results show that scattering of plane wave by dielectric surface with sine-varied relief results in formation of a set of field maxima near hillocks of the relief (Fig. 3). We show that local amplification of field amplitude near the surface differs much for two linear polarizations: it is higher for TE polarization (reaches maximum value of 2.7) and it is about 1.1 for TM polarization. Fig. 3 depicts one of instant space distributions of scattered radiation for TE incident polarization

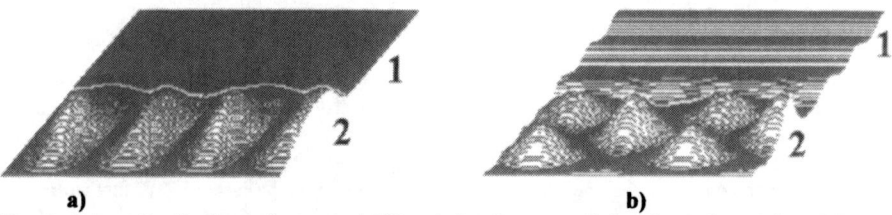

a)   b)

Fig. 3. Space distribution of scattered TE polarization near dielectric surface (glass BK-7, refractive index n_0 =1.50, absorption was assumed to be zero) with sine roughness (space period $d = 0.69\lambda$, λ - radiation wavelength in vacuum, roughness amplitude 0.1λ). There are depicted instant space distributions of total (**b**) and pure scattered field (total field minus reflected and refracted fields) (**a**) squared electric field. Area 1 - field distribution in air, area 2 - field distribution in glass layer.

Field amplification inside material depends critically on ratios of roughness period and amplitude to radiation wavelength (Fig. 4, 5) and single resonance can appear at certain value of the ration.

It is also shown that in case of light scattering by single scratches field maxima are located near edges of the scratches. That is the case practically for all sizes of the scratches. Amplification of electric field amplitude does not exceed 1.6 for all the considered cases. Investigation of enter and exit surfaces shows field amplification to be about 30% less at enter surface than at exit one. Thus, laser radiation of below-threshold fluence can result in local melting of surface scratches and their smoothening. On the other hand, using of too long pulses or rather high fluences can result in formation of ripple surface structures and damage of the surface. This shows that certain optimal parameters of laser pulse must exist. It is also shown critical influence of laser wavelength on local field amplification in case of multiple scratches: it is connected with possible large field amplification at periodic surface roughness.

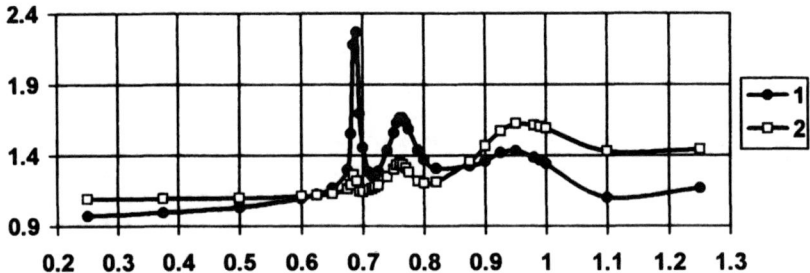

Fig. 4. Dependence of amplification on ratio of roughness spatial period to radiation wavelength d/λ for TE polarization of incident plane wave scattered by glass sample (BK-7, refractive index $n_0=1.50$) with sine roughness of amplitude $A=0.1\ \lambda$. 1 - Amplification of field in silica (transmitted radiation), 2 - in air (reflected radiation). Resonance with maximum amplification of field amplitude 2.269 appears at $d=0.69\lambda$.

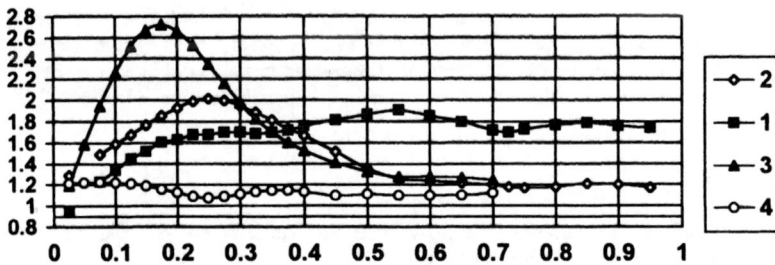

Fig. 5. Dependence of field amplification for TE polarization on roughness amplitude for sine surface roughness of glass sample with off-resonance period $d=\lambda$ (curves 1 and 2) and resonant period $d\ 0.69\lambda$ (curves 3 and 4). 1, 3 – Amplification of field amplitude in glass layer (transmitted radiation); 2, 4 – field amplification in air (reflected radiation)

Modeling of light scattering by dielectric with nonlinear refraction has shown no sufficient nonlinear perturbations to linear solution to wave equation even for incident field amplitude $E_0=10^6$ V/cm, $n_2=10^{-11}$ esu. Such field can induce about 0.05 addition to refraction. Further increasing of field amplitude results in formation of SEW.

Being scattered at rough surface, high-power laser radiation can induce inhomogeneous distribution of refraction (see Figs. 2-4, 7). This results in appearing of light-induced forces in bulk leading to medium deformation. The force is calculated according to (3). Increasing of roughness amplitude with increasing of field amplification (Fig. 5) shows possibility of formation of the following positive feedback: increasing of field amplitude near the hillocks → increasing of laser-induce refraction gradient → increasing of local field-induced force → increasing of surface roughness due to surface deformation → increasing of scattering. Formation of SEW near the surface can result in fast phase transition from solid to liquid that is similar to melting but this time it is induced by electric field near SEW front. To check this idea we assumed the near-surface layer of the medium to be a viscous liquid so that

167

its motion can be described by equation

$$\frac{d\vec{v}}{dt} = \nu\Delta\vec{v} + \vec{f} \qquad (6)$$

where ν is viscosity. Integrating of (6) allows calculating of displacement of single parts of irradiated material. Modeling of surface roughness evolution induced by TE polarization has shown increasing of roughness amplitude accompanied by increasing of scattered field amplitude. This process stopped as soon as roughness reached 0.6λ amplitude what can be expected from Fig. 5.

CONCLUSIONS

Obtained results of modeling show critical role of nonlinear electrodynamical effects in nonthermal interaction of femtosecond laser pulses with matter. One of the most promising effects is formation of SEW which can result in both ionization of irradiated material and damage of its crystal lattice. In the latter case phase transition similar to melting can take place. Characteristic feature of that transition is absence of heating of surrounding material.

Obtained results of modeling of roughness manipulation are in good qualitative agreement with experimental data, in particular, [1] where authors observed generation of ripples at surfaces of isotropic materials. In particular, observed ripples were oriented parallel to incident-field electric vector and their period was about laser wavelength in free space.

Further development of theoretical investigations in the considered way is connected with more detailed investigations of SEW formation, in particular, dependence of that process on material parameters.

REFERENCES

[1] E.E.B.Campbell, D.Ashkenasi, and A.Rosenfeld, in *Lasers in Materials*, edited by R.P.Agarwal (Trans Tech Publ., 1998), Ch. 5.
[2] J.F.Young, J.S.Preston, H.M.Van Driel, J.E.Sipe, Phys. Rev. B, **27**, p. 1141-1155 (1983).
[3] C.B.Schaffer, E.N.Gleser, N.Nishimura, and E.Mazur, in Proc. of SPIE, v. **3269**, pp. 36-45 (1998).
[4] E.J.Lerner, Laser Focus World, v. **34** (10), pp. (1998).
[5] M.N.Libenson, private communications.
[6] V. E. Gruzdev, A. S. Gruzdeva, in *High-power Laser Ablation*, edited by C.R.Phipps (Proc. of SPIE, v. **3343,** Bellingham, WA, 1998) pp. 305-316.
[7] A.S.Gruzdeva, V.E.Gruzdev, paper U3.11, poster session U3, this MRS Meeting.
[8] P.D.Maker, R.W.Terhune, Phys. Rev., **137** (3A), A801-A818 (1965).
[9] L.D.Landau, E.M.Lifshits, *"Electrodynamics of Continuos Media"*, Moscow, Nauka, 1992.
[10] A.Jeffrey, *J. of Mathematics and Mechanics*, v. **15**, pp. 1-14, 1966.
[11] V.E.Gruzdev, S.I.Vavilov State Optical Institute, "Electromagnetic resonances and instabilities and their role in interaction of high-power laser radiation with transparent materials", 1999.

PRODUCTION OF InSb THIN FILMS THROUGH ANNEALING Sb₂S₃-In THIN FILMS

M.T.S. NAIR, Y. RODRÍGUEZ-LAZCANO*, P.K. NAIR

Centro de Investigación en Energía, Universidad Nacional Autonoma de México, A. P. 34, Temixco, Morelos 62580, MEXICO, mtsn@mazatl.cie.unam.mx
*Permanent address: Faculty of Physics-IMRE, University of Havana, Cuba

ABSTRACT

A method to produce large area indium antimonide thin films through a reaction, $Sb_2S_3 + 2 In \rightarrow 2 InSb + 3 S\uparrow$ is presented. A thin film of Sb_2S_3 with typically 0.2 μm thickness is produced on glass substrate by chemical bath deposition (CBD) at 10°C using thiosulfatoantimonate(III) complex. Subsequently, a thin film of indium is deposited on the Sb_2S_3 film by thermal evaporation. Annealing the thin film stack of Sb_2S_3-In at 300°C in a nitrogen atmosphere produces the InSb thin film. The formation of this film is confirmed by x-ray diffraction studies. We would discuss the optimization of the individual film thickness in the Sb_2S_3-In stack to produce a thin film of single phase InSb or a heterostructure, Sb_2S_3-InSb. The electrical and optical properties of the films are presented.

INTRODUCTION

InSb is important for thermal infrared detection and as a magnetoresistive material in position-sensitive or speed-sensitive- sensors [1]. The material has a room temperature band gap of 0.17 eV and a high electron mobility of about 8 m^2/V s at room temperature [2]. InSb of 20 nm - 20 μm thickness is sufficient for magnetoresistive elements. Plasma-assisted epitaxial growth [3] and elctrodeposition [1] techniques have been reported for the deposition of thin films of InSb.

Chemical bath deposition of different semiconductor thin films has been reported from our laboratory, as reviewed in [4]. Annealing multilayers of thin film semiconductors like Bi_2S_3-CuS and Sb_2S_3-CuS has shown the formation of films of ternary composition, Cu_3BiS_3 [5] and Cu_3SbS_4 [6], respectively. Thermal treatments of evaporated indium film on II-VI semiconductors in turn resulted in heterostructures of the type, MX:In-In_2O_3 [7-9]. In this paper we report, for the first time, the formation of InSb thin film through a solid state reaction of evaporated indium film on a chemically deposited Sb_2S_3. The reaction takes place in a nitrogen atmosphere at 300°C. The properties of the films and the mechanism of formation of the III-V compound from V_2-VI_3-III stack are discussed.

EXPERIMENTAL

Deposition of thin films

Antimony sulfide thin films were deposited using a chemical bath reported in a previous work [6]. Microscope glass slides were used as substrates. These substrates were cleaned using commercial detergent and deionized water and dried in air. The bath was prepared in a 200 ml beaker as follows: 1.3 g of $SbCl_3$ (Baker Analyzed Reagent) was dissolved in 5 ml of acetone. To this was added 50 ml of 1 M $Na_2S_2O_3$ (Reactivo Analitico Monterrey) that was previously cooled to 10°C, followed by the addition of 145 ml of cold deionized water and stirred well. The substrates were placed in the bath, vertically supported on the wall of the beaker. The deposition was made at 10°C. At the end of 4 h of deposition, glass slides covered with specularly reflective orange-yellow thin films of Sb_2S_3 were removed from the bath, washed well with deionized water and dried. Step-thickness measurements using an Alpha Step 100 unit showed that the film thickness was about 0.2 μm. By immersing these films in a second bath, film thickness of 0.3 μm or more could be obtained.

Thin films of indium were deposited on the Sb_2S_3 thin films by thermal evaporation of 20-60 mg of 99.999% In (Alfa Products) in a high-vacuum coating unit after attaining a residual pressure of 10^{-6} mbar. Since the films of indium are very soft, step thickness measurement was not possible. Thus the thickness of In film deposited on the slides was estimated gravimetrically: approximately 0.075 μm of In film was deposited on an Sb_2S_3 coated substrate for an indium source of 20 mg.

The Sb_2S_3-In films were annealed for 1 h each in a vacuum oven (T. M. Vacuum Products Inc., New Jersey) in 50 m Torr of nitrogen.

Characterization

X-ray diffraction (XRD) patters were recorded using Cu-K_α radiation on a Siemens D-500 diffractometer. The optical transmittance (T%) and the near-normal specular reflectance spectra (R%) of the films were recorded on a Shimadzu UV-3101PC UV-VIS-NIR scanning spectrophotometer with air and a front aluminized mirror, respectively, as references. For the electrical measurements, silver paint electrodes were printed on the samples in a coplanar configuration. The dark- and the photo- currents were recorded on a computarized system using a Keithley 619 electrometer and a Keithley 230 programmable voltage source. Photocurrent response measurements were made under 300 W/m^2 intensity of illumination from a tungsten-halogen lamp.

RESULTS AND DISCUSSION

Figure 1 gives the XRD patterns of Sb_2S_3 and Sb_2S_3-In films before and after annealing at 300°C in nitrogen for 1 h. The pattern of as-prepared Sb_2S_3-In sample (bottom curve) shows characteristic peaks due to In at 2θ values of 32.965° and 36.328°, corresponding to reflections from (101) and (002) planes, respectively. The films of Sb_2S_3 before annealing do not present any peak in the XRD pattern [6]. Upon annealing in nitrogen at 300°C, the crystallinity improves in the films. Consequently, well-defined peaks matching those of stibnite (JCPDS 6-0474) appear in the XRD pattern of the annealed film (middle curve). The XRD patterns of the annealed samples of Sb_2S_3-In films show new peaks in addition to the characteristic peaks due to Sb_2S_3. These additional peaks match the peaks due to reflections from the (111), (220) and (311) planes of cubic InSb (JCPDS 6-0209). This suggests that annealing the multilayer films of Sb_2S_3–In results in the reaction in solid state between Sb_2S_3 and In leading to the formation of an InSb film.

The thickness of the two participating films could be varied in such a way that the molar ratio of Sb_2S_3: In is nearly 1:2 as required for the stoichiometric reaction, $Sb_2S_3 + 2In \rightarrow 2InSb + 3S\uparrow$. Typically, for a 0.2 μm film of Sb_2S_3 (mass density, 4.1 g/cm^3, molecular mass 339.7 g/mol), 0.076 μm of In (mass density, 7.3 g/cm^3, atomic mass 114.8 g/mol) is required for complete conversion

The XRD patterns of Sb_2S_3-In films of different thickness annealed at 300°C are given in Fig. 2. The pattern obtained in the case of Sb_2S_3 (0.2μm)-In (0.075 μm) film shows peaks that match the standard pattern of InSb (JCPDS6-0208) and no peak due to Sb_2S_3. This shows that a complete conversion of Sb_2S_3-In films to InSb has taken place at this combination of film thickness, in agreement with stoichiometry calculation. Thickness of InSb (mass density 5.78

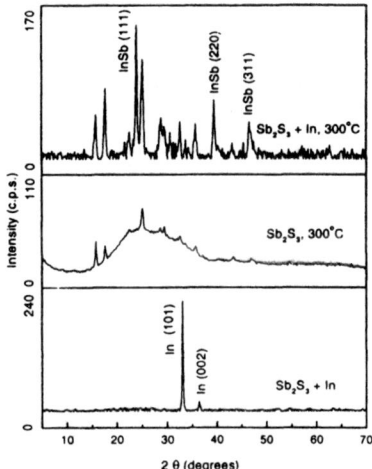

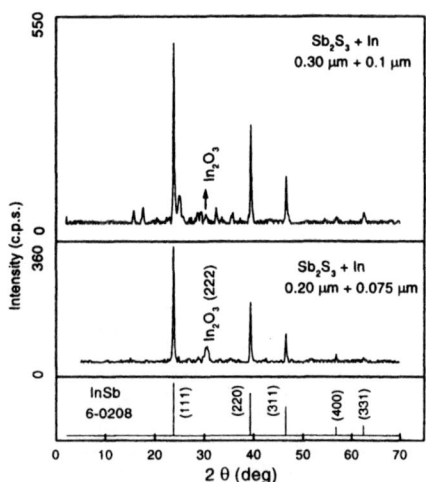

Fig 1. XRD pattern (Cu-K$_\alpha$) of chemically deposited Sb$_2$S$_3$ plus evaporated In films, recorded before (bottom) and after annealing at 300°C in N$_2$.

Fig 2. XRD pattern of annealed (300°C, 1h) Sb$_2$S$_3$–In films for different combination of thin film thickness, indicating total (bottom curve) or partial conversion of Sb$_2$S$_3$ to InSb, along with some quantity of In$_2$O$_3$.

g cm^{-3}, molar mass 236.6) thus obtained, will be 0.1978 μm, approximately the same as the starting Sb$_2$S$_3$ film thickness. Deviation from the stoichiometric reaction can lead to unreacted Sb$_2$S$_3$ at the bottom, illustrated in the top curve in Fig. 2.

The XRD patterns in Fig.2 also show the presence of slight amount of In$_2$O$_3$ in the film. This could arise due to the reaction of In with oxygen that has been liberated from the grain boundaries of chemically deposited Sb$_2$S$_3$ thin films.

The grain size of crystallites in the films are determined using the Scherrer equation, D = 0.9λ/βcosθ (λ, 1.5406 Å for Cu-K$_\alpha$; β, the full width at half maximum of the peak in radian and θ, the Bragg angle). The grain diameter of InSb thin film in the case of samples in Fig. 2 is about 35 nm.

Figure 3 shows the optical transmittance and reflectance spectra of Sb$_2$S$_3$, before and after annealing at 300°C, and of annealed samples of Sb$_2$S$_3$ (0.2 μm)-In (0.075 μm), before and after etching in 1 M HCl for 1 h. XRD analysis has shown that annealing the latter films results in the formation of InSb with some amount of In$_2$O$_3$. Assuming that the In$_2$O$_3$ formed during the annealing is at the surface, the etching will remove the In$_2$O$_3$ film leaving the underlying InSb film. The electrical properties of the films, discussed later, will show that the assumption is valid.

The optical absorption coefficients (α) of the materials were determined from the T% and R% values: $\alpha = (1/d) \ln [(T\%-R\%)/T\%]$, where d is the film thickness. Values of α^2 are plotted against photon energy hv in Fig. 4. The as-prepared Sb$_2$S$_3$ sample shows a direct band gap of 2.43 eV which is higher than that of the annealed sample, 1.76 eV, due to quantum confinement in the nanocrystalline sample [6]. Thin film of InSb show relatively high absorption coefficient

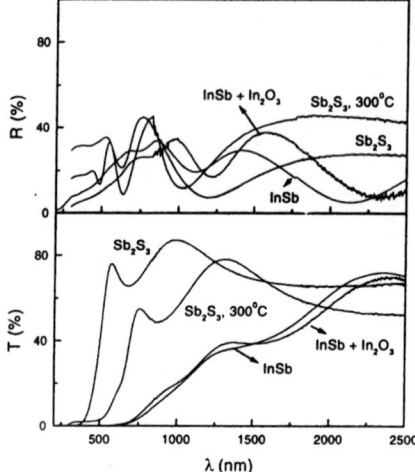

Fig 3. Optical transmittance (T%) and specular reflectance (R%) curves for a 0.2 µm Sb_2S_3 thin film and of the same film recorded after annealing at 300 °C during 1 hour. Also shown are the curves for $InSb+In_2O_3$ thin film produced and for InSb film resulting from the etching of the film for 1 h in 1M HCl.

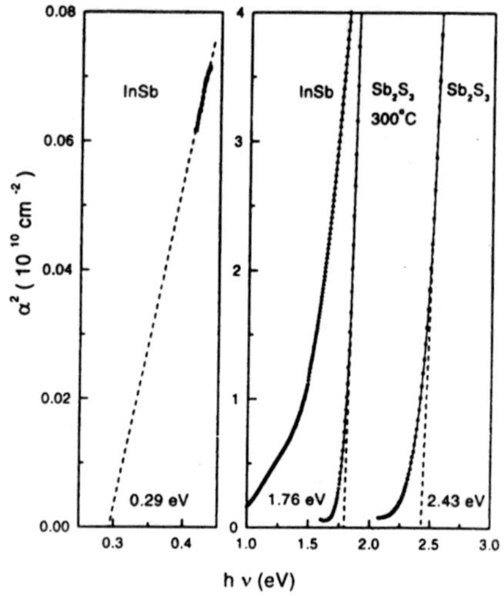

Fig. 4 Square of the optical absorption coefficient (α) plotted against photon energy ($h\nu$) for the samples described in Fig. 3.

in the region 1.5 eV-2 eV. In the low energy region the data may be extrapolated to indicate a direct band gap of less than 0.3 eV. This is higher than the reported band gap value of 0.17 eV, possibly due to the very small crystallites. The order of magnitude of the absorption coefficient is in agreement with the reported values for this material [ref 2, p.751]

Photocurrent response curves of Sb_2S_3-In are shown in Figure 5. A current of 5×10^{-3} A observed at a bias of 10V in the $InSb$-In_2O_3 film indicates a sheet resistance of 2×10^3 Ω/square. This is due to the conductive In_2O_{3-x} layer formed on the surface during the annealing, indicated in the XRD patterns in Fig .2. The conclusion that this layer is located at the top is based on the observation of the decrease in the current by six orders of magnitude during etching the film in dilute HCl. The optical transmittance spectra in Fig. 4 show that the etching does not significantly affect the curves, indicating that the absorber layer is left almost intact during the process. The underlying InSb film has a sheet resistance in the dark of $5 \times 10^9 \Omega$/square, or a dark conductivity of typically, 10^{-5} Ω^{-1}cm^{-1} (film thickness, 2×10^{-5} cm).

The above results lead to the proposal of a scheme in Fig. 6 for the formation of heterostructures, Sb_2S_3-$InSb$-In_2O_{3-x} or $InSb$-In_2O_{3-x}, depending on the relative thickness of the Sb_2S_3 and In thin films. Calculations for complete conversion of Sb_2S_3 to InSb show a thin film thickness of 38 nm In for each 100 nm thickness of Sb_2S_3. The top layer of In_2O_{3-x} formed

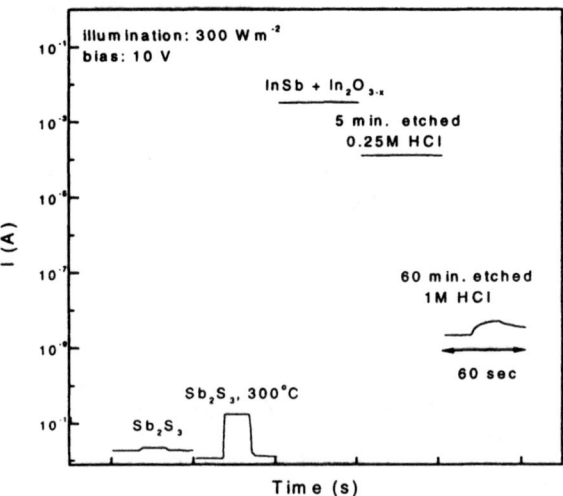

Fig. 5 Photocurrent response curves of the different films recorded using coplanar silver print electrodes (5 mm each at 5 mm separation) for 20 s in the dark, 40 s illumination and 20 s in the dark.

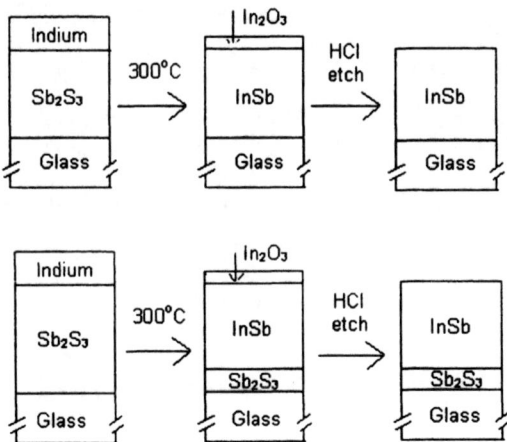

Fig. 6 Schematic diagram for the total or partial conversion of an Sb_2S_3-In stack into InSb during annealing at 300°C in a nitrogen atmosphere.

during the process may be removed through etching, or the heterostructure may be used as such for specific applications.

CONCLUSIONS

In this paper we presented a new technique for the formation of InSb thin film from chemically deposited Sb_2S_3 thin film and evaporated In thin film through heating at 300°C for about 1 h. Large area deposition capability of chemical bath deposition technique is well known [4]. Hence the present method is suitable for producing large area InSb thin films.

ACKNOWLEDGMENT

The authors are grateful to Aaron Sanchez, Oscar Gomezdaza, Juan Carlos Castrejon and Laura Guerrero for their help in the experimental work and to Leticia Baños of IIM, UNAM, for recording the XRD patterns. We acknowledge the financial support received from DGIA-UNAM, DGAPA-UNAM (Projects IN500997 and IN110598) and CONACYT (Mexico).

REFERENCES

1. M. K. Carpenter and M. W. Verbrugge, *J. Mater. Res.*, **9**, p. 2584 (1994).
2. S. M. Sze, *Physics of semiconductor Devices*, Wiley, New York, 1981 p.849 and 751.
3. S. Yamauchi, T. Hariu, H. Ohba, and K. Sawamura, *Thin Solid Films*, **316**, p. 93 (1998).
4. P. K. Nair, M. T. S. Nair, V. M. García, O. L. Arenas, Y. Peña, A. Castillo, I. T. Ayala, O. GomezDaza, A. Sánchez, J. Campos, H. Hu, R. Suárez, and M. E. Rincón, *Solar Energy Materials and Solar Cells*, **52**, p. 313 (1998).
5. P. K. Nair, L. Huang, M. T. S. Nair, Hailin Hu, E. A. Meyers, and R. A. Zingaro, *J. Materials Research* **12**, p. 651 (1997).
6. M. T. S. Nair, Y. Peña, J. Campos, V. M. García and P. K. Nair, *J. Electrochem. Soc.*, **145**, p. 2113 (1998).
7. P. J. George, A. Sánchez, P. K. Nair and M. T. S. Nair, *Applied Physics Letters* **66**, 3624 (1995).
8. V. M. García, P. J. George, M. T. S. Nair and P. K. Nair, *J. Electrochem. Soc.*, **143**, p. 2892 (1996).
9. V. M. García, M. T. S. Nair and P. K. Nair, *Semicond. Sci. Technol.* **14** (1999 at Press).

CHEMICAL VAPOR DEPOSITED TUNGSTEN FILM ON MOLECULAR LAYER EPITAXIALLY-GROWN GaAs AND ITS APPLICATION TO LOW RESISTIVITY CONTACT

Y. OYAMA*, Y.OSHIDA*, J. NISHIZAWA**, F. MATSUMOTO**, P. PLOTKA**
* Dep. of Materials Science, Graduate School of Engineering, Tohoku University, Aramaki02 Aoba-ku, Sendai 980-8579, Japan, oyama@material.tohoku.ac.jp
**Semiconductor Research Institute of Semiconductor Research Foundation, Kawauchi Aoba-ku, Sendai 980-0862, Japan

ABSTRACT

W CVD process was carried out on molecular layer epitaxially-grown GaAs just after in-situ surface treatment under optimized AsH_3 pressure followed by ex-situ photolithography process. The precursor for W CVD used is $W(CO)_6$. The W/GaAs interface is analyzed using SIMS and RBS/channeling technique. Interface structure was directly observed by cross sectional HRTEM. Plan view HRTEM observation was also carried out. From these physical analyses, it is shown that the mixed layer at W/GaAs interface is estimated to be about 2-3 ML and that the deposited W layer is epitaxially aligned with underlying GaAs lattice. The contact resistance in W/GaAs is obtained by the transmission line measurements (TLM) of patterned W on heavily doped GaAs grown by (molecular layer epitaxy) MLE. The dependence of the contact resistance on the surface treatment prior to the W CVD is also studied. Barrier height of W/GaAs structure is measured by the temperature dependence of I-V characteristics. Contact resistance of non-alloyed structure achieved are 3×10^{-7} Ωcm^2 for n-type GaAs:Te and below 5×10^{-8} Ωcm^2 for p-type GaAs:C respectively.

INTRODUCTION

In future novel semiconductor devices, the active region is localized in atomic scale and the thin layered structure is required with atomic accuracy(AA). In such fast devices, the metal/semiconductor contacts limit net operating speed. In addition, the conventional alloyed contact with a few 100 nm spike cannot be applied for such thin layered structures. Therefore, low resistivity metal/semiconductor contact formed at low temperature with atomically flat interface has been urgently required. W/GaAs contacts were found stable up to 500°C. This temperature is higher than that used for selective regrowth with MLE for the 10nm channel GaAs static induction transistor (SIT) [1]. Whereas sputtering was commonly used for W deposition, it results in serious generation of defects in thin active semiconductor layers.

In this paper, novel CVD W suitable for ultra-thin device fabrication is shown. The contact resistance in W/GaAs is shown as a function of surface stoichiometry by using the TLM on heavily doped GaAs grown by MLE. Barrier height of W/GaAs is studied in reference for lowering the contact resistance. The W/GaAs interface and the impurity profiles in MLE GaAs layers are measured by secondary ion mass spectrometry (SIMS). Interface structure and crystalline properties are observed by HRTEM. Rutherford backscattering spectroscopy (RBS) is used to evaluate the structural properties of W/GaAs interface.

EXPERIMENTS

The precursor for the W CVD on GaAs used is $W(CO)_6$. The W layers were deposited in the UHV MLE reactor [2]. Prior to the W CVD, oxides were removed in-situ from GaAs surface in the deposition chamber by exposing AsH_3 at below 480°C. Oxides are chemically reduced with this process rather than physically evaporated [3]. Immediately after this process, $W(CO)_6$ was introduced continuously at 360-400°C with the pressure of 15 mTorr. GaAs surfaces were heated with a halogen lamp located over the wafer.

The impurity profiles in MLE layers were measured by SIMS. Both dynamic SIMS and static SIMS are applied to these kinds of thin structure analysis. A primary sputtering beam of Cs and O ion with 1keV for negative and positive SIMS were used, respectively. W/GaAs interface profiles were analyzed by time of flight (TOF) SIMS, in which the sputtering beams of Ar^+ of 1keV were used to minimize a mixing effect for high depth resolution.

RBS measurements were carried out by the 1.5MeV He^+ irradiation with the scattering angle of 170°. W/GaAs samples used for RBS measurements were the same as those for the barrier height measurements. Random spectra were measured by tilting the crystal surface 7° off to the <100> axis. {110} and {100} Plane channeling was safely avoided. Angular dependence of the backscattering yield was also measured. He^+ dose for each angular step used was 200nC.

Interface structure of W/GaAs was observed by cross sectional HRTEM. Low angle Ar^+ milling is applied for the preparation of TEM sample. Plan view HRTEM observation was also applied for W/GaAs.

The contacts to n-type MLE GaAs layers doped with Te on semi-insulating undoped (100) GaAs were studied. The MLE layers were grown by the alternative injection of triethylgallium (TEG) and AsH_3 precursors. Diethyltellurium (DETe) was used for n-type doping. P-type MLE GaAs layers were doped with Zn from diethylzinc (DEZn) or with C from trimethylgallium [4]. W/GaAs contacts were evaluated by TLM with the SiN patterned structure. By using SiN remote-plasma deposition and two-step etching process. The process-induced defects in W/GaAs interface were safely avoided in TLM fabrication process. Wet etching with $NH_4OH:H_2O_2:H_2O$ was used for patterning of the W layers.

The Schottky barrier heights of W/GaAs contacts were estimated by using temperature dependence of the I-V characteristics. By the following equation, the barrier heights ϕ_b were calculated.

$$Js = A*T^2exp(-\phi_b/kT)$$

where Js is the saturation current density at zero bias voltage, A* is the effective Richardson constant, T is the measurement temperature, k is the Boltzmann constant, ϕ_b is the Schottky barrier height. Therefore, the ϕ_b value at zero bias voltage was obtained in this study. The built-in voltage by C-V measurements at RT was also measured for the reference.

RESULTS AND DISCUSSION

The W deposition was observed only for pressures higher than 1×10^{-3} Torr of $W(CO)_6$ for the entire tested

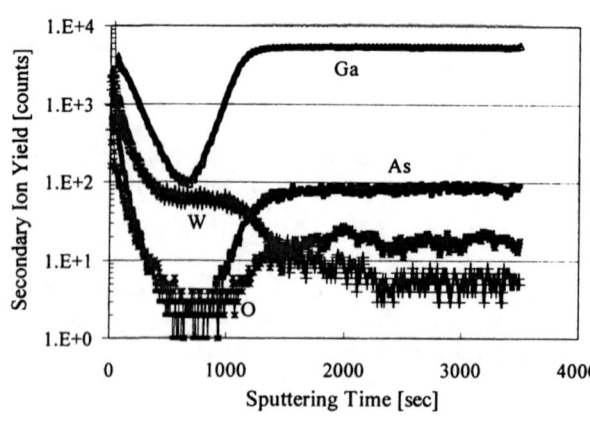

Fig.1
TOF-SIMS depth profile of W/GaAs structure. Sputtering beams of Ar^+ of 1keV were used to minimize a mixing effect for high depth resolution.

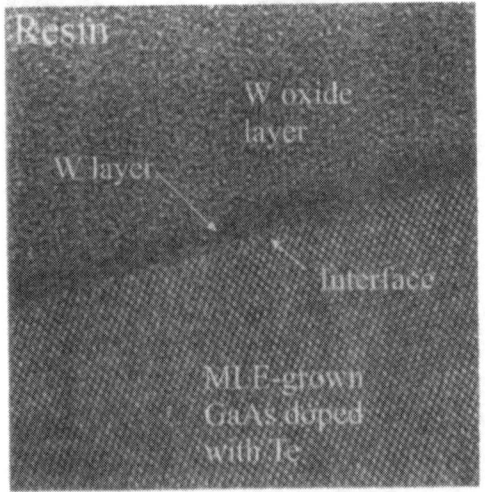

Fig.2
Cross sectional HRTEM image of W/GaAs deposited on MLE-grown GaAs doped with Te with the concentration of about $10^{20} cm^{-3}$. Low angle Ar^+ ion milling was applied for the sample preparation.

range of temperature 360–400°C. Deposition rate was about 3 Å/min on GaAs, like reported for pyrolytic decomposition[5]. However, we can not exclude some photolytic reaction [6]. Although the GaAs substrates were not intentionally illuminated, the light of the halogen heater lamp contained near-UV wavelengths. The W layers on GaAs observed with Nomarski and SEM microscopes appeared mirror-like. AFM observation also shows extremely flat surface with the maximum roughness of about 2nm. The layers on SiN had a grain structure, similar to the one reported previously [5]. The difference in layer morphology between patterned SiN/GaAs edge in plane suggests a catalytic properties of the GaAs clean surface.

Figure 1 shows the TOF-SIMS depth profile of W/GaAs interface. W/GaAs interface was clearly separated, and the mixed layer in the interface was estimated less than 2nm. But strange profile was seen on the surface of W,Ga, As and O piled- up. In addition, O and C yields look to be reduced in W layer. It is possible to explain these results by a surface unstability or a different matrix effect.

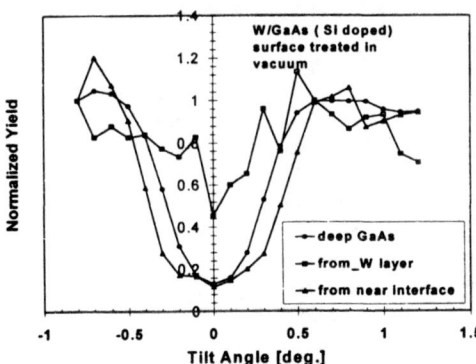

Fig.3 Angular dependence of backscattered 1.5MeV He^+ from W/GaAs. RBS/channeling results show crystalline alignment with underlying GaAs lattice.

Fig.2 shows the cross sectional HRTEM image of the present W/GaAs structure. It is shown that the atomically flat interface between W and MLE-grown GaAs can be achieved with the maximum roughness of about 2-3 ML. As shown in Fig.2, two layered structures were shown in CVD W. TEM-EDX analysis has shown that the composition of light and dark imaged region were W oxide and W respectively. W oxide layer shown at the surface was formed during sample preparation for TEM observation. Many defocused series of cross sectional HRTRM images also have shown almost same extremely flat interface. Therefore, it is concluded from both TOF-SIMS and HRTEM results that the present W/GaAs structure shows atomically flat interface with

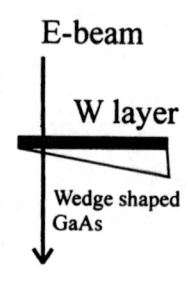

E-beam

W layer

Wedge shaped
GaAs

Fig.4
Plan view HRTEM image
of W/GaAs. Some Moiré
interference image can be
seen as the result of
interference contrast
between CVD W and
underlying GaAs lattice.

the maximum mixed layer of about 2-3 ML thickness.

Figure 3 shows the angular dependence of backscattering yield of RBS from the surface W, W/GaAs interfacial region and the bulk region of GaAs. Dip curve was obtained from the W/GaAs SBD. W layer is deposited on the GaAs surface heat-treated in vacuum for 30 min just prior to the deposition. As shown in Fig.3, the angular dependence of backscattering yield from the surface W shows clear dip almost at the same angle, where the underlying GaAs shows <100> axial channeling. Therefore, the W atoms deposited on GaAs show aligned at least in <100> direction when the GaAs surface is heat-treated in vacuum at 480°C for 30min just prior to the W deposition. The SB height was lowered when the GaAs surface was heat-treated in vacuum at 480°C compared with that heated under AsH$_3$ exposure of 1×10^{-3} Torr. Whereas the atomic structure of deposited W/GaAs is not clear yet, it is considered that the barrier height of W/GaAs structure is closely related with the atomic alignment of W on GaAs.

Figure 4 shows the plan view HRTEM image of W/GaAs structure. This was observed by using the wedge shaped sample. As shown in Fig.4, some Moiré regions were clearly observed in a plan view. It is considered that these Moiré fringes is formed by the interference between restricted W lattice and underlying GaAs crystal lattice. From these results, the RBS results were also confirmed as that the present CVD W layer was partially aligned with the underlying GaAs lattice.

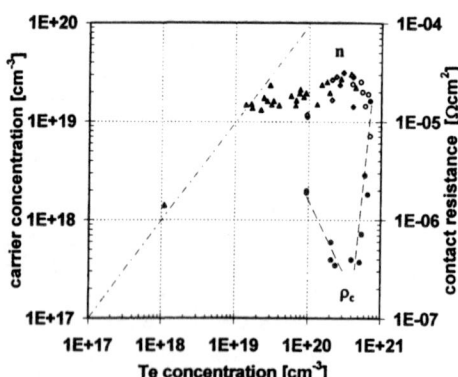

Fig.5 Relation between Te and carrier concentration of MLE GaAs:Te. Resulted ρ_c for n-GaAs is also shown.

By using CVD W layers with the structural properties mentioned above, the contact resistance ρ_c was evaluated as functions of surface treatment conditions and impurity doping levels by MLE method.

Figure 5 shows the relation between the carrier concentration and the Te atomic concentration in MLE-grown GaAs doped with Te. Te atomic concentration was obtained by SIMS. Atomic concentration was calibrated by using a series of bulk GaAs doped with Te of different concentration. It is shown that the carrier concentration gradually increased over 1×10^{19} cm^{-3} Te concentration, and beyond 4×10^{20} cm^{-3}, decreased rapidly. In this figure, the

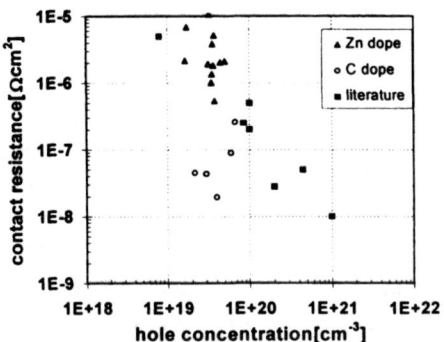

Fig.6 Change of ρ_c of MLE p-GaAs doped with Zn and C as a function of hole concentration. Achieved lowest rc reaches about $10^{-8}\Omega\text{cm}^2$, which is almost the limit of TLM method.

specific contact resistance ρ_c is also shown as a function of Te concentration. ρ_c strongly depended on Te concentration and had a minimum at the peak of the carrier concentration. For heavily doped semiconductors the tunneling becomes predominant in MS conduction and ρ_c is determined by the factor, $\exp(\phi_b /N_D^{1/2})$. The dependence was different from the tunneling theory. It can be related to excess Te atoms adjacent to W/GaAs interface. The lowest ρ_c obtained was about 3×10^{-7} Ωcm^2. This is the lowest reported value for doping concentration in the range of 10^{19} cm^{-3}. Similar values were reported by Patkar et al[7], for molecular beam epitaxial (MBE) GaAs doped with Si as the low temperature grown cap. The MLE layers, reported here, were doped uniformly and no passivation layers were necessary. The fabrication process used was close to that used for device fabrication, including SiN PE CVD and ozone ashing for semiconductor cleaning. Uniformly doped top layers of GaAs has advantages for such an application. In addition, all processes reported here for the CVD W contacts were carried out at the temperature below 480°C, which is required for ultra-thin devices [1] to prevent the thermal degradation of devices. To achieve ρ_c of $3\text{-}4\times10^{-7}$ Ωcm^2, it was also necessary to apply AsH_3 treatment below 480°C for oxide reduction prior to the W CVD. In all experiments, the CVD temperature did not exceed that of AsH_3 treatment. The lowest ρ_c was obtained with the AsH_3 treatment at 450°C for 10 sec.

The average value of the ρ_c for the W contacts to C-doped p-GaAs was 2×10^{-8} Ωcm^2 at acceptor concentration 3×10^{19} cm^{-3} as shown in Fig.6. The low ρ_c value and low mobility of p-type layers result in large uncertainty of this value. The measured values were from almost 0 to 5×10^{-8} Ωcm^2. At that low ρ_c values, the main source of error was an absolute accuracy of distance measurements between the TLM contacts, which was 0.1 μm in our laboratory. The ρ_c extraction accuracy is lost, if a transfer length $L_T=(\rho_c /R_s)^{1/2}$ becomes comparable to the contact distance measurement accuracy, where R_s is a sheet resistance value. The obtained ρ_c for GaAs:C was about 1/10 smaller than that expected from the hole concentration measured in the MLE layer and literature reports. One possible reason for this is oxide free interface of our layers. The results of Stareev [8], who obtained a low ρ_c value for $N_A= 2\times10^{20}$ cm^{-3} after sputtering, in-situ, cleaning and annealing, indicate importance of this factor. The ρ_c value for GaAs:Zn was 10 times larger, 1×10^{-6} Ωcm^2, at the same

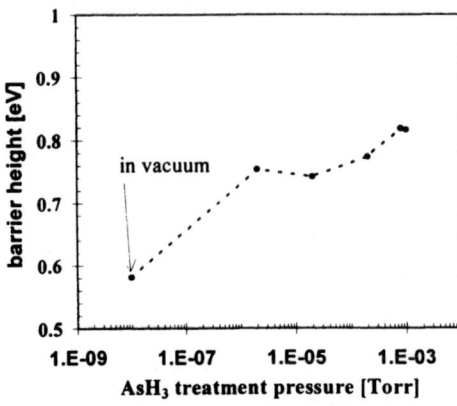

Fig.7 Change of Schottky barrier height ϕ_b of W/GaAs as a function of AsH_3 pressure exposed just prior to CVD W. ϕ_b values were obtained from the results of temperature dependence of I-V characteristics. Φ_b at zero bias voltage was obtained.

acceptor concentration of 3×10^{19} cm^{-3}, than for GaAs:C. The W specific contact resistivity of the Zn-doped MLE layers corresponds well with the reported ρ_c values. To explain the difference between the W contacts to C-doped and to Zn-doped layers requires further investigation.

Fig.7 shows the change of Schottky barrier height ϕ_b as a function of surface treatment condition prior to the W deposition. The surface treatment was carried out at 480°C for 30 min under various AsH$_3$ pressure. W deposition was carried out at 380°C for 30 min. The surface treated without AsH$_3$, in vacuum 5×10^{-9} Torr, gives the lowest ϕ_b 0.58 eV by I-V characteristics. ϕ_b increases up to 0.8 eV with AsH$_3$ pressure at 1×10^{-3} Torr. From the theoretical considerations, it is estimated that this difference in ϕ_b induces large reduction of ρ_c at the same N$_D$ in almost two order in magnitude. The ideal factor of the Schottky barrier diode on the surface treated without AsH$_3$ is 1.05, and that with higher AsH$_3$ pressure, 1.01, indicates that the interface crystal quality depends on the surface treatment prior to the W deposition. From experiments on *pin* diodes regrown with MLE on AsH$_3$ treated interfaces, it was also concluded [3] that this treatment affects crystal structure in layers adjacent to the interface. This can explain the dependence of the ρ_c value on the treatment temperature though ϕ_b is as high as 0.75 eV. Possible mechanisms include depletion of the surface layers from dopant atoms by their evaporation or exchange reactions, as well as modification of the barrier height by crystal structure changing.

CONCLUSION

In summary, CVD of W from W(CO)$_6$ precursor on MLE-grown *n*-type GaAs doped with Te shows contact resistance $\rho_c = 3 \times 10^{-7}$ Ωcm^2 and mirror-like surface morphology. The native oxides can be reduced with AsH$_3$ prior to the CVD. The AsH$_3$ treatment and CVD temperatures were below 480°C. The CVD W contacts to *p*-GaAs MLE layers doped with C give ρ_c in the low range of 10^{-8} Ωcm^2, while for the Zn dopant, only $\rho_c = 1 \times 10^{-6}$ Ωcm^2 was obtained. The low contact resistance is obtained for electrically active dopant concentration about one tenth lower than expected from literature reports. The mixed layer in the interface was estimated less than 2nm and well confirmed to be about 2-3ML by cross sectional HRTEM results. Deposited W layer was not completely amorphous but has shown some lattice alignment with underlying GaAs lattice. The conditions used for GaAs MLE and W CVD are suitable for self-aligning constructions of ultra-thin devices with regrown epitaxial layers.

ACKNOWLEDGEMENT

Authors would like to acknowledge Dr. T.Kurabayashi for TOF-SIMS measurement, and Mr. H.Kikuchi, Mr. T.Hamano, Mr. M.Henmi and Mr. K.Ito for MLE layer growth, W CVD and sample processing. Part of this work was sponsored by the Materials Science Research Foundation.

REFERENCES

[1] P.Plotka, *et al*, *Applied Surface Sci.*, **82/83**, 91-96 (1994)

[2] J.Nishizawa, *et al*, *J. Electrochem. Soc.* **136**, 478-484 (1989)

[3] Y.Oyama, *et al*, *Applied Surface Sci.*, **82/83**, 41-45 (1994)
    J.Nishizawa, *et al*, *Surface Science*, **348**, 105-114 (1996)

[4] T.Kurabayashi and J.Nishizawa, *Applied Surface Sci.*, **82/83**, 97-102 (1994)

[5] L H .Kaplan and F M .d'Heurle, *J. Electrochem. Soc.*, **117**, 693-700 (1970)

[6] R.Solanki, *et al*, *Appl. Phys. Lett.*, **38**, 572-574 (1981)

[7] M P .Patkar, *et al*, *Appl. Phys. Lett.*, **66**, 1412-1414 (1995)

[8] G.Stareev, *Appl. Phys. Lett.*, **62**, 2801-2803 (1983)

Part V

Electronic Devices and Processing

Novel *in-situ* Ion Bombardment Process for A Thermally Stable (> 800 °C) Plasma Deposited Dielectric

F. Ren[1], J.R. Lothian[2], S.J. Pearton[3], R.G. Wilson[4], J.R. LaRoche[1], J.W. Lee[5], D. Johnson[5], and J. M. Zavada[6]

1 Department of Chemical Engineering, University of Florida, Gainesville, FL 32611
2 Multiplex Inc., South Plainfield, NJ 07080
3 Department of Materials Science and Engineering, University of Florida, Gainesville, FL 32611
4 Charles Evans and Associates, Sunnyvale, CA 94086
5 Plasma Therm inc., St. Petersburg, FL 33716
6 US Army Research Office, Research Triangle Pk, NC27709

Abstract

We demonstrated an *in-situ* dielectric film passivation technique by dividing a thick film deposition into many thin film (<40Å) depositions and incorporating the ion bombardment between the depositions. N_2 was used for the plasma treatment to passivate the SiN_x film and a well passivated and thermally stable SiN_x was achieved with this process. The refractive index of N_2 treated SiN_x film only changed 0.3% when the SiN_x film was heated up to 1000 °C and the film with a continuous deposition showed a 2.5% change. From the results of SIMS analysis, the N_2 treated SiN_x film showed a excellent thermal stability after heat up to 1000 °C. The etch rates of passivated SiN_x film in BOE and diluted HF are ≤40 Å/min which is much slower than that of un-treated SiN_x(135 Å/min).

Introduction

Due to the limited thermal stability of III-V based materials, low temperature (≤ 300 °C) plasma enhanced chemical vapor deposited silicon nitride (SiN_x) is widely used in the III-V based integrated circuits fabrication. During the SiN_x deposition, the Si-H and N-H bonds from the precursors, SiH_4 and NH_3, respectively, can not be completely broken in this relative low temperature operation, [1,2]. Therefore, typical PECVD SiN_x does not have a fixed stoichiometry(Si_3N_4) but rather contains a varying amount of Si, N and H. In some case, the hydrogen content can be as high as 39% (atomic) and is strongly dependent on the deposition conditions. The ratio of Si-H to N-H depends on the deposition conditions[3]. The SiN_x film deposited without load-lock system can also contain more than 5% oxygen when low deposition temperature is used (<250 °C).

Hydrogen bonds, free hydrogen, and oxygen in the SiN_x film can diffusion or rearrange which will create the traps, passivate the dopants, and cause the instability in devices during subsequent high temperature IC fabrication steps or high bias voltage device operation conditions. Thermal energy will enhance the diffusion and in a high electric field condition, the hot electrons can inject into the dielectric film which will release the bonded H therefore create electron traps[4-7].

Recently, electron cyclotron resonance (ECR) system has been successfully used for dielectric deposition[8-10]. ECR system can provide high ion density for very low temperature deposition. The high density ion bombardment helps to break the hydrogen bonds in the precursors at low temperature deposition (<100 °C) and enhance the surface species migration, which improves the step coverage and reduces the surface roughness. Nitrogen can be used as the N-precursor to replace NH_3 for ECR SiN_x deposition, therefore, hydrogen content in the SiN_x film can be further reduced.

In this work, we investigated the impact of the plasma treatment on the properties of dielectric film by using an in-situ nitrogen ion bombardment to passivate the SiN_x in a Plasma Therm ECRCVD system. SIMS, refracitve index measurement, and wet chemical as well as dry etching were used to characterize the dielectric films.

Experimental

SiN_x films were deposited with nitrogen and 2% silane diluted in nitrogen. All deposition processes were carried out in a load-luck Plasma Therm SLR 770 remote plasma system which has two chambers, one is the ECR plasma generation chamber and another is the sample position chamber where deposition occurs. The samples sit on an rf-powered, He backside cooled chuck, which was biased at 2W. The nitrogen was introduced into the ECR chamber and the diluted silane was directly employed into the deposition chamber through a gas distribution ring 3 cm above the samples. The deposition temperature was kept at 50 °C and the process pressure was maintained at 2 mTorr. A 400 W ECR power was employed for the deposition and deposition rate was ~60 Å/min. Multi-layer dielectric depositions with the incorporation of N_2 plasma treatment to each thin SiN_x (~40 Å) deposition was performed in a computer controlled automatic manner. The deposition sequence is following: 40 sec. for SiN_x deposition, 1 min for pumping out the system, 1 min N_2 plasma treatment, 1 min for pumping out the system, and continue the deposition cycle until the desired thickness of SiN_x film is achieved. The energy of the rf power applied to the sample during the N_2 treatment was 10 W and the ECR power was kept at 400W and the self-bias voltage was ~ -10V. With a high pumping speed of the equipped turbo-pump, the deposition chamber background pressure was reduced to 2×10^{-6} torr within 1min after switching off the gases. The total deposition time of the multi-layer deposition was about six times of the continuous film deposition.

The wet chemical etching characteristics of SiN_x films were performed in buffered HF and diluted HF (10 : 1 = H_2O : HF) solution. The thermal stability of SiN_x films was studied with SIMS analysis and refractive index measurements.

Results And Discussion

The bonding energies of Si-H and N-H are 3.9 and 4.5 eV, respectively and the self-bias voltage applied on the sample in this study was ~ -10V. Therefore with this self-bias voltage and high ion density bombardment to enhances the surface species migration, a thermally stable film with less dangle bond and hydrogen content was expected. From the SIMS depth profiles of SiN_x film prepared with in-situ nitrogen ion bombardment and normal deposition, the hydrogen content (~15%) in the nitrogen ion bombarded SiN_x film was not significantly reduced as compared to that of normal deposition. Higher bias voltage, longer nitrogen plasma exposure, or thinner SiN_x film for each treatment cycle may be needed to further reduce the hydrogen content.

Besides the hydrogen incorporation, there is ~1.8% oxygen incorporated throughout the nitrogen treated SiN_x film. Although, the ECRCVD is equipped with load-lock, there is still water on the sample or on the deposition chamber-wall. During the deposition, high density ion

184

bombardment desorbed the water molecules from the chamber, sample, and sample carrier and oxygen got incorporated in the SiN$_x$ film. As for the SiN$_x$ film obtained from the continuous deposition, the oxygen incorporation was more serious. Besides the ion bombardment effect, there was a significant heat generated during the continuous deposition which further assisted water molecules desorbing from the system and incorporating the oxygen in the dielectric film. This is clear showed in the SIMS result of continuous deposited SiN$_x$ film that oxygen almost linearly increases toward to the surface of the SiN$_x$ film. The oxygen peak in the SiN$_x$/Si substrate is the oxygen left on Si substrate after the sample preparation.

Fig. 1 shows the SIMS profiles of both film annealed at 600 °C for 1 min. There was no noticeable change in the nitrogen treated SiN$_x$ film, however, there was a significant oxygen movement in the continuous deposited film. This could be resulted from the dangling bonds and defects in the continuous deposited film.

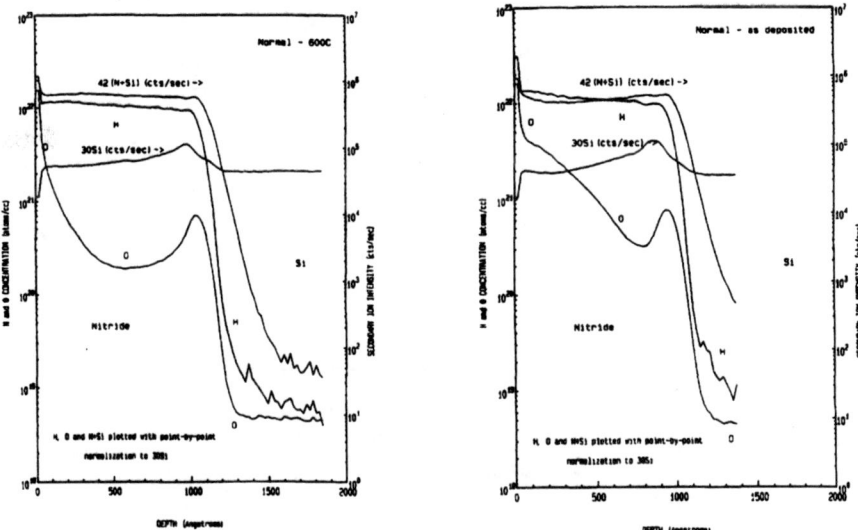

Fig. 1. The SIMS depth profiles of SiN$_x$ with the normal deposition method (left) as deposit (right) annealed at 600 °C for 1min.

As shown in Fig. 2, after heating up the samples to 800 °C for 1 min., the nitrogen treated SiN$_x$ film is still very stable and there is no movement detected. On the contrary, the hydrogen on the surface of continuous deposited film started to diffuse out of the film after the annealing. This instability trend of the continuous SiN$_x$ film is consistent with the refractive index measurements.As illustrated in Fig. 3, there is a change of 2.5% for the continuous deposited SiN$_x$ film. On the contrary, the SiN$_x$ prepared with in-situ nitrogen bombardment showed an excellent thermal stability. The change of refractive index for the N$_2$ treated SiN$_x$ is only 0.3% after been heated up to 1000 °C for 1 min. This further confirms that the dangling bonds in the N$_2$ treatment have been minimized.

185

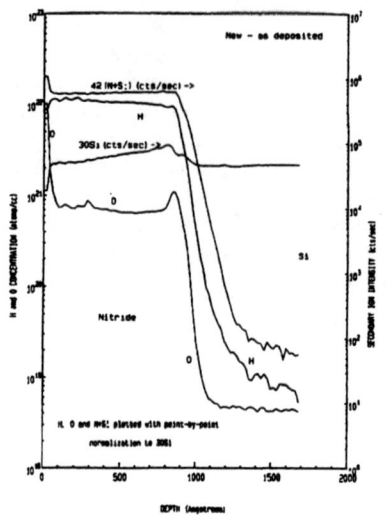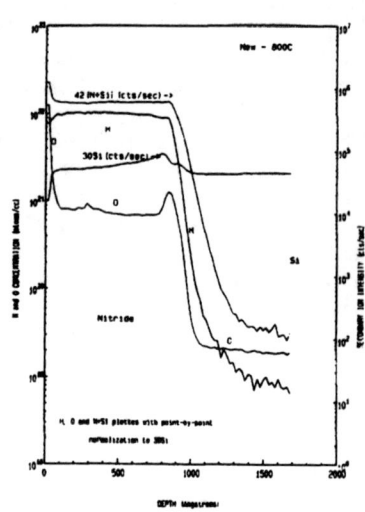

Fig. 2. The SIMS depth profiles of SiN$_x$ with the normal deposition method (left) as deposit (right) annealed at 800 °C for 1 min.

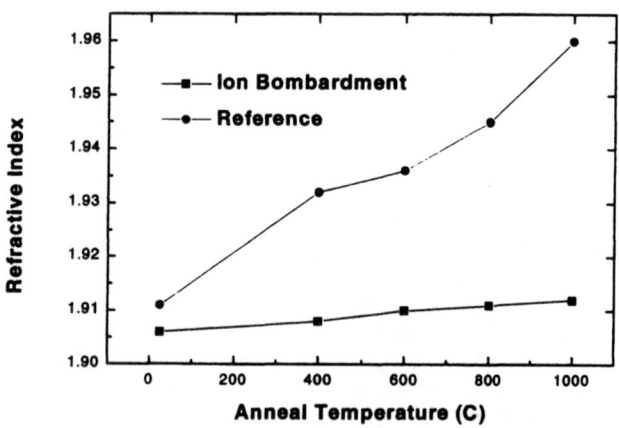

Fig. 3. The refractive index of N$_2$ treated and normal deposition methods as a function of annealing temperature.

The stability of the nitrogen treated SiN$_x$ film was also validated in the etching studies. Table I shows the etch rates of both films in wet chemical and plasma etchings. The plasma etch rates are very similar for both SiN$_x$ films. Since the etch rate strongly depends on plasma

186

conditions and composition of the dielectric materials. The atomic compositions for these two as deposit SiN$_x$ films are very close, therefore the etch rates should be close. However, the wet chemical etch rate depends on the quality of the film, such as dangling bonds, defect density, and atomic composition. The wet chemical etch rates are quite different for these two SiN$_x$ films. The etch rate of N$_2$ treated SiN$_x$ are much slower in BOE is much lower than that without N$_2$ treatment (35 Å/min verses 135 Å/min in BOE) and this demonstrated the effectiveness of the passivation for the N$_2$ ion bombardment process. This is the slowest etch rate of low temperature plasma deposited SiN$_x$ ever reported. This results is consistent with the SIMS profile of both films annealed at 800 °C. The summary of dry and wet chemical etch rate is illustrated in Table I.

Table I Wet chemical and plasma etch rates of nitrogen treated and continuous deposited SiN$_x$ films.

Etch Method	Etch Rate for N$_2$ treated SiN$_x$	Etch Rate for Reference SiN$_x$
CF$_4$/O$_2$ Plasma (Å/min)	1180	1230
BOE (Å/min)	35	125
HF Solution (Å/min)	40	350

Conclusion

We have demonstrated a novel *in-situ* dielectric film passivation technique by dividing a thick film deposition into many thin film (<30Å) depositions and incorporating the ion bombardment between the depositions. The SiN$_x$ film can stand high temperature annealing and there was only 0.3% of refractive index change after 1000 °C anneal. The slowest wet chemical etch rates ever reported was also achieved with this methods. Further studies with different ion treatment(O$_2$ or D$_2$), self-bias voltage, treating time, and thinner dielectric layer are under investigating.

References
1. W. A. Lanford and M. J. Rand, J. Appl. Phys., **49**, 2473(1978).
2. H. J. Stein, V. A. Wells, and R. E. Hampy, J. Electrochem. Soc., **126**, 1750(1979).
3. R. Chow, W. A. Landoed, K. M. Wang, and R. Rosler, J. App. Phys., **53**, 5630(1982).
4. D. G. Esaev, V. M.Efimov and A. A. Shklyaew, Thin Solid Films, **221**, 160(1992).
5. G. V. Gadiyak, J. Appl. Phys., **82**, 5573(1997).
6. S. T. Sah, Solid-State Electron., **33**, 147(1990).
7. R. C. Sun, J. T. Clemens and J. T. Nelson, in Proc. IEEE Rel. Phys., 244(1980).
8. Lee JW, MacKenzie K, Johnson D, Shul RJ, Pearton SJ, Ren F, Solid State Electronics, **42**, 1031(1998).
9. "Effect of ECR Plasma in the Luminescence Efficiency of InGaAs and InP," F. Ren, D. Buckley, K. Lee, S.J. Pearton, C. Constantine, W.S. Hobson, R.A. Hamm and P.C. Chao, Solid State Electronics **38**, 2011(1995).
10. F. Ren, J. LaRoche, T. Anderson, S. J. Pearton, J. W. Lee, D. Johnson, J. R. Lothian, J. Lin, R. J. Shul, and C. S. Wu, J Electrochemical Soc. Lett., **1**, 279-281(1998).

LOW ENERGY, HIGH DENSITY PLASMA (ICP) FOR LOW DEFECT ETCHING AND DEPOSITION APPLICATIONS ON COMPOUND SEMICONDUCTORS

J. ETRILLARD, H. MAHER, M. MEDJDOUB, J.L. COURANT and Y.I. NISSIM
OPTO+, Groupement d'Interet Economique, Route de Nozay, F-91460, Marcoussis, France
France Telecom – CNET - yves.nissim@cnet.francetelecom.fr

ABSTRACT :

The use of a low ion energy of an extremely dense plasma has been studied as a dry etching as well as a thin film deposition tool (same source, two different reactors) for InP and GaAs device processing. Under these working conditions it is expected to control well the etch depth or in the case of deposition to obtain high deposition rates. In all cases minimun ion damages are induced on the processed substrate. Both technologies are presented here from the point of view of material analysis as well as device processing demonstration. For etching, the gate recess of an InP-based HEMT has been addressed as one of the key technological step that requires such properties for good device performances. InGaAs/InAlAs HEMT like structures have been grown and the recess of the InGaAs layer has been conducted with a 13eV $SiCl_4$ inductively coupled plasma (ICP). DLTS and AFM measurements made on the exposed AlInAs surface after InGaAs removal indicate that device quality on its electrical and structural properties are achieved. Passivation of fully processed HEMT devices with a ICP enhanced chemical vapor deposition (ICPECVD) silicon nitride film is being studied.

I. ICP ETCHING

I.1 Introduction

Dry etching of InP and related compounds is an important technological process to develop uniform and reproducible optoelectronic devices and circuits. Since the materials of the InP family are fragile, it is important to reduce the structural and electrical damages produced by chemically or physically active species in the plasma. Among all the plasma setting parameters, most of the time the damages are attributed to the high ion energy [1]. Several plasma etching techniques have been proposed to minimize the energy and still keep anisotropy and acceptable etching rates. The Inductively Coupled Plasma (ICP) that has been utilized in this study is a good candidate, since it can provide extremely high ion density independently of the ion energy. Using a hydrogen free chemistry with $SiCl_4$ and very low ion energy, the problem of the gate recess in the InGaAs/InAlAs/InP HEMT has been addressed since the electrical and structural properties of the surface of the AlInAs barrier after recess are extremely important for the performances of the HEMT. Furthermore the use of a dry process for this technological step is a warrant of uniform threshold voltage throughout a complete wafer and thus open the field of digital electronics for the InP HEMT[2].

I.2 ICP Process

The experiments are performed in an ALCATEL METLAB ICP source operating at 13.6MHz. The ICP etching process on InP has been recently studied [3], with a $SiCl_4$ chemistry to avoid hydrogen diffusion and dopant compensation as well as sidewall polymer redeposition that are observed with the more conventional CH_4/H_2 chemistry. The surface stoechiometry of InP has been studied with Auger spectrometry and the electrical properties with photoluminescence. The variation of these properties are shown in figure 1 as a function of ion energy (the plasma potential is not accounted for) for a fixed ion density of around 10^{10} cm^{-3}. As can be seen from this figure, the structural and electrical disorders are located in a surface region that is only 3 nm thick when the ion energy is set at 30 eV. Most of the work that has been reported to date on ICP process utilizes bias voltages on the order of 30 to 50 V combined with high process pressure (over 1 Pa) in order to obtain high etch rates. Actually etch rates as high as 1μm/mn have been reported on InP [3]. In this work the lowest bias voltage that can be set on the plasma has been used in order to slow down the rate and thus obtain a controlable etching depth with a minimun amount of ion induced damages.

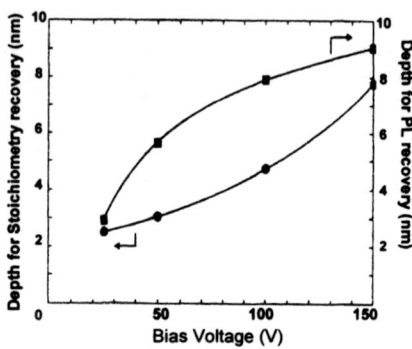

Figure 1: surface stoichiometry and photoluminescence recovery as a function of ion energy of a SiCl$_4$ ICP plasma on an InP substrate.

A HEMT like structure was designed with a 10 nm GaInAs cap layer and a 800nm AlInAs barrier layer MOVPE grown on a S doped InP substrate. The recess of the 10nm thick cap layer has been attempted with an ICP process where the ion energy was reduced to 13 eV. The ion density was kept at around 10^{10} cm^{-3} , and the substrate held at room temperature. Since the gate recess of the HEMT requires a reproducible etch of a 10nm InGaAs layer, the base pressure of the reactor was kept as low as possible (less than 10^{-6} Torr) with a reactor plasma cleaning procedure prior to the gate recess. With these settings two different etching times were utilized : 75 seconds in order to reach the InGaAs/InAlAs interface and 90 seconds in order to penetrate into the InAlAs. As a standard, a chemical selective etch in a citric acid solution was also prepared.

I.3 DLTS and AFM Measurements

Ti-Au Schottky diodes (320 μm in diameter) were fabricated in the recess area, and the back side of the sample was contacted. The gate leakage current of the diodes was measured as a function of reverse bias on the gate. The resullts are shown in figure 2, for a chemical etch and for two different durations of ICP process. As can be seen from this figure a comparable leakage current is obtained for the chemical etch and the 75 second ICP process. This indicates that the InAlAs layer has been reached in the ICP process since the leakage current of a diode fabricated on InGaAs would have been larger due to a smaller metal/semiconductor barrier height. It indicates also that the amount of ion induced damage after 75 seconds is comparable with the amount obtained after a chemical selective etch. On the contrary a strong leakage current is observed when the plasma interaction reaches 90 seconds. For this duration part of the AlInAs layer has been etched away and a higher density of surface damage is obtained.

DLTS measurements were performed on these Schottky diodes. The reverse bias of the diode and the filling pulse votage were set to explore either the surface of the AlInAs (figure 2A) or the bulk close to the surface (figure 2B). Three traps are usually evidenced in AlInAs [4 , 5] with activation energies close to 0.2eV (E_1), 0.42 eV (E_2) and 0.62eV (E_3). E_1 is not observable in our experiment as is also the case in molecular beam epitaxial InAlAs. The DLTS intensity of this level is also very small on InAlAs surfaces that have been etched with a more conventional CH_4/H_2 RIE process [6]. On the contrary the E_2 and E_3 traps are very sensitive to surface treatments. Figure 3A shows that the ICP process increases the concentration of the E_2 traps at the surface of the AlInAs while it reduces drastically the deeper traps E_3. This behaviour has already been observed with RIE processes [6] although with the ICP process, the density of E_2 traps is largely reduced.

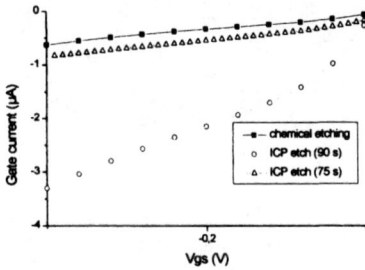

Figure 2 : variation of the gate current of the Schottky diodes with the gate voltage for 3 different recess processes : a chemical etch , and two ICP etches with different duration

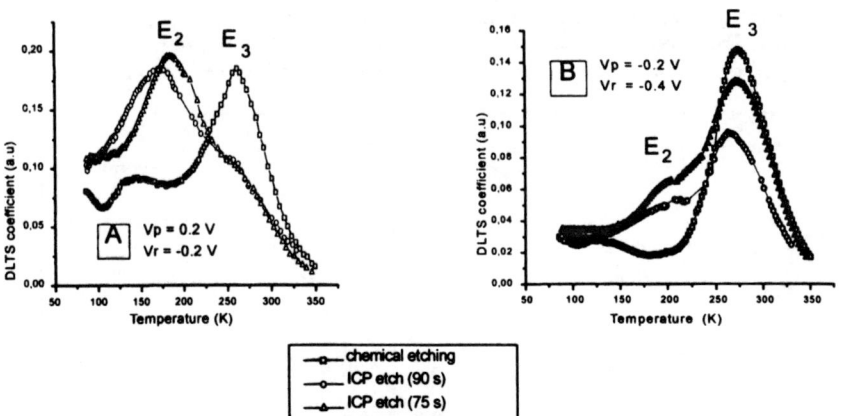

Figure 3: DLTS spectra of AlInAs surfaces revealed with 3 different recesses of the InGaAs cap layer: A chemical etch, and two ICP etching processes with different duration. The choice of the reverse bias voltage (Vr) and the voltage pulse (Vp) allows to explore the surface (A) or the bulk (B) of the AlInAs.

191

The E_3 traps are dominant in the wet chemical recess as well as in the as-grown material. With a good choice of reverse bias voltage (Vr) and voltage pulse (Vp), it is possible to explore the bulk of the AlInAs material just beneath the surface. The DLTS coefficient of this bulk material is shown in figure 3B. As can be seen, the ICP process reduces progressively the E_3 density as it etches away the surface of AlInAs. This result combined with the gate current results suggests that there should be an optimun etching depth for defect removal (E_2 and E_3) without formation of excess gate leakage current.

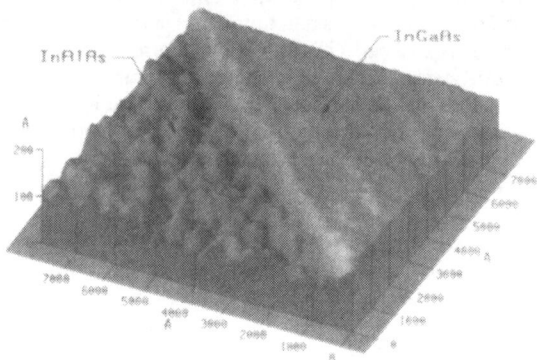

Figure 4: AFM micrograph of an ICP recess made on an InGaAs/AlInAs HEMT like structure. The depth of the recess is about 8 nm and the rms value of the roughness on the AlInAs surface is only 1.2nm.

These good electrical results are associated with the good morphology of the recess surface observed by AFM measurements . Figure 4 shows the edge of a recessed area. The rms value of the roughness on the as-grown structure is 4Å, while it is 12Å on the etched AlInAs surface. These results are indicative of the quality of the MOVPE grown structure. They indicate also that the use of an extremely low ion energy combined with a high density of the

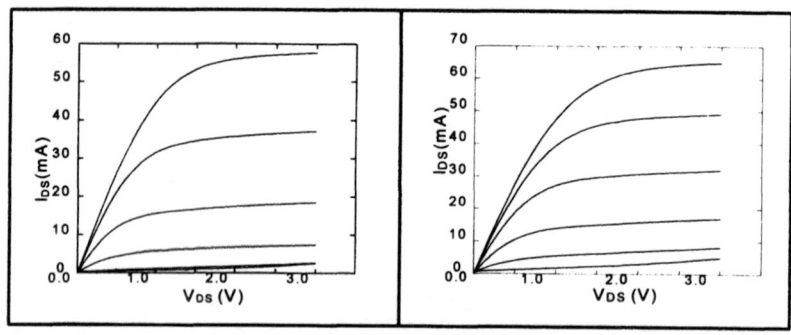

WET RECESS (CITRIC ACID)   DRY RECESS (SiCl₄ ICP)

Figure 5 : Static characteristics of two 0.8µm gate length HEMTs processed either with a wet or dry gate recess.

plasma results in a surface after etching structurally adequate for device processing. HEMT structures have been processed to verify this assumption and static device characteristics are presented in figure 5. As can be seen from this figure well behaved static I(V) characteristics are obtained with even higher drain current and lower gate leakage current for the ICP recess gate transistor as compared to the wet gate recess transistor. The comparison between the two processes will not be pushed further since the two barrier thicknesses in the two processes are probably slightly different (the wet chemical etch is perfectly selective while the dry one is not). Nevertheless a full device has been processed with an ICP recess gate. This demonstrates that low power ICP is a dry technology that leaves treated surfaces with device grade electronic properties.

I.4 Conclusions

The electrical and structural quality of the AlInAs surface revealed with a low ion energy SiCl$_4$ ICP etch of an InGaAs cap layer have been studied. The quality of the Schottky diode obtained in this process is comparable to the quality of a diode fabricated with a conventionnal chemical etch. The ion induced damages on InAlAs are not important, although they have a different signature than the damages measured on a wet etch surface. Finally the rms value of the roughness of the AlInAs surface after this process is only a few monolayers and thus is compatible with a sensitive device interface. All these results allow to perform a gate recess of a InP HEMT without any additional wet etch.

II. LOW TEMPERATURE ICP ENHANCED CVD (ICPECVD) OF SiN

II.1 Introduction

Classical Plasma-Enhanced Chemical Vapour Deposition (PECVD) systems need a relatively high deposition temperature to ensure a good quality of the deposited material (over 400°C). For III-V technology, it is necessary to reduce the power density injected in the plasma, and the temperature to avoid the degradation of the surface. The quality of the dielectric is drastically reduced (Si$_3$N$_4$ is porous and H-rich) and thus poor passivating characteristics are obtained. To avoid these problems, different reactor geometries have been designed with some success. Microwave plasmas with a remote plasma source, have given very good results with SiO$_2$ and Si$_3$N$_4$ layers. Especially with a microwave excitation, the high dissociation rate and the quasi room temperature (<100°C) deposition capabilities permit to obtain a good stoechiometric material, and good interface properties with InP and related materials [7,8,9]. If the low deposition rate in such equipment is well adapted for the control of the interface quality, it had to be improved to answer the needs of technological layers.

Similarly, UVCVD reactors have been developped to provide high quality electrical passivation on III-V materials. Good results in terms of electrical behaviour have already been reported for both SiO$_2$ and Si$_3$N$_4$ films. With the absence of electric power excitation, to avoid ion bombardment, and the capability to deposit film at temperature lower than 300°C, the electrical results were good enough to be used as passivation film in a wide range of devices (Field Effect Transistors, PIN photodiodes, avalanche photodiodes) [10].

The requirements for InP technology and for the bulk dielectric quality are contradictory. The different deposition technics have advantages and drawbacks each impeding their industrial development. The "ideal "equipment would have the following characteristics. Extremely low ion energy bombardment as in UVCVD or DECR systems, but with higher deposition rate and no limitation of maximum thickness; controllable deposition temperature between 0°C and 300°C, to avoid semi-conductors degradations but keeping in mind that at such low temperature good material cannot be obtained without a highly dissociated plasma as it has been shown in DECR reactors; good control of the SiN-semiconductor interface, with the capability to provide an in-situ surface treatment.

Reactors with an ICP source are a possible answer to the conflicting requirements. The ability of such source to provide a highly dissociated plasma [11,12], should allow a low temperature deposition with a high quality material. The high density plasma, which has demonstrated an extremely high etching rate (several micron per minute) in dry etching of Si and III-V material [3,13], should provide very high deposition rates. The configuration of the reactor, with a very low ion energy is compatible with fragile materials, as already demonstrated in InP etching experiments [14]. The wide range of working pressure is an important advantage to explore a wide range of deposition conditions, and accurately control the interface between dielectric and semi-conductor.

the first results obtained with an ICP equipment using the same plasma source than theetching equipment described above are presented here.

II.2 ICP reactor configuration:

The ICP equipment designed for deposition (ALCATEL 601D) is composed as followed: the source is built ground an alumna cylinder. The top closing piece is equipped with nitrogen content gas injector (N_2 or NH_3) . An external circular antenna provides the RF (13.56 MHz) power, in the range between 0 W and 2000W. The inductively coupled plasma generates a highly dissociated plasma from N_2 and/or NH_3. At the lower part of the source body, an external solenoid improves the coupling between antenna and plasma. This bottom part of the source is opened to the diffusion chamber. This diffusion chamber is designed with 14 permanent magnets, to direct the actives species toward the sample and to improve the uniformity of the plasma. In this chamber, a ring shaped injector , placed just above the sample, supplies the SiH_4 in a post-decharge position. SiH_4 is dissociated by the N active species, outside the source. The substrate holder, is placed just below the diffusion chamber. The sample is clamped on the chuck, and its temperature can be controlled between 20°C and 300°C.

The process chamber is equipped with a turbomolecular pump. The pressure in the chamber is regulated with a laminar valve.

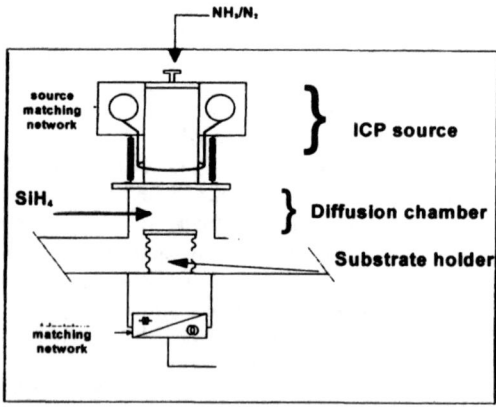

Figure 6: Schematic diagram of the ALCATEL 601D ICP reactor.

The sample is introduced in the reactor through a load-lock system. After pumping down to 10^{-3} Pa, the gases are introduced, the pressure is stabilised at the desired value, prior to plasma ignition. The complete process sequence, introduction of the sample, gas flow and pressure and temperature regulation, ignition of the plasma and sample exit are completely automated and computer controlled.

II.3 Results obtained with a substrate at room temperature

In this section the first results obtained for deposition at room temperature, and are given together with the impact of the different chemistries (SiH_4 + NH_3 and SiH_4+N_2) on growth behaviour. Si_xN_y layers have been deposited on silicon with no heat or bias on the substrate. The sample temperature stays below 40°C for any plasma power used. The thickness and refractive index have been measured at λ=633 nm with a variable angle ellipsometer. The etch rate speed was measured in buffered HF (NH_4F: HF, 87.5:12.5).

The deposition rate of SiN obtained with NH_3 plasma is shown on fig. 7. As expected, the values are very high, with a maximum of more than 500Å/min, and they are strongly dependant on the SiH_4 flow. At low power (500W), the refractive index and deposition rate are constant when the SiH_4:NH_3 ratio is higher than 20%.

At low RF power the limiting factor seems to be the dissociation rate of the NH_3. Without any pressure regulation (at maximum pumping speed), the chamber pressure increases with the plasma power, (e.g. from 0.23 Pa without plasma to 1.86 Pa at 1000W), and does not increase anymore above 1200W. This could explain the limited variations of deposition rate and refractive index with reactant ratio. In such conditions, the growth is limited by the N flow, and the dissociation seems too low to improve the absorption of active species on the film.

At higher power, the growth rate and characteristics of the layers are controlled by the SiH_4 flow. Both deposition rate and refractive index can be varied over a wide range of values (index is increasing from 1.72 to 2.1 with the increase of SiH_4 and thus the increase of Si incorporation in the layer).

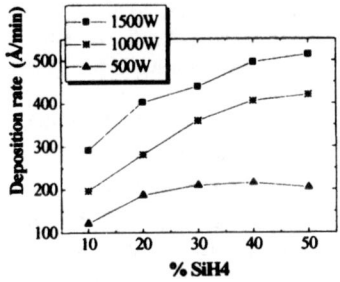

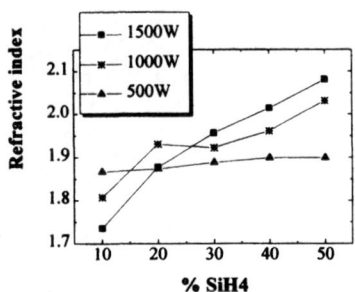

Fig.7a : Deposition rate vs SiH_4/NH_3 flow ratio, on a Si substrate at room temperature for 3 different RF power values.

Fig.7b : Refractive index measured at 633 nm, vs SiH_4/NH_3 flow ratio, on aSi substrate at room temperature for 3 different RF power values.

The deposition rate of SiN obtained when using N_2 and SiH_4 as reactant is shown on figure 8. The deposition rate appears to be much higher with this mixture. The behaviour is the same as in the previous case, the characteristics are controlled by the SiH_4 flow, but with a higher range of values particularly for the refractive index (from 1.6 to 2.3 at $\lambda = 633$ nm).

Whether using NH_3 or N_2, the refractive index is low probably du to the low substrate temperature. The increase of n with an increase of the SiH_4 flow is the result of the highest amount of Si incorporated in the layer. The ratio between the deposition rate with N_2 and NH_3 is closed to a factor 2. The value of the deposition rate seems highly dependant on the incoming [N] concentration for any gas used but not dependant on the dissociation energy of molecules. The high dissociation of the gas seems to lead to a growth mechanism driven only by the flow of N atoms.

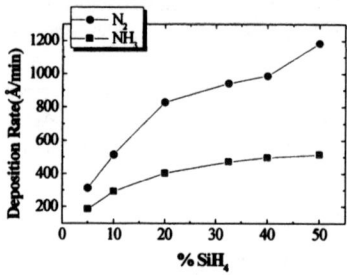

Fig. 8a : Comparison of the deposition rate between NH_3 and N_2 plasmas (deposition at room temperature and RF power P=1500W)

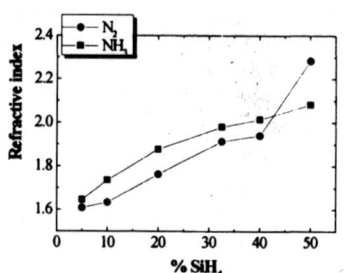

Fig. 8b: Comparison of refractive index at 633nm between NH_3 and N_2 plasmas (deposition at room temperature and RF power P=1500W)

Measurement of Infra-red transmission has been used to evaluate the structural properties of the films (SiN bonds) and the amount of hydrogen in dielectric films (SiH and NH bonds). FTIR (Fourier Transform Infra-Red) characterisation have been performed on Si substrate. The amount of hydrogen was measured by integration of SiH and NH absorption peaks, and calculated with the Landolph method [15,16].

Over the range of parameters investigated, the [H] concentration is evaluated at about 2 to $3\ 10^{22}\ cm^{-3}$. But one should notice that the hydrogen is not incorporated the same way for the N_2 and NH_3 films (fig. 9a and 9b). The NH bond ($\sim 3300\ cm^{-2}$) peak is quite strong at low SiH_4/NH_3 ratio (with a weak SiH peak at 2200 cm^{-3}) and weak for a high SiH_4/NH_3 ratio. A low deposition rate seems to enhance the reaction of hydrogen at the surface of the growing layer with N atoms. Still we are lacking evidence of better or worse material overall characteristics, depending incorporation of H as NH or SiH.

The SiN stretching position is an important value to assess the quality of the film. The peak maximum absorption position is between 820 and 835 cm^{-1} in the $SiH_4 + NH_3$ chemistry, a value closed to the reference for LPCVD material (835 cm^{-1}). With the $SiH_4 + N_2$ chemistry, the stretching position is closed to 835 cm^{-1}, except for low SiH_4 flow, for which a much higher value is observed (885 cm^{-1}).

These results show the possibility to deposit films with a good structural quality at very low temperature. The relatively high amount of hydrogen (as compared with LPCVD) is still remarkably low for a room temperature deposited material. Futhermore, it has already been shown that H-rich film deposited at low temperature could exhibit good passivation properties. The amount of Si in the layer can be the source of electrical traps leading to a degradation of the electrical behaviour of passivated devices.

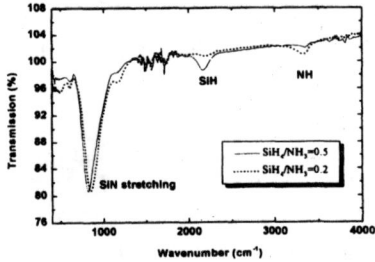
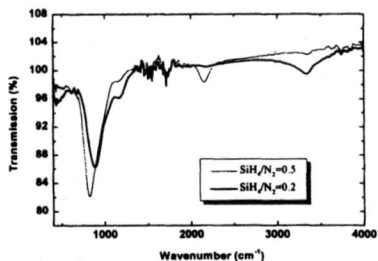

Fig.9a: Infra-red absorption spectra of films deposited at 2 flow ratios (SiH₄/NH₃=0.5 and 0.2)

Fig.9b: Infra-red absorption spectra on films deposited at 2 flow ratios (SiH₄/N₂=0.5 and 0.2)

The etch rate of layers has been measured in a solution of NH_4F and HF (87.5:12.5). The etch rate is strongly dependant on the structural quality of the material. The porosity, the non-stoechiometry, will be revealed by a higher etch rate. The LPCVD material (deposited at 900°C) is etched at about 0,3 nm/min. The etch rate of ICP layers shown is figure 10 is strongly dependant of the SiH_4 /NH_3 ratio. But with very low SiH_4 content in gas phase, the etch rate is quite low (30 nm/min), over a large range of SiH_4 /NH_3 ratio in the chamber. As previously mentionned, the hydrogen content is relatively constant, but the stoechiometry Si/N in the film seems to vary strongly , and appear as the main factor controlling the etch rate.

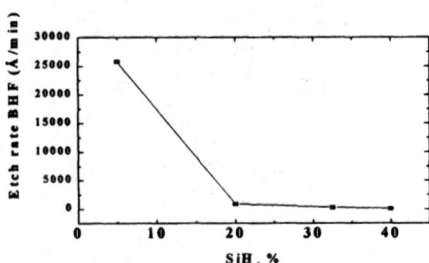

Fig.10: Etch rate of SiN thin film in buffered HF (87.5:12.5 NH_4F:HF)

Obviously low internal stress of thin films is essential, to avoid unsticking of layer during fabrication process, such as annealing, diffusion, even during deposition of thick films. Bowing of silicon wafers has been measured prior and after deposition of a 200 nm layer by an optical interferometry. The difference in bowing is used to calculate the internal stress of films [17,18].

In all cases, the ICPECVD films present low stress (below 300 MPa decreasing with the SiH_4/NH_3 ratio). This result is in agreement with all previous studies [19] : the incorporation of Si in the dielectric film reduces the intrinsic stress.

197

Figure 11b plots the intrinsic stress measured on films deposited from a mixture of SiH_4 and $(NH_3 + N_2)$ vs the NH_3/N_2 ratio. The material deposited with pure N_2 is more strained but the stress stay under 300 MPa. The addition of NH_3 in the plasma permits to reduce the intrinsic stress. This result has already been observed on films deposited with the DECR technique, but lower values are obtained with ICP technique (270MPa for ICP vs 1000 MPa with DECR). This has also been observed on UVCVD films. This result could be explained by a higher Si content in NH_3 film, than in N_2 films. This is presently investigated through the analysis of film stoechiometry.

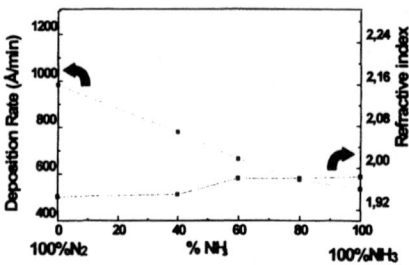

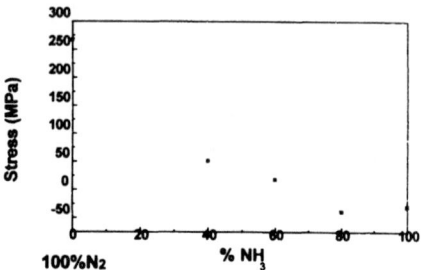

fig.11a Refractive index and deposition rate of SiN layers from plasma of SiH_4 and N_2+NH_3 (plasma power P=1500W and SiH_4 flow are kept constant)

fig.11b Stress measurement on layer obtained with a plasma SiH_4 and N_2+NH_3 (plasma power P=1500W and SiH_4 flow are kept constant)

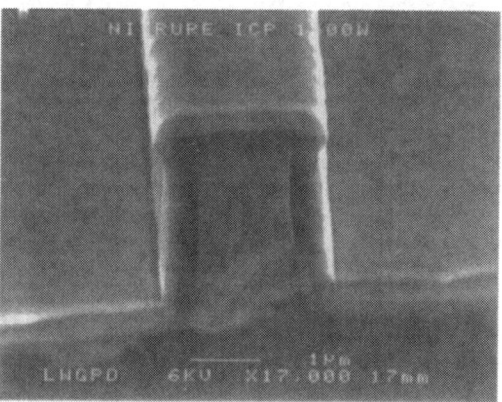

Figure 12 : SEM view of a Si_3N_4 layer deposited on a patterned InP, without substrate bias.

The conformity, i.e. the ability for a layer to cover vertical edges as well as horizontal ones, is important in applications as spacers, or anti-reflect coating. Figure 12 shows a scanning electron microscope view of a SiN film deposited (without substrate bias) on a patterned InP wafer. The thickness of the film is about 8000Å on the flat top surface. On the vertical walls of the structure, the thickness estimated with SEM is about 6000Å. The reason of this difference seems to be the shape of the etched InP structure with an overhanging edge leading to an effect of shadowing of deposition on the sides of the ridge. Nevertheless, the SEM picture shows a good conformity as one could expect without substrate bias.

II.4 Results obtained with the substrate held at 180°C

To improve the material quality, SiN layers were deposited at 180°C. As one could expect the refractive index is increased, but the deposition rate is slightly lower. The desorption of active species is a more efficient process than their reaction with surfaces atoms of the growing layer. The increase of refractive index can be the result of a more compact material or of a higher incorporation of Si favored by the lower absorption of N atoms.

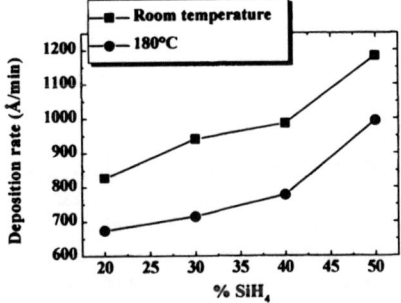
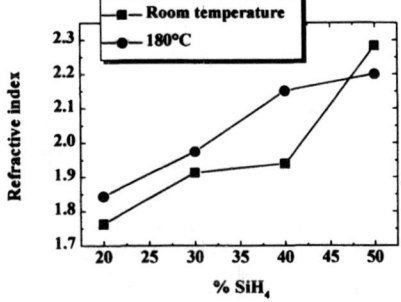

fig.13a Comparison of deposition rate for layers deposited at 180°C and without heating (<50°C).

fig.13b Comparison of refractive index (λ=633nm) for layers deposited at 180°C and without heating (<50°C).

Metal Insulating Semiconductor (MIS) diodes have been fabricated to address the passivation properties of these films. After dielectric deposition (thickness in the range 600Å-700Å) on n-type InP (n=8.10^{15} cm^{-3}), Ti-Au electrodes were evaporated (500µm diameter). C(V) characteristics between +10V and -10V were then recorded at 1MHz as well as leakage current. The values of hysteresis, resistivity and breakdown field were extracted from the recorded curves. The C(V) characteristics, of as deposited films and after a 200°C anneal, are shown on figure 14. In both cases the curves exhibit conventionnal shapes, with a large modulation from accumulation to depletion and a low hysteresis (about 2V). These results show that the InP surface has not been degraded by the deposition. The interface state density N_{ss} is relatively high, in the 10^{13} cm^{-2}eV^{-1} range after annealing.

The breakdown voltage is in the 5-6 MV/cm range and the resistivity is about 10^{14} Ω.cm. These values are sufficiently high to permit a suitable device passivation.

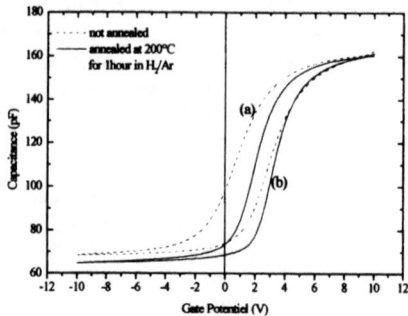

Fig 14: C(V) characteristics measurement at 1MHz on a MIS diode on InP substrate (as deposited and with a 200°C anneal).

II.5 Conclusions

The Inductive Coupled Plasma (ICP) source provides a high density plasma that produces high dissociation ratio and low bombardement energy in a thin film deposition reactor. Soft deposition and high deposition rates (for Si_3N_4 thin film : over 500Å/min with NH_3, and 1500A/min with N_2) without limitation on the maximum thickness layer are the main properties of this merging technology. The study reported here show that it allows to control the quality of the film over a wide range of deposition rates and refractive index with low induced stress on III-V semiconductors. Finally the electrical properties of the deposited films are very promising for passivation purpose since they display low leakage current, and strong breakdown field.

CONCLUSIONS

The two technologies presented here use the same high density plasma source. The results obtained demonstrate that efficient processes with high rates are obtain with amazingly low induced damages. These properties were expected for III-V semiconductors device processing. To go one step further, one would wonder if such technologies would be integrable in the same industrial equipment. Crosscontamination between etching and deposition would be the an important problem to address. In situ cleaning procedure can certainly be worked out with the dense plasma. This leaves us avenues for more work to be done.

ACKNOWLEDGEMENTS

The authors would like to thank Aline Falcou for devive processing, Dr J.F. Bresse for surface analysis. They particularly acknowledge the expertise of Dr J.M. Moison for the AFM analysis They also would like to acknowledge A.Madouri from L2M CNRS laboratory for infrared measurements

REFERENCES

1. Y.H. Lee, J. Vac. Sci. Technol. A 10, 1318 (1992)

2. T. Enoki, H. Ito, K. Ikuta and Y. Ishii, IPRM Proceedings, p.81, Sapporo, (1995)

3. J. Etrillard, P. Ossart, G. Patriarche, M. Juhel, J.F. Bresse, and C. Daguet, J. Vac Sci. Technol. A 15, 626, (1997)

4. W.P. Hong, P.K. Bhattacharya, J. Singh, Appl. Phys. Lett., Vol. 50, (1987)

5. J.K. Luo, H. Thomas, S.A. Clark, R.H. Williams, J. Appl. Phys., Vol. 74, 6726, (1993).

6. M. Achouche, A. Clei, and J.C. Harmand, J. Vac. Sci. Technol. B 14, 2555 (1996).

7. F.Delmotte et al. J.Vac.Sci.Technol. B 15(6), p 1919, Nov/Dec 1997

8. F.Plais, B.Agius, N.Proust, S.Cassette, G.Ravel, M.Puech, Appl. Phys. Lett., Aug 12,., 59(7), p.837 (1991)

9. B.Lescaut, Y.I. Nissim and J.F. Bresse, Proceedings of IPRM, p 319, (1996)

10. L.S. How Kee Chun J.L. Courant , P.Ossart and G.Post. Proceedings of IPRM, p.412 (1996)

11. R.W.Boswell, A.J.Perry, M.Emami, J.Vac.Sci.Technol. A 7(6), 1989 p. 3345-3350

12. A.Durandet, C. A.Davis, and R.W.Boswell, Appl. Phys. Lett.,April 7, 70(14), p. 1814 (1997)

13. R. J. Shul, G. B. McClellan, R. D. Briggs, and D. J. Rieger et al., J. of Vac. Sci. Technol. A 15,(3), May 1997, pp. 633-637

14. M.Achouche: Thesis, Université PARIS VII, 05/12/1996.

15. R.Chow, W.A. Lanford, W.Ke-Ming, R.S.Rosler, J.Appl.Phys., Vol 53, No.8, 1982, p. 5630-5633

16. W.A. Lanford, M.J. Rand, J.Appl.Phys., Vol 49, No.4, 1978, p. 2473-2477

17. E.I. Bromley et al. J.Vac.Sci.Technol. B 1(4), Oct/Dec 1983 p 1364

18. L.K. Nanver, P.J. French, E.J.G. Goudena H.W. Van Zeijl. Materials science and technology, 1995, 11, 1995, p. 36-40.

19. L.Shi A.M. Steenbergen et al., J.Vac.Sci.Technol. A 14(2), Mar/Avr 1996 p 471

WET OXIDATION OF HIGH-Al-CONTENT III-V SEMICONDUCTORS: IMPORTANT MATERIALS CONSIDERATIONS FOR DEVICE APPLICATIONS

CAROL I. H. ASHBY
Sandia National Laboratories, Albuquerque, New Mexico 87185-0603

ABSTRACT

Wet oxidation of high-Al-content AlGaAs semiconductor layers in vertical cavity surface emitting lasers (VCSELs) has produced devices with record low threshold currents and voltages and with wall-plug efficiencies greater than 50%. Wet oxidation of buried AlGaAs layers has been employed to reduce the problems associated with substrate current leakage in GaAs-on-insulator (GOI) MESFETs. Wet oxidation has also been considered as a route to the long-sought goal of a III-V MIS technology. To continue improving device designs for even higher performance and to establish a truly manufacturable technology based on wet oxidation, the effect of oxidation of a given layer on the properties of the entire device structure must be understood. The oxidation of a given layer can strongly affect the electrical and chemical properties of adjacent layers. Many of these effects are derived from the production of large amounts of elemental As during the oxidation reaction, the resultant generation of point defects, and the diffusion of these defects into adjacent regions. This can modify the chemical and electrical properties of these regions in ways that can impact device design, fabrication, and performance. Current understanding of the problem is discussed here

INTRODUCTION

Wet oxidation of AlGaAs with Al mole fractions in excess of 85% is an enabling technology for high-efficiency vertical cavity surface emitting lasers (VCSELs) [1]. Threshold currents below 10 μA and threshold voltages only 50 meV above the emitted photon energy have been obtained [2] and greater than 50% wall-plug efficiency can be routinely achieved [3]. Oxide layers generated by heating in a H_2O-saturated N_2 ambient have been employed in gallium-arsenide-on-insulator (GOI) metal semiconductor field effect transistors (MESFETs) [4] and metal-insulator-semiconductor field effect transistors (MISFETs) [5]. To date, such applications have been largely restricted to laboratory demonstrations. For such devices to achieve widespread commercial availability, the many factors that influence the extent of oxidation in a particular device and the influence of wet oxidation on relevant device properties must be understood. Oxidation rates depend to first order on Al mole fraction, but this dependence can be markedly changed by the details of a particular structure, including layer thickness and the composition of adjacent layers. The generation of elemental As during the oxidation process can produce significant changes in the chemical, electronic, and optical properties of devices. Important considerations derived from the chemical processes involved and their effect on important device parameters are discussed here.

EXPERIMENT

Two different sample geometries have been employed in our study: lateral and surface oxidation. Before lateral wet oxidation of Al-III-V heterostructures, the Al-containing layer is exposed by etching a mesa structure to permit direct access of water vapor to the layer to be oxidized. Wet oxidation is typically performed with a water-saturated nitrogen stream flowing through a tube furnace heated to a temperature between 375 and 500 °C [6]. Oxidation begins at the exposed edges and proceeds inward. The oxidation rate is strongly temperature dependent, so a three-zone furnace is especially desirable for well-controlled oxidation profiles. Water temperature and gas flow can be varied to control the supply of H_2O reactant to the exposed layer edges. Typically, water temperatures between 75 and 95 °C are employed. The oxidized length can be determined either optically (for sufficiently deep oxidations) or using scanning electron microscopy (SEM). Lateral oxidation results presented here were obtained with 80 ±0.5 °C water and a flow rate of 3.0 slm through a 4"-diam. 3-zone tube furnace.

Raman spectroscopy was used to identify intermediate species formed during the oxidation process. Raman samples consisted of planar 2-μm-thick $Al_{0.98}Ga_{0.02}As$ layers on GaAs that were oxidized from the surface down rather than laterally from an exposed edge [7]. Prior to wet oxidation, a 300-Å GaAs cap was selectively removed using a citric acid/peroxide mix [5:1 of (1g citric monohydrate/1g H_2O):30% H_2O_2]. Samples were heated to the reaction temperature (400<T<455 °C) in dry nitrogen. The nitrogen flow (0.4 slm, 2-in. diam. tube) was then switched to bubble through 80 ± 1 °C water. The reaction was terminated by switching to a dry nitrogen flow. Raman spectra were measured in the x(y',y'+z')$\bar{x}$ backscattering configuration (y' and z' parallel to (110) planes) using 514.5-nm light at < 85 W/cm^2.

RESULTS AND DISCUSSION

Some of the key intermediate species formed during wet oxidation of AlGaAs have been identified using Raman spectroscopy. Spectra of partially oxidized planar AlGaAs structures always show the presence of a significant amount of elemental As that has been liberated during the oxidation process (Fig. 1). Dominant features include crystalline As peaks at 198 and 257 cm^{-1} and a broad feature between 200 and 250 cm^{-1} peaking near 227 cm^{-1} due to amorphous As [8]. Resonant enhancement produces a detection limit on the order of a 10-20 Å layer [9]. At higher reaction temperatures, significant amounts of As_2O_3 are also detected. The broad feature centered at 475 cm^{-1} is due to amorphous a-As_2O_3 [9]. This species, if present, is below the

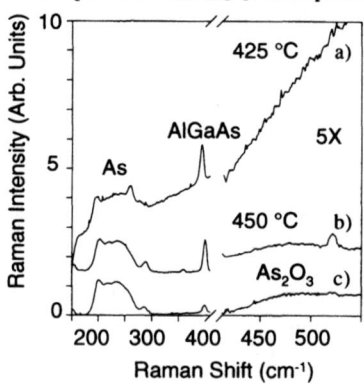

Fig. 1. Raman spectra of partially oxidized, 2-μ layers of $Al_{0.98}Ga_{0.02}As$

Raman detection level for oxidations performed at 400 and 425 ° (Fig. 1a), but it is observed at relatively constant levels throughout the oxidation at 450 °C (Figs. 1b and 1c). As the oxidation front advances into AlAs or AlGaAs films, one observes a relatively constant Raman signal from

As and a-As_2O_3 while the intensity of the AlAs-like phonon steadily decreases as the AlGaAs is converted to oxide [10]. The presence of both As_2O_3 and As intermediates in the wet oxidation process can be explained as follows.

Water has been observed to adsorb dissociatively on AlAs to produce Al-O, Al-OH, and As-H type species on the AlAs surface [11]. Literature values for the thermodynamic quantities of these surface species are unavailable so we have chosen to model the process using molecular analogs [12]. This leads to the possible initial reaction products described by Eqn. 1a and 1b.

$$2\ AlAs + 3\ H_2O_{(g)} = Al_2O_3 + 2\ AsH_3 \qquad (1a)$$

$$2\ AlAs + 4\ H_2O_{(g0)} = 2\ AlO(OH) + 2\ AsH_3 \qquad (1b)$$

$$2\ AsH_3 = 2\ As + 3\ H_2 \qquad (2)$$

$$2\ AsH_3 + 3\ H_2O = As_2O_3 + 6\ H_2 \qquad (3)$$

$$As_2O_{3(l)} + 3\ H_2 = 2\ As + 3\ H_2O_{(g)} \qquad (4)$$

$$2\ AlAs + 3\ H_2O_{(g)} = Al_2O_3 + 2\ As + 3\ H_2 \qquad (5a)$$

$$2\ AlAs + 4\ H_2O_{(g)} = 2\ AlO(OH) + 2\ As + 3\ H_2 \qquad (5b)$$

$$2\ AlO(OH) = Al_2O_3 + H_2O \qquad (6)$$

It is likely that both AlO(OH) and Al_2O_3 are present during the initial phases of oxidation. SIMs measurements have shown significant amounts of hydrogen to be present after $Al_5Ga_{0.5}As$ has been oxidized at 450 °C [13]. Formation of both AlO(OH) (boehmite) and Al_2O_3 (γ-alumina) are highly favored at typical oxidation temperatures. Conversion of boehmite to γ-alumina is favored above 176 °C. The mixed character of the Al-oxide can explain the less-than-expected shrinkage that occurs upon oxidation. If one wishes to minimize stress in an oxidized structure, conditions that minimize total dehydration may be desirable. Although the water-rich ambient probably retards the AlO(OH)-to-Al_2O_3 conversion during the wet oxidation process, crystallites of γ-alumina are more prominent farther back from the oxidation front under some oxidation conditions and voids appear at the oxide/semiconductor interfaces that are parallel to the sample surface [14]. Extended oxidation times of surface-oxidized samples can produce crystallization and delamination of 2-• m films of oxidized $Al_{0.98}Ga_{0.02}As$ [7], presumably due to increased film shrinkage as it becomes more extensively dehydrated.

At 425°C, As-H can decompose directly to give As and H_2, as suggested by the Raman spectra (Eqn. 2). As-H can also react with water to produce As_2O_3, but the free energy change is quite small (Eqn. 3), e.g., 22 kJ/mole at 425 °C. It is likely that the observed As_2O_3 results from the combination of reactions like Eqns. 1a, 1b, and 3. Both AsH_3 and As_2O_3 can be readily converted to elemental As for removal from the reacting layer (Eqns. 2 and 4), finally resulting in a porous Al-oxide layer, as described by Eqns. 5a and 5b. Gibbs free energies for Eqns. 1a-6 at 425 °C are -451, -404, -153, -22, -131-, -604, -557, and -35 kJ/mole, respectively.

The relative importance of Eqns. 3 + 4 vs. Eqn. 2 appears to be temperature dependent. This is manifested in the temperature dependence of the As_2O_3-associated peaks in Fig. 1. At temperatures of 425 °C and below, any As_2O_3 is present below the Raman detection limit.

However, at 450 °C, a nearly constant As_2O_3 signal is present for almost the entire course of the reaction. The presence and buildup of this product under some reaction conditions can explain the variation in time dependence of the oxidation process.

Although the oxidation rate is determined, to first order, by the Al mole fraction in the oxidizing layer [6], the time dependence of the reaction rate for samples with identical Al mole fraction has been observed to vary depending on processing conditions. Many workers operate in a reaction-rate-limited regime (linear time dependence) to facilitate more precise control of oxidized depth in device structures and others report diffusion-limited behavior (parabolic time dependence). Extrapolations of oxidized depth vs. time curves often fail to pass through the origin, with deviations in both directions having been reported. This complicates mechanistic interpretations. However, sufficient information is available about the temperature and composition dependence of wet oxidation rates to support the following understanding [15] of the important dynamic characteristics that determine the time dependence.

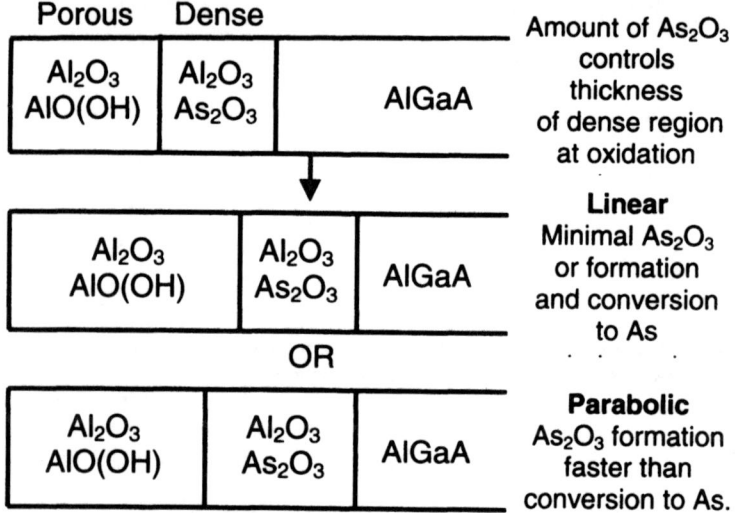

Fig. 2. Schematic of effect of increase in rate of As_2O_3 formation relative to its reduction to As for removal to leave behind a porous $AlO(OH)/As_2O_3$ film.

The Deal and Grove model for oxidation [16] describes the temporal dependence of an oxidation process as the sum of a linear and a parabolic term,

$$\frac{d^2}{k_{diff}} + \frac{d}{k_{rxn}} = t \qquad (7),$$

where the linear term dominates when the oxidation rate is reaction-rate limited and the parabolic term dominates when the rate is diffusion limited. A range of dependencies on time have been reported for wet oxidation of AlGaAs. The oxidation of one of the more common device compositions, $Al_{0.98}Ga_{0.02}As$, display a linear time dependence between 380 and 440 °C [6], while the oxidation of AlAs has been reported to have a parabolic dependence from 370 to 450 °C [17]. Another study of AlAs has reported a linear time dependence at 356 °C , a parabolic depencence at 516 °C, and a mixed linear/parabolic dependence at intermediate temperatures [18]. Yet another study of AlAs has shown linear dependence at T ≤ 350 °C and parabolic behavior at T ≥ 375 °C [19].

The apparent shift between reaction-rate-limited and diffusion-limited regimes can be explained by the relative importance of the As_2O_3 intermediate in the process, as illustrated in Fig. 2. Partially oxidized planar AlGaAs structures always show strong As-associated Raman peaks due to the production of a significant amount of elemental As during the oxidation process. Only at higher reaction temperatures are significant amounts of As_2O_3 detected. Below a certain temperature (e.g., 425 °C for $Al_{0.98}Ga_{0.02}As$), very little, if any, As_2O_3 is present near the oxidation front. Arsenic loss under these conditions leaves a porous Al-oxide region that does not impede reactant or product transport to the oxidation front. Hence, the linear time dependence characteristic of a reaction-rate-limited process is obtained. During the 450 °C oxidation of a 2-μm layer of $Al_{0.98}Ga_{0.02}As$, As_2O_3 formation (Eqn. 3) is approximately balanced by the subsequent reduction to elemental As (Eqn. 4) so a progressively thickening dense oxide layer does not form. As long as the dense As_2O_3-containing region at the oxidation front does not appreciably increase in thickness as the oxidation proceeds to the desired depth (typically <50 μm in device structures), the observed time dependence will remain linear. This is the most desirable condition for device fabrication since the linear time dependence facilitates more precise control of the final depth of an oxidized layer. If the importance of As_2O_3 formation increases relative to its reduction to As, a progressively thickening dense oxide will form at the oxidation front and the time dependence will shift toward parabolic. The existence of a thin, dense, amorphous region of a few nanometers thickness has been observed at the oxidation front by transmission electron microscopy (TEM) [20]. Behind this dense region is a less dense region of amorphous Al-oxide that extends back to the exposed mesa edge.

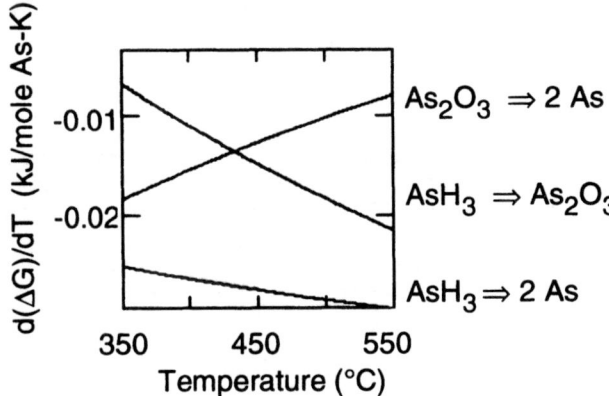

Fig. 3. Temperature dependence of d(ΔG/DT) for Eqns. 2-4.

The transition from reaction-limited (linear) to diffusion-limited (parabolic) behavior for a particular AlGaAs composition will depend on the relative temperature dependencies of Eqns. 1a-5b. For AlAs, increasing temperature causes the linear-to-parabolic shift [18,19]. Calculation of the temperature-dependent change in the Gibbs free energy, dΔG/dT (Fig. 3) shows that the favorability of As_2O_3 formation increases with increasing temperature while the favorability of its subsequent reduction decreases. This is consistent with the increase in the As_2O_3 Raman intensity with increasing temperature. Addition of Ga to form AlGaAs decreases the total oxidation rate and appears to shift the linear-to-parabolic transition to higher temperature, since $Al_xGa_{1-x}As$ with $x \leq 0.98$ displays linear time dependence at 425 °C, where AlAs oxidation is either parabolic [17,19] or mixed linear/parabolic [18] at that temperature. This suggests that Eqn. 3 becomes less dominant as Ga is substituted for Al and the total oxidation rate decreases. Experimental results to date suggest that Eqn. 3 becomes increasingly important for higher Al contents and at higher reaction temperatures.

This is clearly illustrated in the time-dependence of the oxidation of 450-nm layers of $Al_{0.98}Ga_{0.02}As$ and $Al_{0.94}Ga_{0.06}As$ within the same sample (Fig 4). While the time dependence of the 94% material remains linear at both temperatures, the deviation from linear dependence is clearly seen for the 98% material as reaction times and total penetration of the oxidized region into the sample increases.

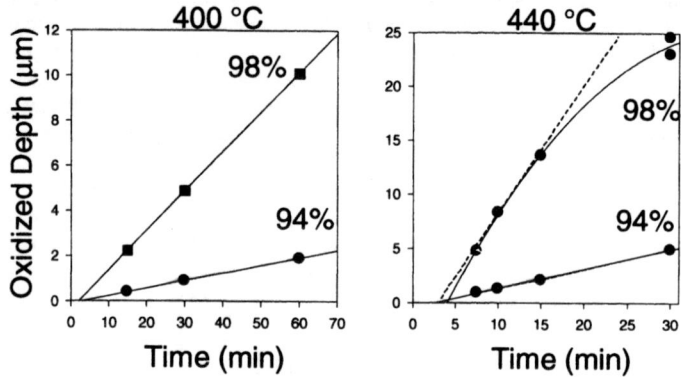

Fig 4. Temperature dependence of the oxidation of 450-Å layers of $Al_{0.09}Ga_{0.02}As$ and $Al_{0.94}Ga_{0.06}As$ within the same sample.

Since As is a key product of wet oxidation, an As-rich ambient is present at the surfaces adjacent to the oxidation fronts, both at the homointerface where the oxidation is proceeding deeper into a given layer and along the heterointerfaces parallel to the oxidation direction. Such As-rich conditions result in the generation of numerous defects in the regions adjacent to the oxidized layer. The defects generated by excess As [21] include As interstitial, As_i, As antisites, As_{III}, group III vacancies, V_{III}, and even group III interstitials, III_i, as described by the following equations:

$$As^0 \leftrightarrow As_i$$
$$As^0 \leftrightarrow As_{III}$$
$$As^0 \leftrightarrow GaAs + V_{III}$$
$$As^0 \leftrightarrow As_{III} + III_i$$

The increased concentrations of these point defects can have profound effects on the chemical reaction rates of adjacent layers and the optical and electronic properties of devices made using wet oxidation.

The oxidation rate of a lower-Al-content layer can be appreciably increased when it is in close proximity to a faster-oxidizing layer. The effect of such close-proximity faster-oxidizing layers [10] is shown in Fig.5 for 45-nm layers of undoped $Al_{0.94}Ga_{0.06}As$ separated from 45-nm layers of undoped $Al_{0.98}Ga_{0.02}As$ by thin layers of undoped GaAs (0.7, 1.5, 3, 6, 9, and 12 nm) oxidized at 400 °C for 90 min. Although no enhancement of the reaction of the $Al_{0.94}Ga_{0.4}As$ layer is observed after this reaction time with 9 and 12 nm barriers, there are progressively greater enhancements of the oxidation of the $Al_{0.94}Ga_{0.06}As$ layers that are separated for the already oxidized $Al_{0.98}Ga_{0.02}As$ by 6, 3, 1.5, and 0.7 nm GaAs layers. All $Al_{0.94}Ga_{0.06}As$ layers exhibit rate enhancements at longer reaction times. The oxidation of the $Al_{0.94}Ga_{0.06}As$ layers can be modeled as the sum of the unenhanced rate plus a contribution derived from the diffusion of something into the layer from the already oxidized adjacent layers [22]. The activation energy for some reaction-enhancing species diffusing through the barrier and the diffusion constant are consistent with those reported for column III vacancies [23]. The close-proximity enhancement can also be significant in the absence of barriers. Structures with graded compositions result in profiles that show appreciably deeper oxidation than expected from compositional effects alone [24]. This proximity enhancement is especially important since compositional grading is used to fabricate tapered oxide current apertures for optimum VCSEL performance [6,25] and to minimize stresses in the oxidized devices.

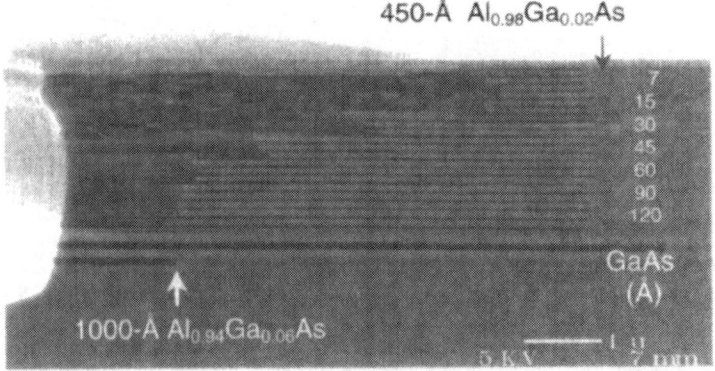

Fig 5. Cross sectional SEM of a partially oxidized (400 °C, 90 min) sample of alternating 450 Å layers of $Al_{0.98}Ga_{0.02}As$ and $Al_{0.94}Ga_{0.06}As$ separated by GaAs barriers with thicknesses varying from 120 Å near the substrate to 7 Å near the surface. Dark regions are oxide, light regions are unoxidized.

One indication of the formation of large quantities of point defects during wet oxidation is the appearance of As precipitates quite remote from the original oxidation. Such As precipitates have been observed widely dispersed through the adjoining layers and even segregated at the GaAs surface of a wet-oxidized sample [26]. Precipitates have also been reported in the InGaAs channel of a GOI MESFET that had a 10 nm $Al_{0.25}Ga_{0.75}As$ layer interposed between the channel and the oxidized layer [27].

The photoluminescence from quantum wells that are near wet-oxidized regions exhibits changes that are typically seen in the presence of excessive numbers of point defects. The photoluminescence is decreased for quantum wells (QWs) situated 20 and 40 nm from oxidized

AlAs [28]. The time and temperature dependence of this degradation is consistent with the presence of oxidation-generated point defects that have a diffusivity comparable to As in GaAs. Post-oxidation hydrogen treatment partially restores the photoluminescence intensity from GaAs quantum wells separated from the oxidized layer by 25-nm AlGaAs spacers. [29] While oxidation reduces the PL intensity to 6% of the original intensity, post-oxidation H-annealing restores the intensity to 67% of its original level. Such PL restoration is to be expected from hydrogen passivation of As-associated deep levels, e.g., As_{Ga}, that were formed during oxidation by indiffusion of excess As.

Intermixing at quantum well interfaces by heating is seen to be greater near oxidized AlGaAs regions than near unoxidized regions of the same device. However, even the QW emission from the regions that are not immediately above or below oxidized material exhibit a blue-shift. Such behavior is expected if there is increased mobility of column III atoms in the presence of excess As [30].

In addition to the observation of As precipitates in the InGaAs channel of a GOI MESFET, a deterioration of both the mobility and the free carrier concentration is obtained following oxidation. This deterioration was minimized by decreasing oxidation temperature and/or oxidation time [27]. Both process alterations would reduce the diffusion of As-derived defects into the device, thereby minimizing device degradation.

SUMMARY

Elemental As is always an important product of the wet oxidation of high-Al-content AlGaAs. The loss of As from the oxidized layer leads to the formation of a porous, relatively As-free mixture of Al-oxides, such as AlO(OH) and Al_2O_3. Under some reaction conditions, As_2O_3 forms at appreciable levels near the oxidation front. The time dependence of the oxidation reaction depends on the thickness of the dense, As_2O_3-containing region at the oxide/semiconductor interface. The shift from a reaction-rate-limited (linear) behavior toward a diffusion-limited (parabolic) time dependence is favored by increasing temperature or increasing Al mole fraction. The elemental As that is produced in the reaction can serve as a source of point defects. These can affect important chemical and electrical properties of structures. Since formation and loss of elemental As is essential to produce the porous layer that permits rapid, reaction-rate-limited oxidation, wet oxidation should prove very useful for applications that do not depend strongly on concentrations of point defects in the material. Those devices that are highly defect sensitive are less promising candidates for this fabrication approach.

ACKNOWLEDGMENTS

The author wishes to acknowledge her Sandia colleagues working in the oxidation area: Andrew A. Allerman, Albert G. Baca, Olga Blum, Monica M. Bridges, Kent D. Choquette, David M. Follstaedt,,Kent M. Geib, Michael J. Hafich, B.E. Hammons, Hong Q. Hou, Nancy A. Missert, Paula P. Newcomer, John P. Sullivan, and Ray D. Twesten.

Sandia is a multiprogram laboratory operated by Sandia Corporation, a Lockheed Martin Company, for the United States Department of Energy under Contract DE-AC04-94AL85000.

REFERENCES

1. J. M. Dallesasse, N. Holonyak, Jr., A. R. Sugg, T. A. Richard, and N. El-Zein, Appl. Phys. Lett. **57**,2844 (1990).
2. K. D. Choquette, R.P. Schneider, Jr., K. L. Lear, and K. M. Geib, Electroni. Lett. **30**, 2043 (1994).
3. K.L. Lear, J.D. Choquette, R.P. Schneider, Jr., S.P. Kilcoyne, and K. M. Geib, Electron. Lett. **31**, 208 (1995).
4. P.A. Parikh, P. M. Chavarkar, and U.K. Mishra, IEEE Electron. Device Lett. **18**, 111 (1997).
5. E.I.Chen, N.Holonyak, Jr., and S.A Maranowski, Appl. Phys. Lett. **66**, 2688, (1995).
6. K. D. Choquette, K. M. Geib, C. I. H. Ashby, R. D. Twesten, O. Blum, H. Q. Hou, D. M. Follstaedt, B. E. Hammons, D. Mathes, and R. Hull, IEEE J. Select. Topics in Quant. Electron. **3**, 916 (1997).
7. C. I. H. Ashby, J. P. Sullivan, P.P. Newcomer, N. A. Missert, H. Q. Hou. B. E. Hammons, M. J. Hafich, and A. G. Baca, Appl. Phys. Lett. **70**, 2443 (1997).
8. G.P. Schwartz, B. Schwartz, D. DiStefano, G.J. Gualtieri, and J.E. Griffiths, Appl. Phys. Lett. **34**, 205 (1979).
9. G.P. Schwartz, G.J. Gualtieri, J.E. Griffiths, C.D. Thurmond, B. Schwartz, J. Electrochem. Soc. **127**, 2488 (1980).
10. C. I. H. Ashby, J. P. Sullivan, K. D. Choquette, K.M. Geib, and H. Q. Hou, J. Appl. Phys. **82**, 3134 (1997).
11. W. J. Mitchell, C.-H. Chung, S.I. Yi, E.L. Hu, and W.H. Weinberg, J. Vac. Sci. Technol. **B15**, 1182 (1997).
12. Thermochemical data used in calculations found in O. Kubaschewski, C. B. Alcock, P. J. Spencer, "Materials Thermochemistry", Pergamon Press, UK, 1993.
13. F.A. Kish, S.J. Caracci, N. Holonyak, Jr., K.C. Hsieh, J.E. Baker, S. A. Maranowski, A.R. Sugg, J.M. Dallesasse, R.M. Fletcher, C.P. Kuo, T.D. Osentowski, and M.G. Craford, J. Electron. Mater. **21**, 1133 (1992).
14. Z. Liliental-Weber, O. Richter, W. Swider, M. Li,G.S. Li, C. Chang-Hasnain, and E.R. Weber, in "Semiconducting and Insulating Materials", edts. Z. Liliental-Weber and C. Miner, IEEE, 1999 in press.
15. C.I.H. Ashby, M.M. Bridges, A.A. Allerman, B.E. Hammons, and H.Q. Hou, to be published in App. Phys. Lett.
16. B. E. Deal and A. S. Grove, J. Appl. Phys. **36**, 3770 (1965).
17. T. Langenfelder, St. Schröder, and H. Grothe, J. Appl.Phys. **82**, 3548 (1997).
18. M. Ochiai, G. E. Giudice, H. Temkin, J. W. Scott, and T. M. Cockerill, Appl. Phys. Lett. **68**, 1898 (1996).
19. S. A. Feld, J. P. Loehr, R. E. Sherriff, J. Miemeri, and R. Kaspi, IEEE Photon. Technol. Lett. **10**, 197 (1998).
20. R. D. Twesten, D. M. Follstaedt, and K. D. Choquette, "Vercital-Cavity Surface Emitting Lasers, K. D. Choquette and D. G. Deppe, eds., Proc. SPIE-The International Society for Optical Engineering Proceedings 3003, 55 (1997).
21. T. Y. Tan, Mater. Sci. and Engin. **B10**, 227 (1991).
22. O. Blum, C. I. H. Ashby, and H. Q. Hou, Appl. Phys. Lett. **70**, 2870 (1997).
23. S.F. Wee, M.K. Chai, K.P. Homewood, and W.P. Gillin, J. Appl. Phys. Lett. **82**, 4842 (1997).
24. O. Blum, K. L. Lear, H. Q. Hou, and M. E. Warren, Electron. Lett. **32**, 1406 (1996).
25. R. L. Naone, E. R. Hegbloom, B. J. Thibeault, and L. A. Coldren, Electron. Lett. **33**, 300 (1997).
26. Z. Liliental-Weber, S. Ruvimov, W. Swider, J. Washburn, M. Li, G.S. Li,

and C. Chang-Hasnain, and E.R. Weber et al., SPIE--The International Society for
Optical Engineering Proceedings **3006**, 15 (1997).

27. P.A. Parikh, P.M. Chavarkar, L. Zhao, J. Ibbetson, J.S. Speck, U.K. Mishra,
"Effect of oxidation of AlxGa1-xAs on Adjacent Semiconductor Layers : Hall (Electrical) and
TEM (Structural Characterization)", 1997 EMC Proceedings, pp 40.

28. S. S. Shi, Ph.D. thesis, 'Hydrogen Passivation of Native Oxides in GaAs-based III-V
Devices', September 1997.

29. S. S. Shi, E. L. Hu, J.-P. Zhang, Y.-I. Chang, P. Parikh, and U. K. Mishra, Appl. Phys. Lett.
70, 1293 (1997).

30. C.-K. Lin, X. Zhang, P. D. Dapkus, and D. H. Rich, Appl. Phys. Lett. **71**, 3108 (1997).

STUDIES OF SURFACE STATE DENSITIES OF SEMICONDUCTORS BY ROOM-TEMPERATURE PHOTOREFLECTANCE

J. S. Hwang, and G. S. Chang

Dept. of Physics, National Cheng Kung University, Tainan, Taiwan, R. O. C.,

jshwang@ibm60.phy.ncku.edu.tw

ABSTRACT

In this study, we develop a novel approach to determine the surface Fermi level and the surface state densities of semiconductors. The built-in electric field and thus the surface barrier height are evaluated from the Franz-Keldysh oscillations in the PR spectra. Based on the thermionic-emission theory and current-transport theory, the surface state density as well as the pinning position of the surface Fermi level can be determined from the dependence of the surface barrier height on the pump beam intensity. Even though this method is significantly simpler, easier to perform, and time efficient compared with other approaches, the results obtained agree with the literature.

INTRODUCTION

There exists at many semiconductor surfaces an appreciable density of uncontrolled electronic states, known as surface states, whose origin is still not clear[1]. Since these states tend to pin the surface Ferim level at these surfaces in the bandgap, high surface state density may result in low gain, high loss, and low efficiency in semiconductor devices[3]. Therefore the determination of surface state densities of semiconductor surfaces has become an interest topic of studies. Recently, contactless and nondestructive technique of photoreflectance spectroscopy has become an important technique in the characterization of semiconductors[2]. Hwang *et al.* in their previous reports[4,6] have studied, using photoreflectance(PR), the surface state density D_s and surface state distribution function of InAlAs SIN[+] structures by measuring the surface barrier height V_b as a function of the thickness of the toplayer. Yin *et al.*[2] calculated the surface state density of GaAs by fitting V_b as a function of temperature. Both methods are very time consuming and tedious.

This study used room temperature PR[5] spectra to investigate the surface Fermi levels V_F and the surface state densities of semiconductors by measuring V_b as a function of pump beam intensity P_m. The built-in electric field and then the surface barrier height as a function of P_m were evaluated using the Franz-Keldysh oscillations[7] in the PR spectra. Based on the current-transport theory and thermionic emission theory, D_s as well as the V_F position can be obtained.

Mat. Res. Soc. Symp. Proc. Vol. 573 © 1999 Materials Research Society

THEORY

In PR, the surface electric field is modulated through the photo-injection of electron-hole pair via chopped incident laser beam. The line shape of the PR signal, $\Delta R/R$, is directly related to the perturbed complex dielectric function. For a moderate electric field, the PR spectrum exhibits a series of oscillations (FKO) originating from the electric field F in the samples. The nth extremum E_n of FKO occurs when[8,9]

$$n\pi = (4/3)[(E_n-E_g)/\hbar\Theta]^{3/2} + \chi \tag{1}$$

where E_g is the energy gap, χ is an arbitrary factor and $\hbar\Theta = (e^2\hbar^2F^2/8\mu)^{1/3}$ with e being the electron charge, and μ the reduced effective mass. A plot of $(4/3\pi)(E_n-E_g)^{3/2}$ vs. the index number n will yield a straight line with slope $(\hbar\Theta)^{-3/2}$. Therefore, the electric field (F) can be obtained directly from the period of the FKOs.

The surface barrier height is related to the built-in electric field by[2,10]

$$V_b = Fd + kT/e + \varepsilon F^2/2eN \tag{2}$$

where d, k, T, ε and N represent the undoped layer thickness, the Boltzmann constant, temperature, dielectric constant and doping concentration, respectively. The surface Fermi level V_f, measured from the conduction band at the surface, is equal to V_b+V_s, where V_s is the photovoltage induced by both pump and probe beam. If the probe beam is defocused on the sample and maintained at sufficiently low intensity, the photovoltage induced by the probe beam can be neglected. The photovoltage and hence the surface barrier height V_b, at a constant temperature, is a function of P_m only. The photocurrent density J_{pc} consists of the drift and diffusion currents and can be written as[11]

$$J_{pc} = [eP_m\gamma(1-R_0)/\hbar\omega]\{[1-exp(-\alpha L)] + \alpha L_d exp(-\alpha L)/(1+\alpha L_d)\} \tag{3}$$

where γ is the quantum efficiency, R_0 is the reflectivity of sample surface and $\hbar\omega$ is the photon energy of the pump beam, L is the depletion width approximately equivalent to the toplayer thickness d, α is the absorption coefficient and L_d is the diffusion length of the minority carriers. For cases in which the diffusion length is much larger than the penetration depth, i.e. $\alpha L_d \gg 1$, Eq.(3) reduces as follows

$$J_{pc} = eP_m\gamma(1-R_0)/\hbar\omega \tag{4}$$

According to current-transport theory, the photovoltage V_s can be expressed as[12]

$$V_s = (\eta kT/e)ln(I_{pc}/I_0 + 1) \qquad (5)$$

where η is an ideality factor[3,12], I_{pc} equals J_{pc} times the surface area A_{pc} simultaneously illuminated by both pump and probe beams, and $I_0(T)$ is the saturation current. The saturation current depends on the dominant current flow mechanism and is equal to the saturation current density $J_0(T)$ times some effective area A_0 , which contributes to the current mechanism. For samples of SIN$^+$ structure, thermionic-emission and diffusion are the major contributions to $J_0(T)$, therefore, J_0 can be expressed as[13,14]

$$J_0(T) = [A^*T^2/(1+BT^{1/2})]exp[-eV_F(T)/kT] \qquad (6)$$

where A^* is the modified Richardson constant defined as[13] $m^*ek^2/(2\pi^2\hbar^3)$ and $B = (k/2\pi m^*)^{1/2}(300/v_o)$, where m* is the effective mass of the electron. By substituting Eqs.(4) and (6) into Eq.(5) with $I_{pc} = A_{pc}J_{pc}$ and $I_0 = A_0J_0$, the surface barrier height is then

$$V_b = V_F - (\eta kT/e)ln[1 + eP_m\gamma(1-R_o)(1+BT^{3/2}) exp(eV_F/kT) /\hbar\omega\, rA^*T^2] \qquad (7)$$

where the geometry factor[2] $r \equiv A_0/A_{pc}$ is the fraction of the surface that has surface states.

At constant temperature, the only variable in Eq.(7) is the P_m. When V_b as a function of P_m is least-squares fitted to Eq.(7), V_F , η and r can be obtained from the fitting parameters. By assuming one surface state per atom at the surface, the surface state density D_s can be calculated from rN_o , where N_o is the number of atoms per unit area of the surface.

EXPERIMENT

GaAs and InAlAs SIN$^+$ structures were grown by conventional molecular beam epitaxy. These heterostructures possesse a common structure consisting of 1000Å undoped layer on top of 1μm of a Si-doped, n-type buffer layer grown on a semi-insulating (100) GaAs substrate for GaAs SIN$^+$ structure or on a Fe-doped semi-insulating InP for InAlAs SIN$^+$ structure. In InAlAs SIN$^+$ structures, both the undoped and the buffer layers share the same AlAs mole fraction. The doping concentration in the buffer layer was approximately 1×10^{18} cm^{-3} for all samples studied. A standard PR apparatus was used in this study[8]. A He-Ne laser served as the pump beam. The probe beam was defocused on the sample and its intensity was kept at $0.1\mu W/cm^2$ to reduce the photovotaic effect. PR spectra were measured at room temperature with the intensity of the pump beam, varying from 0.05 to 270μW/cm^2 .

RESULTS

Figure 1 displays the PR spectra of the GaAs SIN$^+$ structure with various pump beam intensities. All the spectra exhibit a large number of Franz-Keldysh oscillations (FKO) originates from the uniform built-in electric field[15] in the undoped region. Figure 2 depicts V_b as a function of the P_m. The solid line in Fig. 2 is a least-squares fit of the experimental data to Eq. (6). For GaAs sample, where $A^* = 8.0A/cm^2K^2$, $B = 3.3\times10^{-4}K^{-3/2}$, $\gamma \cong 1$, $N_0 = 6.3\times10^{14}cm^{-2}$, and $R_0 = 0.34$, the fitting parameters obtained are $V_F = 0.70 \pm 0.02eV$, $\eta = 0.80 \pm 0.05$ and $r = 0.040 \pm 0.005$. Assuming one state per atom on the (100) GaAs surface[2], the surface state density calculated from $r\,N_0$ is $(2.52\pm0.32)\times10^{13}cm^{-2}$. These results are listed in Table. I. Yin et al.[2] measured V_b as a function of temperature at fixed P_m. By fitting the measured $V_b(T)$ as a function of temperature to Eq. (6), they obtained $V_F = 0.77 \pm 0.02eV$, $\eta = 0.93 \pm 0.05$, and $D_s = (1.26\pm0.63) \times10^{13}cm^{-2}$ for GaAs. Their results are also included in Table. I.

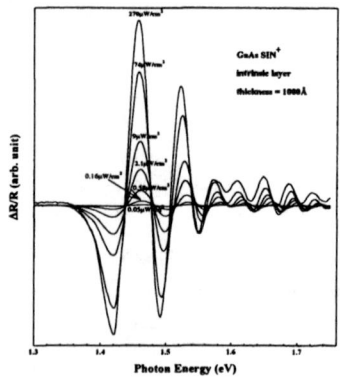

FIG. 1. The PR spectra of the GaAs SIN$^+$ structure with the pump beam intensity varying from 0.05 to 270μW/cm^2.

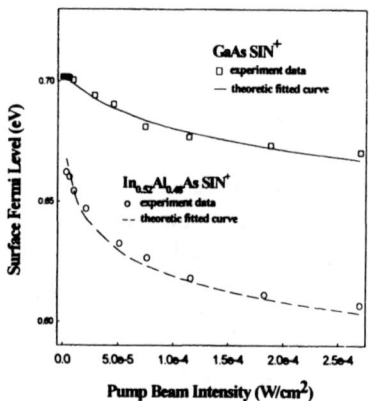

FIG. 2. The measured surface barrier heights V_b of the GaAs and InAlAs SIN$^+$ structures as a function of the pump beam intensity. The solid and dash lines are least-squares fits of the data of V_b to Eq. (6).

Figure 3 displays a series of PR spectra of an In$_{0.52}$Al$_{0.48}$As SIN$^+$ structure with 1000Å toplayer under various P_m. V_b is also plotted as a function of P_m in Fig. 2. Again, the dashed line represents a least-squares fit to Eq. (6). For InAlAs, where $A^* = 8.5A/cm^2K^2$, $B = 3.2\times10^{-4}K^{-3/2}$, $\gamma \cong 1$, $N_0 = 2.9\times10^{14}cm^{-2}$, and $R_0 = 0.30$, the fitting parameters obtained are $V_F = 0.66 \pm 0.02eV$, $\eta = 0.80 \pm 0.05$, and $r = (2.0\pm1.0) \times10^{-3}$. The density of occupied surface states estimated from D_s is $(5.8\pm2.9)\times10^{11}cm^{-2}$ with $N_0 = 2.9\times10^{14}cm^{-2}$. In our previous studies[13], we found that the surface Fermi level of In$_{1-x}$Al$_x$As SIN$^+$ structures is not pinned at midgap over aluminium concentration of 0.42~0.57. For each Al composition there exists certain ranges of

216

top layer thickness within which the surface Fermi level is weakly pinned. From the dependence of the electric field and surface Fermi level on the top layer thickness, we concluded that the surface states distribute over two separate regions within the energy band gap and the densities of surface states are as low as $(1.36\pm0.15)\times10^{11}$cm^{-2} for the distribution near the conduction band (U) and $(4.38\pm0.50)\times10^{11}$cm^{-2} for the distribution near valence band (L). In this study, the toplayer thickness of the sample is 1000Å. Our previous studies[5] showed that the surface Fermi level is pinned within the lower distribution. Therefore the surface state density obtained in this study is the surface state density of the lower distribution occupied by electrons and is comparable with the result obtained in our previous study.

Table. I. The surface Fermi level, ideality factor η, geometry factor r, and the surface state denties of GaAs and In$_{0.52}$Al$_{0.48}$As SIN$^+$ structures obtained in various studies.

	(300k)	V_F (eV)	η	r	D_s (cm^{-2})
GaAs	Present work	0.70±0.02	0.80±0.05	0.040±0.005	$(2.52\pm0.32)\times10^{13}$
	Yin et al.[a]	0.77±0.02	0.93±0.05	0.02±0.01	$(1.26\pm0.63)\times10^{13}$
In$_{0.52}$Al$_{0.48}$As	Present work	0.66±0.02	0.80±0.05	$(2.0\pm0.5)\times10^{-3}$	$(5.8\pm1.45)\times10^{11}$
	Hwang et al.[b, c]	0.62±0.01			$(3.31\pm0.05)\times10^{11}$

[a]Reference 2, [b]Reference 6, [c]Reference 7.

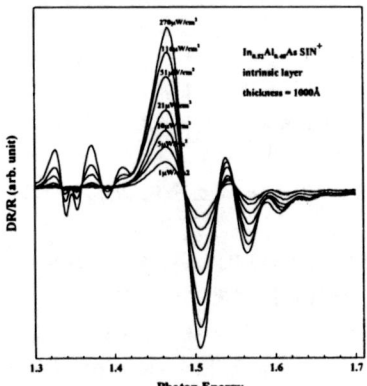

FIG. 3. The PR spectra of an In$_{0.52}$Al$_{0.48}$As SIN$^+$ structure with 1000Å undoped toplayer under various pump beam intensities.

CONCLUSIONS

In conclusion, we have introduced a new approach to investigate the surface state density and the Fermi level pinning position of semiconductor by simply changing the intensity of the pump beam in photoreflectance experiment. Since the intensity of the pump beam can be adjusted simply by inserting a neutral density fitter in the beam path, it is a simple and efficient method to investigate the surface state densities of various semiconductor surfaces or interfaces. The surface state densities of GaAs and $In_{0.52}Al_{0.48}As$ obtained using this technique are in good agreement with those obtained from other procedures, which provides further support of its validity. This work was supported by National Science Council of the Republic of China under Contract No. NSC 88-2112-M-006-015.

REFERENCES

(1) J. M. Woodall, P. D. Kirchner, J. L. Freeouf, D. T. McInturff, M. R. Melloch, and F. H. Pollak. Trans. R. Soc. London **344**, 521 (1993).

(2) X. Yin, H.-M. Chen, F. H. Pollak, Y, Cao, P.A. Montano, P. D. Kirchner, G. D. Pettit, and J. M. Woodall, J. Vac. Sci. Technol. A **10**, 131 (1992)

(3) S. M. Sze, in Semiconductor Devices Physics and Technology, (Wiley, New York, 1985), p. 113.

(4) J. S. Hwang, S. L. Tyan, W. Y. Chou. M. L. Lee, D. Weyburn, Z. Hang. H. H. Lin, and T. L. Lee, Appl. Phys. Lett. **64**. 3314 (1994).

(5) J. S. Hwang, W. Y. Chou. S. L. Tyan, H. H. Lin, and T. L. Lee, Appl. Phys. Lett. **67**. 2350 (1995).

(6) J. S. Hwang, W. C. Hwang, G. S. Chang, and Y. J. Cheng, "A photoreflectance study of the surface state density and distribution function of InAlAs", submitted to J. of Appl. Phys.

(7) C. Van Hoof, K. Deneffe, J. DeBoeck, D. J. Arent, and G. Borghs, Appl. Phys. Lett. **54**, 608 (1989).

(8) R. N. Bhattacharya, H. Shen, P. Parayanthal, F. H. Pollak, T. Coutts, and H. Aharoni, Phys, Rev. B **37**, 4044 (1988).

(9) H. Shen, F. H. Pollak, and J. M. Woodall, J. Vac. Sci. Technol. B **8**, 413 (1990).

(10) H, Shen, and M. Dutta. J. Appl. Phys. **78**, 2151 (1995).

(11) T. Kanata, Matsunaga, H. takaura, Y. Hamakaea, and T. Nishino, in Proceedings of the Society of Photo-Optical Instrumentation Engineer (SPIE, Bellingham, 1990), Vol. **1286**, p. 56.

(12) H. Hovel, in Semiconductor and Semimetals (Academic, New York, 1975), Vol. **11**, p. 59.

(13) E. H. Rhoderick, in Metal-Semiconductor Contacts (Clarendon, Oxford, 1980), p. 101.

(14) M. Hecht, Phys. Rev. B **41**, 7918 (1990).

(15) Fred H. Pollak, H. Shen, Materials Science and Engineering, R10, Nos. 7-8, Oct. 1, 1993, pp. 275-374.

ADVANCES IN GaAs MOSFET'S USING $Ga_2O_3(Gd_2O_3)$ AS GATE OXIDE

Y. C. WANG, M. HONG, J. M. KUO, J. P. MANNAERTS, J. KWO, H. S. TSAI,
J. J. KRAJEWSKI, J. S. WEINER, Y. K. CHEN, A. Y. CHO

Bell Laboratories, Lucent Technologies, Murray Hill, NJ 07974, yuchiwang@lucent.com

ABSTRACT

In this article, we review the recent progress on GaAs MOSFET's using *in-situ* MBE-grown $Ga_2O_3(Gd_2O_3)$ as the gate dielectric. Both depletion-mode (D-mode) and inversion-mode (I-mode) GaAs MOSFET's with negligible drain current drift and hysteresis are demostrated. The absence of drain current drift and hysteresis indicates that the excellent stability of the oxide and low oxide/GaAs interface state density have been achieved. The drain current density and transconductance are about one order of magnitude higher than the best previous reported data in the literature for an inversion-mode GaAs MOSFET. Excellent high frequency and power performances were also measured from the depletion-mode devices. These improvements are attributed to the excellent $Ga_2O_3(Gd_2O_3)$ oxide properties and novel processing techniques.

INTRODUCTION

GaAs metal-oxide-semiconductor field effect transistors (MOSFET's) potentially have advantages over Si-based MOSFET's for high-speed low-power logic integrated circuits (IC's) and monolithic microwave integrated circuits (MMIC's), due to the five times higher electron mobility in GaAs and the availability of semi-insulating GaAs substrates. In contrast to GaAs metal-semiconductor FET's (MESFET's) and high electron mobility transistors (HEMT's), which exhibit small forward gate voltages limited by the Schottky barrier heights, GaAs MOSFET's feature a much larger logic swing which gives a greater flexibility for digital IC designs. Furthermore, the current gain cutoff frequency (f_T) for depletion-mode MOSFET's is greater than that of MESFET's, due to the much smaller input capacitance C_{gs} for the insulating-gate devices [1]. In addition, the C_{gs} of MOSFET's is less RF drive dependent, resulting in a better linearity (lower intermodulation distortion) [1], [2]. Essentially, GaAs MOSFET's have the combined advantages of Si MOSFET's and GaAs MESFET's.

Extensive efforts have been devoted to the realization of a technologically viable GaAs MOSFET. However, the drain current drift and hysteresis hindered the deployment of GaAs MOSFET's due to the difficulty in fabricating an insulating film with insignificant bulk trapped charges and a low interfacial density of states on GaAs [3], [4]. Research efforts using an oxide mixture of $Ga_2O_3(Gd_2O_3)$ as the gate dielectric have led to the demonstration of inversion-mode *p*- and *n*-channel GaAs MOSFET's [5], [6], the attainment of a midgap interfacial density of states less than 10^{11} $cm^{-2}eV^{-1}$ [7], as well as undetectable hysteresis and negligible drain current drift [8]. In this article, we present the recent progress on GaAs MOSFET's using *in-situ* MBE-grown $Ga_2O_3(Gd_2O_3)$ as the gate dielectric.

Mat. Res. Soc. Symp. Proc. Vol. 573 © 1999 Materials Research Society

OXIDE GROWTH AND DEVICE FABRICATION

Depletion-Mode N-Channel GaAs MOSFET's

The $Ga_2O_3(Gd_2O_3)$/GaAs MOSFET structure was grown in a multi-chamber ultra-high vacuum (UHV) MBE system. After a 110-nm-thick silicon-doped ($2\times10^{17}cm^{-3}$) GaAs channel layer was grown on a (100) semi-insulating GaAs substrate, the wafer was then *in-situ* transferred to an arsenic-free chamber for oxide deposition under UHV. The $Ga_2O_3(Gd_2O_3)$ with thickness of 38-52nm was then deposited at a substrate temperature of 520°C using e-beam evaporation from a $Ga_5Gd_3O_{12}$ single crystal source. The interfacial roughness of as-grown $Ga_2O_3(Gd_2O_3)$/GaAs is typically 0.3-1.0nm. Further details of $Ga_2O_3(Gd_2O_3)$ film growth and micro-structural properties of $Ga_2O_3(Gd_2O_3)$/GaAs interfaces have been reported earlier [9], [10].

The fabrication process started with device isolation, which was achieved by oxygen implantation to define the active device area. The activation annealing was performed at 500 °C under helium gas ambiance. After the removal of oxide on the source and drain regions, ohmic contacts were formed by e-beam deposition of Ge/Ni/AuGe/Ag/Au and a 400 °C anneal under helium gas ambiance. Finally, the conventional Ti/Pt/Au metals were e-beam evaporated sequentially as the gate electrodes. A cross-sectional view of the depletion-mode *n*-channel GaAs MOSFET is schematically shown in Fig. 1(a).

Inversion-Mode N-Channel GaAs MOSFET's

First, a 300-nm-thick Be-doped ($5\times10^{16}cm^{-3}$) GaAs was grown and followed by the same oxide growth procedure for the D-mode devices. Si was used as the implanted dopant for the source and drain ohmic contacts. After the dopant activation annealing was performed, Ge/Ni/AuGe/Ag/Au were e-beam evaporated and annealed at 400 °C under helium gas ambiance to form ohmic contacts. Finally, the Ti/Pt/Au metals were deposited sequentially as the gate electrodes. The cross-sectional view of an inversion-mode *n*-channel GaAs MOSFET is schematically shown in Fig. 1(b).

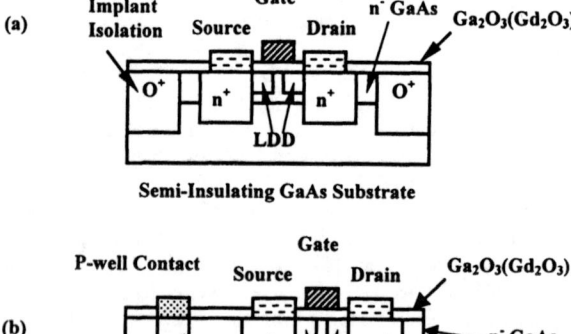

Fig. 1   Cross-sectional view of (a) depletion-mode *n*-channel GaAs MOSFET and (b) inversion-mode *n*-channel GaAs MOSFET.

DEVICE CHARACTERISTICS AND DISCUSSION

Depletion-Mode N-Channel GaAs MOSFET's

The drain *I-V dc* characteristics of a 20 μm×100 μm device is shown in Fig. 2(a). The device shows a clean pinch-off behavior at a threshold voltage of –2V with drain-source breakdown voltage higher than 25V. Fig. 2(a) also clearly demonstrates that the device can be operated at the accumulation mode up to +3V gate voltage. Higher forward gate voltages for accumulation mode operation have been achieved for devices with larger gate length. The device shows a maximum drain current (I_{max}) of 20 mA and a peak extrinsic transconductance (g_m) of 50 mS/mm. The flat transconductance profile in accumulation, as shown in Fig. 2(b), reveals the advantage of MOSFETs for linearity consideration.

Fig. 3 shows the *dc I-V* characteristics of a sub-micron gate length (0.8 μm×60 μm) device. The gate *I-V* characteristics of the 0.8 μm×60 μm device shows a symmetric breakdown behavior of the oxide gate, as shown in the insert. The drain-source breakdown voltage is as high as 24 V, which corresponds to a breakdown field of 6.3 MV/cm. The output characteristics measured with a curve tracer from the same device shows no substantial *I-V* hysteresis and drain current drift (Fig. 4). There is no significant disagreement between *I-V* curves measured under quasi-*dc* condition and measured by curve tracer operating at 120 Hz as reported in Reference [11]. In addition, the drain *I-V* characteristics are not sensitive to light, either. These phenomena indicate that insignificant bulk oxide trapped charges and low interfacial density of states have been achieved. The low doping concentration (4×10^{17} cm^{-3}) in the channel layer has caused a high source resistance, thus resulting in high on-resistance (r_{on}). The calculated effective channel mobility is 1100 cm^2/V·sec by use of the following formula [12]

$$\mu_{eff} = \frac{g_m L}{V_{d0} C_{ox} W},$$

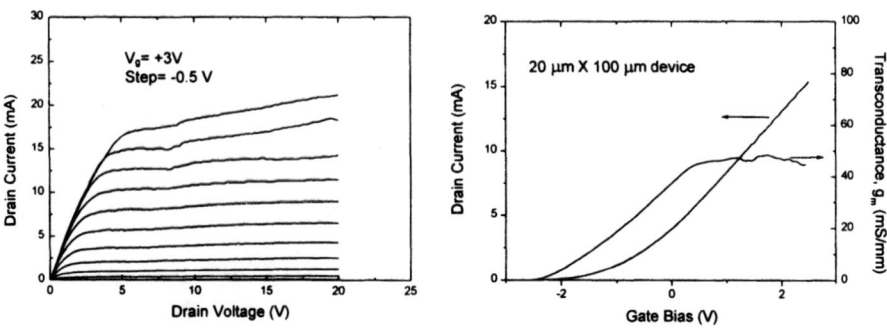

Fig. 2 (a) The drain *I-V* dc characteristics of a 20 μm×100 μm device. (b) The drain current and transconductance as a function of gate bias for a 20 μm×100 μm device.

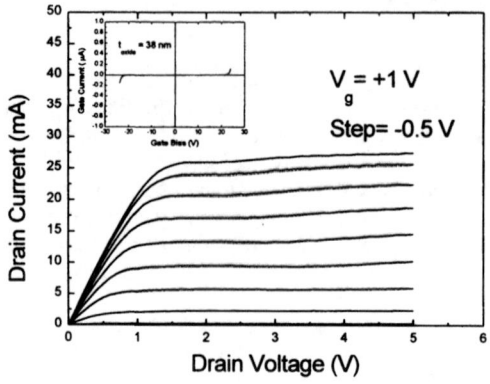

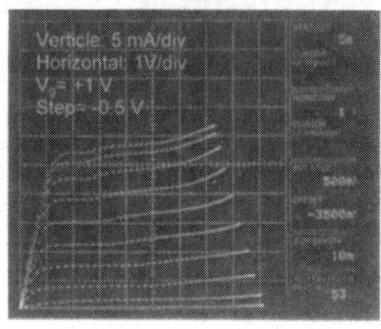

Fig. 3 The *dc I-V* characteristics of a 0.8 μm×60 μm Ga₂O₃(Gd₂O₃)/GaAs MOSFET. Insert shows the gate *I-V* characteristics of this device.

Fig. 4 The output characteristics measured by a curve tracer operating at 120 Hz of a 0.8 μm×60 μm GaAs MOSFET.

where L and W are the channel length and width, respectively, C_{ox} is the oxide capacitance per unit area, and V_{d0} is the drain voltage at saturation. The oxide dielectric constant of 14 was used in the calculation. The high oxide dielectric constant (~3.6 times higher than that of SiO₂) also reveals an advantage of using Ga₂O₃(Gd₂O₃) as the gate insulator because the larger current drive capability (higher g_m) may be achieved. The maximum drain current density (I_{max}) and the peak extrinsic transconductance (g_m) of the 0.8-μm gate-length device are 450 mA/mm and 130 mS/mm, respectively, as shown in Fig. 5. This figure also shows the drain current density as a function of gate bias in both forward and reverse sweep directions. The device shows negligible hysteresis indicating low mobile charge density and no charge injection.

The microwave results are shown in Fig. 6. The short-circuit current gain cutoff frequency (f_T) and the maximum oscillation frequency (f_{max}) are 17 GHz and 60 GHz, respectively.

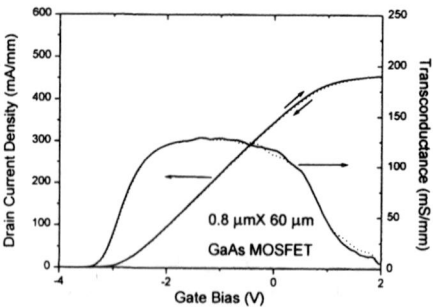

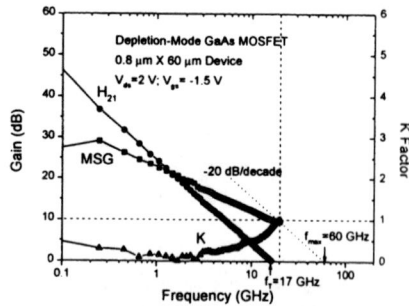

Fig. 5 The drain current density and transconductance as a function of gate bias for a 0.8 μm×60 μm device.

Fig. 6 Microwave performance of a 0.8 μm×60 μm Ga₂O₃(Gd₂O₃)/GaAs MOSFET measured at a drain voltage V_{ds} of 2 V and a gate voltage V_{gs} of −1.5 V.

Fig. 7 shows the long-term drain current drifting behavior of the MOSFET when the devices were biased at an extreme stress condition of V_{ds} = 4 V and V_{gs} = +1 V. No detectable short-term current drift was observed in a period shorter than 1 second after the device was turned on. The long-term drain current drift is less than 1.5% during operation for a period of over 150 hours. This result compares favorably with the best data reported previously [13]. The reference disclosed a 22% decrease of drain current over a period of 10^4 seconds (<3 hours) when the device was biased at V_{ds}=4 V and V_g=+0.5 V.

Power GaAs MOSFET's

Fig. 8 shows the drain current-voltage characteristics for a fabricated 1μm×2.4mm GaAs MOSFET measured from a curve tracer operating at 120Hz. The power transistor shows no significant hysteresis for such a high current operation. The low doping concentration in the channel layer gives high contact resistance, which in turn results in the high on-resistance r_{on} (5.1Ω·mm).

The power measurements were performed on a load-pull system with automatic tuners to measure the optimum load impedance for maximum output power tuning. When measured at 850MHz under 3V operation tuned for maximum output power, a peak power-added efficiency (PAE) of 45% was obtained from the 1μm×2.4mm device, as shown in Fig. 9. Under the same condition, a linear gain G_L = 20dB and a saturated output power P_{sat} = 23dBm were measured. When the drain bias increased to 5V the maximum PAE, G_L, and P_{sat} are 56%, 20dB, and 26.5dBm (power density=186mW/mm), respectively, as shown in Fig. 10. It is expected that the PAE can be further improved by reducing the source and drain contact resistances as described earlier. Larger devices (gate periphery up to 2cm) were also fabricated. However, the load impedances of those devices are too small (<10 Ω), which are beyond the tuning range of our load-pull system for optimum matching. Further improvement on PAE is expected by reducing the contact resistance and optimizing the layer structure.

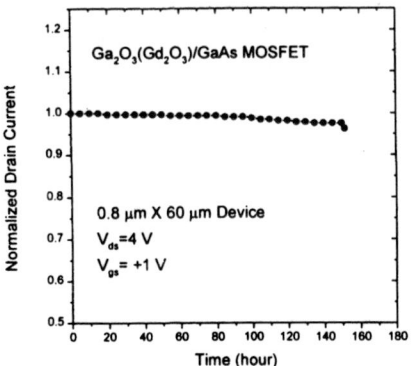

Fig. 7 The long-term drain current drifting behavior of a 0.8 μm×60 μm GaAs MOSFET.

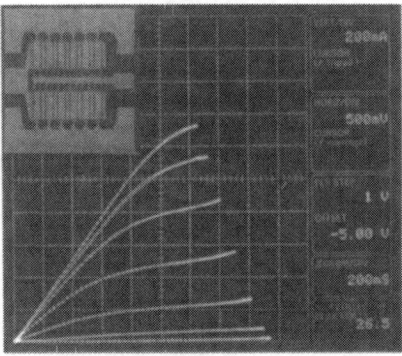

Fig. 8 The drain current-voltage characteristics for a 1μm×2.4mm Ga$_2$O$_3$(Gd$_2$O$_3$)/GaAs MOSFET measured from a curve tracer operating at 120Hz. The insert shows the photograph of a finished 1μm×2.4mm device.

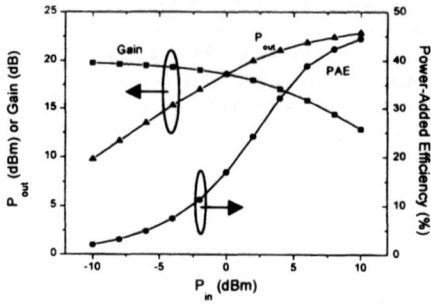

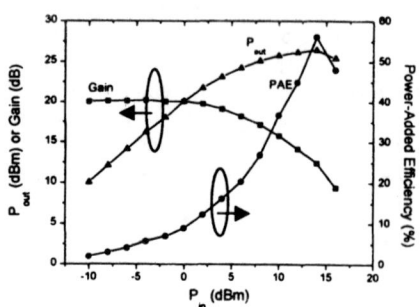

Fig. 9 Output power, power-added efficiency, and power gain against input power measured at 850MHz for a 1μm×2.4mm Ga$_2$O$_3$(Gd$_2$O$_3$)/GaAs MOSFET under 3V operation.

Fig. 10 Output power, power-added efficiency, and power gain against input power measured at 850MHz for a 1μm×2.4mm Ga$_2$O$_3$(Gd$_2$O$_3$)/GaAs MOSFET under 5V operation.

Inversion-Mode N-Channel GaAs MOSFET's

Fig. 11 shows the typical drain current vs drain voltage characteristics of the fabricated inversion-mode n-channel GaAs MOSFET's with gate channel length of 1μm. The gate voltage is varied in steps of 1V from 0 to 7V. The maximum current density and extrinsic transconductance are 30mA/mm and 4mS/mm, respectively. When biased at V_{ds}=4V and V_g=+3V, the f_T=1.2 GHz and f_{max}=2.6 GHz were measured from a 1 μm×200 μm device, as shown in Fig. 12. Currently, we believe the device performance is limited by the oxide/GaAs interface damage due to the high temperature activation annealing process (typically 800-850 °C for Si).

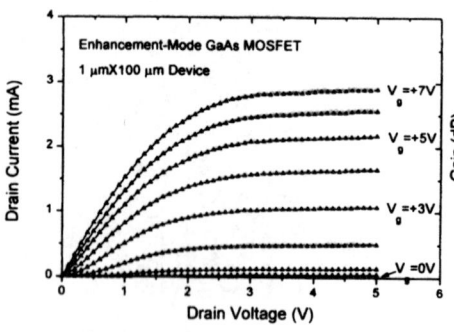

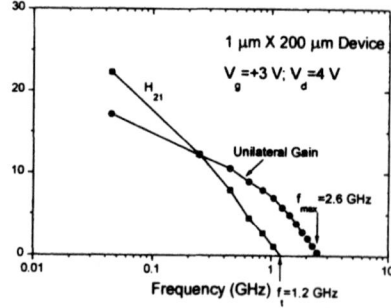

Fig. 11 Drain current vs drain voltage characteristics of the fabricated inversion-mode n-channel GaAs MOSFET's with gate channel length of 1μm.

Fig. 12 Microwave performance of a 1 μm×200 μm inversion-mode MOSFET measured at a drain voltage V_{ds} of 4 V and a gate voltage V_{gs} of +3 V.

224

CONCLUSIONS

We have successfully fabricated depletion-mode GaAs MOSFETs with undetectable hysteresis and negligible drain current drift I-V characteristics. For the time interval between 1 second and 150 hours, the device shows a drain current drift of less than 1.5%, indicating the excellent stability of the oxide and the oxide/GaAs interface. We also demonstrate the power performance at 850MHz cellular frequency of a 1μm×2.4mm GaAs MOSFET, which shows the power-added efficiency of 45% and 56% under 3V and 5V operation, respectively. Preliminary results on inversion-mode GaAs MOSFET's are reported. High current density (30 mA/mm), transconductance (4 mS/mm), and cutoff frequencies (f_T=1.2 GHz and f_{max}=2.6 GHz) have been achieved using *in-situ* MBE-grown $Ga_2O_3(Gd_2O_3)$ as the gate insulator.

REFERENCES

1. K.Yamaguchi, and S. Takahashi, IEEE Trans. Electron Devices, **ED-28**, p. 581 (1981).

2. P. D. Gardner, and S. Y. Narayan, SPIE Indium Phosphide and Related Materials for Advanced Electronic and Optical Devices, **1144**, p. 186 (1989).

3. T. Mimura and M. Fukuta, IEEE Trans. Electron Dev., **ED-27**, p. 1147 (1980).

4. D. S. L. Mui, Z. Wang, and H. Morkoç, Thin Solid Films, **231**, p. 107 (1993).

5. F. Ren, M. Hong, W. S. Hobson, J. M. Kuo, J. R. Lothian, J. P. Mannaerts, J. Kwo, Y. K. Chen, and A. Y. Cho, Tech. Dig. Int. Electron Devices Meet., p. 943 (1996).

6. F. Ren, M. Hong, W. S. Hobson, J. M. Kuo, J. R. Lothian, J. P. Mannaerts, J. Kwo, S. N. G. Chu, Y. K. Chen, and A. Y. Cho, Solid-State Electronics, **41**, p. 1751 (1997).

7. M. Passlack, M. Hong, and J. P. Mannaerts, Appl. Phys. Lett., **68**, p. 1099 (1996).

8. Y. C. Wang, M. Hong, J. M. Kuo, J. P. Mannaerts, J. Kwo, H. S. Tsai, J. J. Krajewski, Y. K. Chen, and A. Y. Cho, IEDM Technical Digest, San Francisco, CA, December 6-9, p. 67 (1998).

9. M. Hong, F. Ren, J. M. Kuo, W. S. Hobson, J. Kwo, J. P. Mannaerts, J. R. Lothian, and Y. K. Chen, J. Vac. Sci. Technol. B, **16**, p. 1398 (1998).

10. M. Hong, M. A. Marcus, J. Kwo, J. P. Mannaerts, A. M. Sergent, L. J. Chou, K. C. Hsieh, and K. Y. Cheng, J. Vac. Sci. Technol. B, **16**, p. 1395 (1998).

11. S.-J. Kim, J.-W. Park, M. Hong, and J. P. Mannaerts, IEE Proc.-Circuits Devices Syst., **145**, p. 162 (1998).

12. H. Becke, R. Hall, and J. White, Solid-State Electronics, **8**, p. 813 (1965).

13. Y. H. Jeong, K. H. Choi, and S. K. Jo, IEEE Electron Dev. Lett., **15**, p. 251 (1994).

Passivation of InGaAs/InP heterostructures

R. DRIAD, Z. H. LU[*], W. R. McKINNON, S. LAFRAMBOISE, S. P. McALISTER, P.J. POOLE, S. RAYMOND and S. CHARBONNEAU
Institute for Microstructural Sciences, National Research Council of Canada, Ottawa, Canada K1A 0R6.
*Department of Materials Science, University of Toronto, Toronto, Canada M5S 3E4.

ABSTRACT

In this study we report different surface treatments and device designs that can be used to improve the performance of InGaAs/InP heterostructure devices. The surface properties of InGaAs (100) after sulfur or UV-ozone passivation were investigated by photoluminescence and high energy-resolution X-ray photoelectron spectroscopy. The base leakage current and the dc current gain of InGaAs/InP heterostructure bipolar transistors (HBTs) have been used to evaluate the efficiency of the passivation treatments. Although these treatments successfully passivated large area HBTs, the improved device characteristics degraded after a dielectric was deposited by plasma enhanced chemical vapor deposition (PECVD) or even just with time. Nevertheless, we found a combined surface treatment that is successful even under PECVD deposition – a UV-ozone treatment that produces a sacrificial oxide that is then removed by HF. This approach will be contrasted with a different method based on an optimized HBT layer structure with a thin InP emitter. In this case, the thin layer of depleted InP from the emitter left on the extrinsic base passivates the surface, and no treatment is required.

INTRODUCTION

During the last decade various approaches, using wet or dry treatments, for the passivation of III-V compound semiconductor surfaces have been studied [1-5]. These surfaces are highly reactive to oxygen species; reaction with oxygen gives large surface recombination velocities and high densities of surface states. This is detrimental to device performance and reliability. Therefore, surface passivation of III-V compound semiconductors remains a critical issue for the fabrication of advanced micro-electronic and opto-electronic devices and integrated circuits.

Because of their high sensitivity to surface recombination, minority carrier devices such as lasers and heterojunction bipolar transistors (HBTs) can be used as test vehicles for passivation studies [2,6]. In compound semiconductor HBTs the base-emitter surface is crucially important for good device performance. Surface states and associated recombination centres can degrade the performance of HBTs, especially when the emitter-base junction area is small (emitter-size effect [7]).

InGaAs-based devices are cited to be better than their GaAs counterparts, because InGaAs has a lower surface recombination velocity than GaAs. Even so, problems of surface recombination have been also reported in InGaAs/InP heterostructures [8-11]. Surface currents have been correlated with leakage paths between the InP emitter side-wall and the base contact [11]. Using a low temperature deposition technique (ECR-PECVD), Fukano et al. [10] found that although the surface currents could be suppressed with a treatment of HF before dielectric deposition, annealing at 200°C regenerated the surface currents, indicating poor thermal stability of the oxide. Recently, Kikawa et al. [12] reported that the Fermi level of an InP

227

surface after SiO_2 deposition shifts closer to the conduction band edge, resulting in a surface leakage path. Thus it appears that the degradation mechanisms in dielectric-coated electronic devices are not yet fully understood.

In this study, we discuss the passivation of InGaAs (100) surfaces and its application to InGaAs/InP heterostructures. We illustrate the effects of different treatments on the semiconductor surfaces through high energy-resolution x-ray photoelectron spectroscopy (XPS) and photoluminescence (PL) measurements. First, we discuss the passivation effects of $(NH_4)_2S$ on InGaAs (100) surfaces. This treatment has been demonstrated to reduce the surface recombination velocity and enhance the photoluminescence intensity of III-V surfaces [13,14]. Sulfur treatment is now being implemented in the fabrication of various devices [15]. Then we discuss the effectiveness of using a simple dry UV-ozone process to passivate the InGaAs surfaces. This process has been reported to produce uniform thin stoichiometric native oxides [16], and results in good electrical characteristics in MIS diodes [17].

We have integrated the passivation treatments in the fabrication process of InGaAs/InP HBTs, and observed a dramatic increase in the current gain of large-area devices. Unfortunately, the improvement in device characteristics degrades after PECVD dielectric deposition or even just with time. However, we found a surface treatment that works even under PECVD deposition—using the UV-ozone treatment to produce a sacrificial oxide, then removing that oxide by HF just before dielectric deposition. Finally, we consider an alternative to these surface treatments—leaving a depleted InP layer (ledge) on the extrinsic base to passivate the surface.

EXPERIMENT

The chemical compositions of undoped InGaAs (100) surfaces before and after sulfur or UV-ozone treatment were studied using XPS. The experiments were carried out in a PHI 5500 system equipped with monochromated Al K_α source and a hemispherical electron-energy analyzer. As 3d, In 3d5/2, Ga 2p3/2, and S 2p core levels were recorded and analyzed. The effect of the passivation was also investigated using PL measurements performed at room temperature with a 632.8 nm He-Ne laser. The luminescence signal was analyzed using a FTIR (Fourier transform infra-red) spectrometer coupled with an InGaAs detector. To avoid the dominance of bulk recombinations, the PL measurements were done on thin ($\sim$ 50 nm) undoped InGaAs epilayers grown by chemical beam epitaxy (CBE) on (100) oriented semi-insulating InP substrates. Tri-methylindium and tri-ethylgallium were used as group-III precursor sources using high purity hydrogen as a carrier gas. The group V source materials were pure arsine and phosphine. The growth temperature was $490^\circ C$ as measured by an optical pyrometer.

Layer structures of InGaAs/InP for HBTs were grown by gas source molecular beam epitaxy (GSMBE) by Tutcore Ltd. on (100) oriented semi-insulating InP substrates (Table I). The main difference between the two designs is the thickness of the InP layer in the emitter (150 nm in structure A, 30 nm in structure B). Both designs have a Be-doped InGaAs base layer, a (p^+/n^+) dipole at the InGaAs/InP collector heterojunction, and a thin InP collector. The dipole-doped composite-collector is used to prevent the electron-blocking effect induced by an abrupt InGaAs/InP heterojunction, and to improve the current collection [18]. The n- and p-type dopants were Si and Be, respectively.

Conventional mesa structure transistors were fabricated using selective wet chemical etchants ($H_3PO_4:H_2O_2:H_2O$ and $H_3PO_4:HCl$) and lift-off processes. Non-alloyed TiPtAu metallization was used simultaneously for both n- and p-type ohmic contacts for devices fabricated from structure A. PtTiPtAu metallization was used for devices from structure B, since this metallization is better for contacting the base through a thin InP layer. For the contribution of the passivation treatments to be separated from the effects of the PECVD deposited dielectrics, large area devices allowing direct probing of the transistors were fabricated both with and without dielectric films. The emitter-base junction areas were typically in the range 10x10 - 70x90 μm^2. The effect of PECVD-dielectric deposition was also investigated on smaller devices that were coated with SiO_2 and contacted by high frequency pads through via holes in the oxide.

Table I: Layer structures for standard (A) and thin-emitter (B) InGaAs/InP HBTs

(A)

Layer	Material	Thickness (nm)	Doping (cm^{-3})	Dopant
Cap	InGaAs	100	2E19	Si
	InP	60	2E19	Si
Emitter	InP	90	3E17	Si
Set-back	InGaAs	10	Undoped	-
Base	InGaAs	40	1-3E19	Be
Spacer	InGaAs	40	5E15	Si
p-Dipole	InGaAs	10	1E18	Be
n-Dipole	InP	10	1E18	Si
Collector	InP	290	5E15	Si
Subcoll.	InGaAs	450	5E18	Si
Buffer	InP	50	Undoped	-
Substrate	InP		S.I	

(B)

Layer	Material	Thickness (nm)	Doping (cm^{-3})	Dopant
Cap	InGaAs	100	2E19	Si
	InGaAs	70	3E17	Si
Emitter	*InP*	*30*	*3E17*	*Si*
Set-back	InGaAs	10	Undoped	-
Base	InGaAs	40	5E19	Be
Spacer	InGaAs	40	5E15	Si
p-Dipole	InGaAs	10	1E18	Be
n-Dipole	InP	10	1E18	Si
Collector	InP	290	5E15	Si
Subcoll.	InGaAs	450	5E18	Si
Buffer	InP	50	Undoped	-
Substrate	InP		S.I	

For sulfur passivation, samples were dipped in the $(NH_4)_2S$ solution for different periods of time at room temperature, then rinsed in acetone, and washed in methanol. The UV-ozone exposures were carried out at room temperature for times from 5 to 60 minutes. After treatment, the samples were then transferred immediately to characterization chambers or to sample holders for device measurements.

RESULTS

Sulfur passivation

In this section, we report the improved surface properties of undoped InGaAs epilayers and the resulting performance of InGaAs/InP HBTs after $(NH_4)_2S$ passivation. (No capping layers of PECVD oxide were used in this aspect of the work.)

Figure 1 shows various core levels recorded on undoped InGaAs samples with and without a 30 sec $(NH_4)_2S$ treatment. The data show that the air-exposed surface (shown as upper curves) is covered with a few nm thick layer of mixed As, Ga and In oxides. These native oxides are known to produce non-radiative recombination defects. The core level spectra recorded from the S-treated surface (shown in lower curves) indicate that these surface oxides were completely removed after the S-treatment. The peak shapes of As 3d, Ga 2p3/2, and In 4d5/2 core levels recorded from S-treated InGaAs (100) surface are virtually identical to those recorded from S-treated GaAs (100) and InP (100) surfaces. We concluded in our previous studies [19] that S forms ordered Ga-S-Ga and In-S-In bridge bonds along [011] on GaAs (100) and InP (100) surfaces, respectively. The present XPS data thus suggest that S is bonded to both Ga and In with possible bond configurations such as Ga-S-In, Ga-S-Ga and In-S-In.

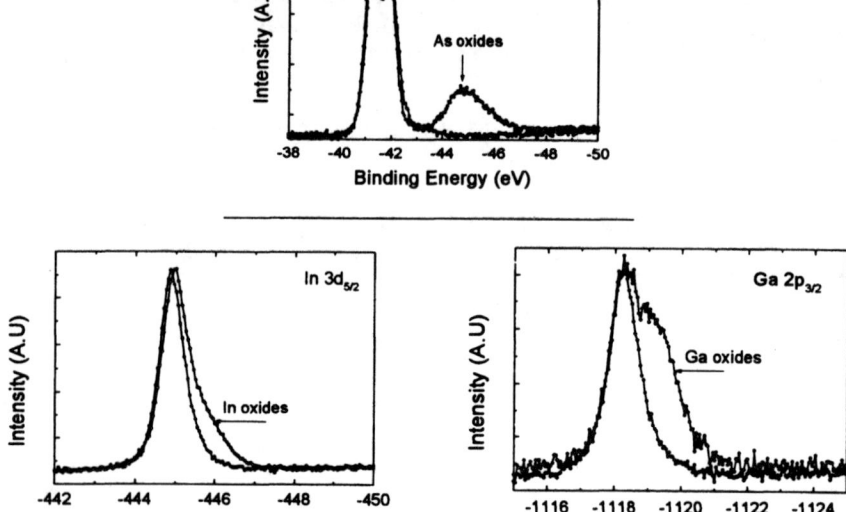

Fig. 1: As3d, In3d5/2 and Ga2p3/2 core level XPS spectra recorded from untreated (upper curves) and sulfur-treated (lower curves) InGaAs surfaces. This shows the elimination of the surface oxides after sulfur passivation.

The beneficial effect of sulfur passivation was also confirmed by room temperature PL measurements. Figure 2 shows the PL peak intensity for undoped InGaAs epilayers before and after sulfur treatment for different times. In agreement with previous studies [14], a dramatic increase in integrated luminescence intensity was observed after sulfur passivation. Compared to an untreated sample, the PL intensity increased by ~40 times after a treatment of 1 minute. The luminescence improves further with treatment time, but saturates after 2 minutes. The improvement in PL intensity is attributed to the elimination of native surface oxides after sulfur treatment, as shown by our XPS data. According to Skromme et al. [14] this treatment substantially reduces the density of pinning and/or non-radiative recombination centres on exposed surfaces.

The mechanism of surface passivation by sulfur treatment can be summarized as follows: a slight etching of the semiconductor surface removes the native oxides as well as the surface and near-surface defects; then the surface states are passivated by S atoms. Given these results, we expect the sulfur treatment to improve the device performance of InGaAs/InP HBTs. Figure 3 shows typical variation of the current gain β as a function of collector current (I_C) for unpassivated (dotted line) and sulfur-passivated (solid line) InGaAs/InP HBTs with an emitter junction area of 20x30 μm^2. As expected, the sulfur treatment improves β over several decades of I_C, by an order of magnitude at low I_C. For the sulfur-passivated devices, β is above 50 for a wide range of I_C, and it depends less on I_C than for the unpassivated devices.

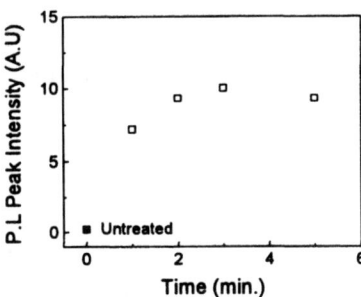

Fig. 2: Photoluminescence peak intensity from InGaAs samples before (untreated) and after different sulfur duration treatments.

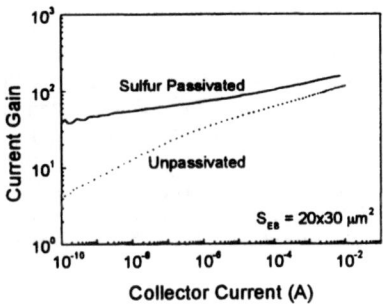

Fig. 3: Current gain as a function of collector current of InGaAs/InP HBTs with (solid line) and without (dashed line) sulfur passivation.

To investigate the emitter size-effect in the sulfur passivated InGaAs/InP HBTs; we measured β as a function of collector current density J_C for various devices with different emitter-base junction area. The relationship between β and the emitter size can be expressed approximately as [20]:

$$1/\beta = \{(J_{Bulk} + J_{Bscr} + J_{Bp}) + K_{Bsurf}.(P/A)\}/J_C \qquad (1)$$

where J_{Bulk} J_{Bscr} and J_{Bp} are the base bulk recombination current density, the base-emitter space charge recombination current density and the base-to-emitter back-injected current density, respectively. K_{Bsurf} is the surface recombination current density divided by emitter periphery P, and A is the emitter area.

Figure 4 shows $1/\beta$ versus P/A for sulfur-passivated and unpassivated InGaAs/InP HBTs at $J_C = 100$ A/cm^2. From equation (1), the slope of $1/\beta$ versus P/A is K_{Bsurf}/J_C. The value of K_{Bsurf} from Fig. 4 is considerably higher in unpassivated devices, and this, together with the strong variation of β with I_C (Fig. 3), reflects the larger surface recombination current in unpassivated devices. In contrast, $1/\beta$ for sulfur passivated HBTs show almost no variation with P/A, demonstrating that surface recombination current is negligible. This is important for the fabrication of small size HBTs for good high frequency performance while still maintaining high current gains.

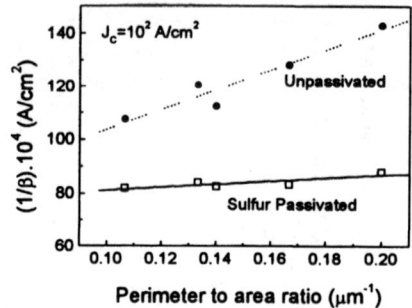

Fig. 4: Variation of (1/β) as a function of perimeter to area ratio (P/A) for unpassivated (solid circles) and sulfur passivated (open squares) InGaAs/InP HBTs with different emitter-base junction area.

The reduction in leakage current of the sulfur-treated devices indicates that the electronic properties of the surface have been modified and the number of non-radiative recombination centers reduced [14]. Unfortunately, this treatment, like all chemical passivation treatments, has limitations of uniformity and reproducibility. Since device reliability, as well as reproducibility, and uniformity of device characteristics over a wafer, are important requirements for advanced integrated circuits (ICs), dry passivation treatments are preferred and actively investigated [21,22]. In the second approach of this study, we present a simple UV-ozone process for the passivation of InGaAs and InGaAs/InP heterostructures.

<u>UV-ozone passivation</u>

The combination of UV-irradiation and ozone has been extensively used in device processing and has been also reported to produce good interface electronic properties [16,17,23]. Thus we investigated the chemical structure and the physical thickness of UV-ozone oxidized InGaAs surfaces by XPS. Based on Ga 3d, Ga 2p, As 3d, and In $3d_{5/2}$ core-level studies, we found that the UV-ozone treated InGaAs (100) surface is covered with As_2O_3, As_2O_5, Ga_2O_3, and In_2O_3. [The oxide thickness was also measured by XPS using a methodology described in Ref. 24.]

Fig. 5a shows the thickness of the surface oxide as a function of UV-ozone time, for oxide growths on InGaAs (100), GaAs (100) and InP (100) surfaces. The detailed discussions for GaAs (100) and InP (100) were reported elsewhere [24]. As expected, the growth rate on InGaAs (100) surface falls between that on GaAs (100) and InP (100) surfaces. We also found that the UV-ozone treatment not only produces a surface oxide film, which removes surface and near-surface defects that might be induced during device processing, but also restores the stoichiometric composition at the semiconductor surface. To study this, we investigated surface compositions of GaAs (100) and InP (100) as a function of UV-ozone treatment. The relative sensitivity factors of Ga 3d, As 3d, In 3d, and P 2p were obtained by measuring in-situ cleaved (110) surfaces. Fig. 5b, which shows the bulk Ga/As and In/P ratios as a function of UV-ozone time, indicates that a stoichiometric surface is restored after 10 minutes of treatment. This exposure time results in the formation of about 15-20 Å of UV-oxide; longer exposures (>20 min.) produce non-stoichiometric surfaces.

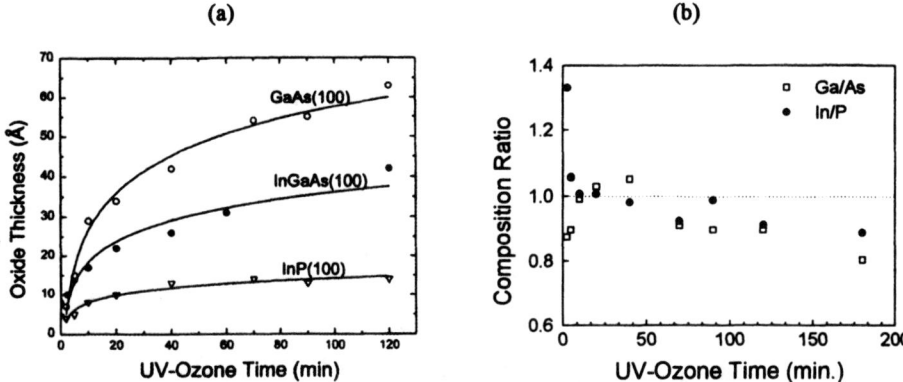

Fig. 5: Surface oxide growth (a) and semiconductor composition ratio near the surface (b) as a function of UV-ozone time.

The UV-ozone treatment was applied to large-area InGaAs/InP HBTs using different exposures. These devices received only the UV-ozone treatment and were not capped with PECVD oxide. In agreement with the XPS data, the best results were obtained for the 10-minute exposure, whereas longer exposures degrade the current gain probably because of the non-stoichiometric surfaces [25]. This process was found to be uniform and highly reproducible. Figure 6 shows the Gummel plots for a large area (20x30 μm^2) InGaAs/InP HBT before (dashed lines) and after (solid lines) a UV-ozone treatment of 10 minutes. The curves for I_C are identical, and are not affected by the UV-ozone treatment. In contrast, I_B is reduced at low bias. This reduction drastically improves β, by almost an order of magnitude at low I_C (from 17 to 100 at 1 nA). The values of the base ideality factor (n_B) decreased from 1.5 to 1.36 after UV-ozone treatment.

In the fabrication process of small area (high frequency) devices, a dielectric deposited by CVD-based techniques is needed for device protection and isolation. However, it has been reported by several authors [10,11,25,26] that the electrical properties of InP-based HBTs degrade drastically when dielectric films (SiO_2, SiN) are deposited by CVD-based techniques. To assess the passivation effect of UV-ozone, we integrated this treatment into the fabrication process of high frequency devices. Unfortunately, we found that the improvements from the UV-ozone are lost when the dielectric is added on the top surface.

Figure 7 shows the Gummel plots of InGaAs/InP HBTs before (dashed line) and after (dotted line) PECVD-SiO_2 deposition, as well as the Gummel plots of a device treated by UV-ozone before SiO_2 deposition (solid line). The curves for I_C are all identical, and are not affected by the dielectric deposition. In contrast, I_B increases at low bias after dielectric deposition (curve labelled as Ref.+SiO_2), drastically degrading β. The sample treated with UV-ozone (Ref.+UV-ozone+SiO_2) shows almost the same behavior. Compared with uncapped HBTs, the devices capped with dielectric show higher values of n_B, close to 2. This degradation is worse, as expected, when the device size is reduced.

233

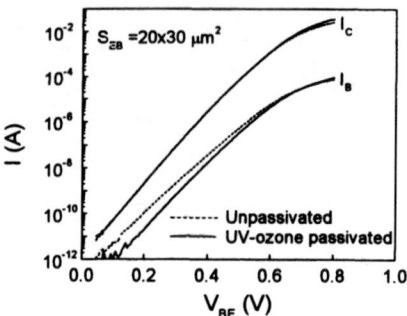

Fig. 6: Gummel plot of InGaAs/InP HBTs with (solid lines) and without UV-ozone treatment (dashed lines).

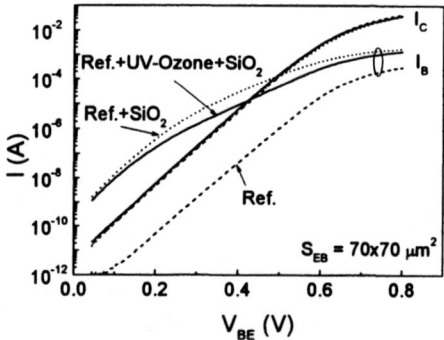

Fig. 7: Gummel plots of InGaAs/InP HBTs before (dashed line), and after SiO₂ deposition (dotted line). The third Gummel plot labelled as Ref.+UV-ozone+SiO₂ (solid line) was obtained for a device including the UV-ozone treatment before PECVD-SiO₂ deposition.

Nevertheless, the degradation process was found to be reversible, and the initial characteristics are fully restored, if not slightly improved, after the dielectric film is removed in a solution of buffered hydrofluoric acid. This may indicate that the SiO₂ deposition is reactivating defect species previously passivated by the UV-ozone treatment. Since the surface of the semiconductor was found stoichiometric after UV-ozone process (Fig. 5b), these reactivated species are presumably in the surface oxide formed by the UV-ozone treatment, rather than in the semiconductor. If so, then removing the surface oxide formed by the UV-ozone should reduce degradation of the device characteristics associated with PECVD, as we verified. Specifically, we found that a UV-ozone process followed by an HF etch (to remove these defects) before the PECVD deposition produces high-frequency InGaAs/InP HBTs with good electrical characteristics even after the PECVD coating. In effect, the oxide from the UV-ozone is being used as a sacrificial layer. Although the HF treatment may have additional benefits, such as decreasing the surface state density by compensating free dangling bonds with hydrogen or fluorine, the UV-ozone step remains an essential part of the process, since HF etching alone followed by PECVD deposition results in poor characteristics.

Figure 8 shows the Gummel plots for small InGaAs/InP HBTs with $5 \times 10 \ \mu m^2$ emitter-base junction area, with (10 min.) and without UV-ozone/HF treatments. After treatment, I_B is an order of magnitude smaller at a collector current I_C of 20 μA, than before treatment. This reduction in I_B of the treated devices, together with the decrease in n_B from 1.7 to 1.5, implies that non-radiative recombination is reduced for the surface states located at the emitter side-walls and on the extrinsic base. Values of β of 10 have been measured on the treated devices at I_C as low as 10 nA. Without the treatment β is less than unity for I_C below 20 μA.

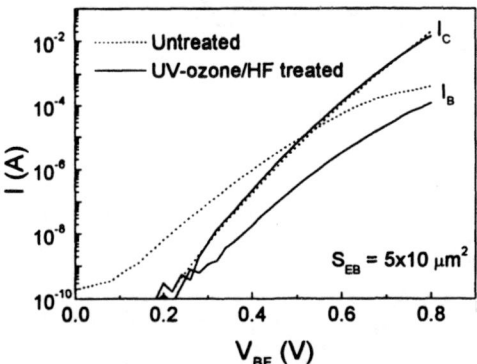

Fig. 8: Gummel plots for small-area high-frequency InGaAs/InP HBTs with (solid lines) and without (dashed lines) a UV-ozone/HF treatment before PECVD-SiO$_2$ coating.

There are, however, some drawbacks to this procedure. The HF etch might cause the metal contacts to corrode, especially Ti-based contacts. This would undoubtedly affect the device reliability. Moreover, to be effective, the process of UV-ozone/HF might have to be repeated several times, depending on the extent of the defects created during device processing. An alternative approach, which avoids exposing the InGaAs base layer to air during processing, and so eliminates any contamination of the base surface, is discussed in the next section.

Ledge and Self-passivation

In this section, we discuss a passivation method that avoids the need for the surface treatments discussed above. In this approach, a fully depleted emitter layer (guard ring) on the extrinsic base acts as a passivation layer to prevent surface recombination on the base. In GaAs-based HBT technology, the depleted emitter ledge has been an effective approach for the passivation of the base surface [7,27]. This approach requires an optimized layer design, since the thickness of the guard ring is hard to control by etching the InP emitter (even harder than in the AlGaAs/GaAs system). All devices discussed in this section were fabricated using the thin-emitter structure (B) described in Table I. Using this structure, the contact to the base can be either through a via hole opened in the InP layer just before metallization, thus creating a ledge around the emitter periphery (self-passivated structure, Fig. 9b), or by a contact diffused through the thin InP layer (self-passivated structure, Fig. 9c).

(a) Unpassivated

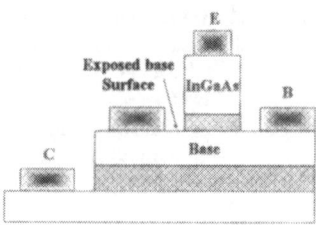

(b) Ledge-passivated

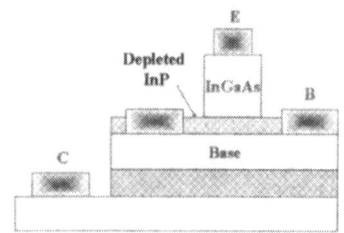

(c) Self-passivated

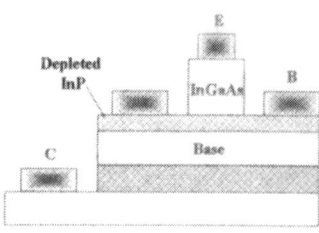

Fig. 9: Schematic diagrams for unpassivated (a), ledge-passivated (b), and self-passivated (c) thin-emitter InGaAs/InP HBTs.

The passivation effect of this approach was first checked by PL measurements on samples of (100) InGaAs with an InP cap layer. A similar sample where the InP cap layer was etched just before the PL measurement was used as a reference. Typical PL spectra are shown in Fig. 10. Etching the InP layer (curve-b) reduces the PL intensity by a factor of 2.5 from the intensity of the passivated sample (curve-a). This indicates that important non-radiative recombinations occur at the InGaAs surface after the InP layer is removed.

To assess the effectiveness of the InP layer in passivation, we fabricated high-frequency devices that include a PECVD deposition step. Figure 11 shows β as a function of I_C for thin-emitter InGaAs/InP DHBTs with an emitter-base junction area of 10×10 μm^2, with and without the extrinsic InP passivation layer. In the unpassivated devices (dashed line), β is degraded after PECVD dielectric film deposition, and drops to less than unity for I_C below 4.5 μA. Even at 10^3 A/cm^2, β of the unpassivated device is only 1/3 the value for the ledge-passivated and self-passivated devices. Moreover, n_B decreases from 1.74 for an unpassivated device to 1.45 for ledge or self-passivated devices).

The improved performance of the passivated devices is associated with a drastic decrease of I_B at low I_C, indicating a reduction of the current from non-radiative recombinations at the states located at the surface of the extrinsic base. This is also supported by the decrease of the PL intensity obtained from uncapped InGaAs as compared to an InP-passivated InGaAs (Fig. 10). These results show that the presence of the InP passivation layer avoids the contamination of the base surface during processing and prevents the interaction between the dielectric film and the InGaAs base.

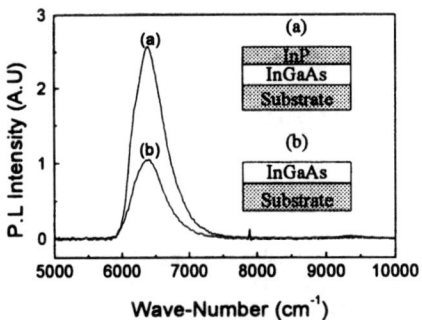

Fig. 10: Photoluminescence spectra from InGaAs samples with (a) and without (b) an InP cap layer.

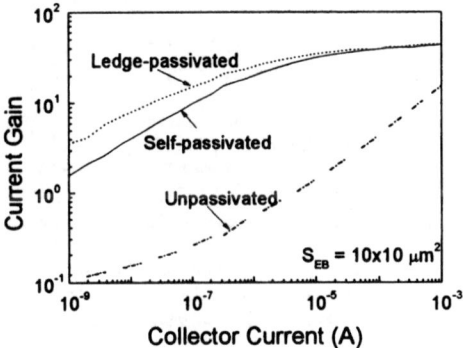

Fig. 11: Current gain variation as a function of collector current for unpassivated, ledge- and self-passivated thin-emitter (structure-B) InGaAs/InP DHBTs. The devices did not receive any treatment before PECVD oxide deposition.

CONCLUSIONS

In summary, we have studied surface passivation of InGaAs/InP heterostructures using different techniques. Sulfur passivation reduces the number of defect centres and surface recombination sites, by eliminating native oxides and slightly etching the surface and also passivates dangling bonds. UV-ozone also reduces recombination sites by restoring the stoichiometry of the semiconductor surface.

We also found that surface recombination caused by PECVD deposition in InP-based HBTs can be reduced substantially by a two-step process combining UV-ozone and HF-etching. The UV-ozone process "traps" surface and near-surface defects in a sacrificial oxide layer, and the HF etch removes the oxide (and thus those defects), before the PECVD deposition. This treatment is an effective, simple cleaning and passivating technique, which can be performed at low temperature with little damage.

Finally, we have shown that a depleted thin ledge of InP between the base contact and the emitter side-wall (guard ring) is an effective technique for the passivation of the extrinsic base surface. This approach also allows the deposition of dielectric films using conventional CVD-based techniques, without degrading the electrical device characteristics. Since the extrinsic base is passivated without extra treatment, this approach is probably the most effective passivation method for InP-based HBTs.

REFERENCES

1. W.E. Spicer, I. Lindau, P. Skeath, and C.Y. Su: J. Vac. Sci. & Technol. 17 (1980) p.1019
2. C.J. Sandroff, R.N. Nottenburg, J.C. Bischoff, and R. Bhat: Appl. Phys. Lett. 51 (1987) p.33
3. M. Hong, M. Passlack, J.P. Mannaerts, T.D. Harris, M.L. Schnoes, R.L. Opila, and H.W. Krautter: Sol. State Elect. 41 (1997) p.643
4. T. Hashizume, K. Ikeya, M. Mutoh, and H. Hasegawa: Appl. Surf. Sci. 123/124 (1998) p.599
5. H.M. Dauplaise, K. Vaccaro, A. Davis, G.O. Ramseyer, and J.P. Lorenzo: J. Appl. Phys. 80 (1996) p.2873
6. N.K. Dutta, W.S. Hobson, G.J. Zydzik, J.F. de Jong, P. Parayanthal, M. Passlack, and U.K. Chakrabarti: Electron. Lett. 33 (1997) p.213
7. O. Nakajima, K. Nagata, H. Ito, T. Ishibashi, and T. Sugeta: Jpn. J. Appl. Phys. 24 (1985) p. L-596
8. B. Jacobs, M. Emmerling, A. Forchel, I. Gyuro, P. Speier and E. Zielinski: Jpn. J. Appl. Phys. 32 (1993) p. L-173
9. Ren, F., Buckley, D.N., Lee, K.M., Peraton, S.J., Bartynski, R.A., Constantine, C., Hobson, W.S., Hamm, R.A., and Chao, P.C. Solid State Electronics, 38 (1995) p.2011
10. Fukano, Y. Takanashi and M. Fujimoto: Jpn. J. Appl. Phys. 32 (1993) p. L-1788
11. A. Ouacha, M. Willander, B. Hammarlund and R.A. Logan: J. Appl. Phys. 74 (1993) p.5602
12. Kikawa, T., Takani, S., Masuda, H. and Tanoue: T. Int. Conf. on Indium Phosphide and Related Materials, 1998, p.76.
13. R. Iyer, R.R. Chang, and D.L. Lile, Appl. Phys. Lett. 53 (1988) p.134
14. B.J. Skromme, C.J. Sandroff, Y. Yablonovitch, and T. Gmitter, Appl. Phys. Lett. 51 (1987) p.2022
15. U. Mohideen, W.S. Hobson, S.J. Pearton, F. Ren, and R.E. Slusher, Appl. Phys. Lett. 64 (1994) p.1911
16. S. Ingrey: J. Vac. Sci. & Technol. A-10 (1992) p.829
17. A.A. Iliadis: Electron. Lett. 25 (1989) 572 ; S. Loualiche, A. Ginoudi, H. L'Haridon, M. Salvi, A. Le Corre, D. Lecrosnier and P.N. Favennec: Appl. Phys. Lett. 54 (1989) p.573
18. S. P. McAlister, W.R. McKinnon, R. Driad and A.R. Renaud: J. Appl. Phys. 82 (1997) p.5231
19. Z.H. Lu, and M.J. Graham: Phys. Rev. B 48 (1993) p.4604 ; Z.H. Lu, M.J. Graham, X.H. Feng, and B.X. Yang: Appl. Phys. Lett. 60 (1992) p. 2773
20. N. Hayama, and K. Honjo: IEEE Electron. Dev. Lett. 11 (1990) p.388
21. Y.L Chang, I.H. Tan, C. Reaves, J. Merz, E. Hu, and S. DenBaars: Appl. Phys. Lett. 64 (1994) p.2658
22. L.S. How Kee Chun, J.L. Courant, A.Falcou, P.Ossart, and G. Post: Microelectron. Eng. 36 (1997) p.69
23. J.R. Vig: J. Vac. Sci. & Technol. A-3 (1985) p.1027
24. Z.H. Lu, B. Bryskiewicz, J. McCaffrey, Z. Wasilewski and M.J. Graham: J. Vac. Sci. & Technol. B-11 (1993) p.2033
25. R. Driad, Z.H. Lu, S. Laframboise, D. Scansen, W.R. McKinnon and S.P. McAlister: To be published in Jpn. J. Appl. Phys. 38 (1999)
26. D. Caffin, L. Bricard, J.L. Courant, L.S. How Kee Chun, B. Lescault, A.M. Duchenois, M. Meghelli, J.L. Benchimol and P. Launay: Int. Proc. of InP and Related Materials, Massachusets (1997) p.637
27. Malik, R.J., Lunardi, L.M., Ryan, R.W., Shunk, S.C., and Feuer, M.D.: Electron. Lett. 25 (1989) p. 1175

Surface Passivation of GaAs Power FETs

Tsuyoshi Tanaka, Hidetoshi Furukawa, Kazuo Miyatsuji, and Daisuke Ueda

Electronics Research Laboratory, Matsushita Electronics Corporation
Yagumonakamachi 3-1-1, Moriguchi, Osaka 570-8501, Japan

ABSTRACT

The surface passivation of GaAs power FET has been investigated. Intermodulation distortion of GaAs power FET was found to be affected by frequency dispersion which originates from electron trap at the surface in the vicinity of the gate. There are two ways to suppress the frequency dispersion. One is reducing electron trap itself by using surface passivation, the other is making surface insensitive to the surface trapping effect. We found the FET with undoped InGaP layers on the n-GaAs channel is free from surface trapping effects. The undoped InGaP layer acts as an ideal passivation layer for the channel, since it shows only 2% frequency dispersion of drain current at 1MHz compared to DC condition.

INTRODUCTION

Digital communication systems are gaining in popularity since they can support more subscribers by using TDMA and CDMA than analog one. In the handsets for these systems, GaAs FETs are widely used as the power amplifier. Because they can achieve inherently higher efficiency and higher gain than silicon bipolar transistors or MOSFETs. In view of designing power FETs for these digital systems, high linearity that can achieve low distortion characteristics is the most important feature in order to suppress bit-error-rate. For this purpose, improvement of linearity has been investigated by the optimization of the impurity profile such as pulse doping [1]. However, distortion characteristic has not sufficiently improved as predicted by dc characteristics. They have been reported that the poor distortion characteristic was caused by frequency dispersion which originated from electron trapping at the surface of the gate recess [2-6].

In this paper, we discuss surface passivation of GaAs power FET that can achieve low distortion characteristic and high power performance. First, the frequency dispersion of GaAs FET was analyzed. The Vgs-Ids characteristics at high frequency differed from that at dc. The non-linearity of the Vgs-Ids characteristics became large at high frequency. A simulation was carried out in order to clarify the relationship between the nonlinearity of the Vgs-Ids characteristics and intermodulation distortion. It was found that the increase in nonlinearity of the Vgs-Ids characteristics due to the frequency dispersion is a cause of the increase of the intermodulation distortion. Since the frequency dispersion of GaAs FET originates from the surface trapping states, the intermodulation distortion can be improved by reduction of the surface state. There are two ways to suppress the frequency dispersion. One is reducing electron trap itself by using surface passivation, the other is making surface insensitive to the surface trapping effect. We found the FET with undoped InGaP layers on the n-GaAs channel is free from surface trapping effects. The undoped InGaP layer acts as an ideal passivation layer for the channel, since it shows only 2% frequency dispersion of drain current at 1MHz compared to DC condition. Because of this small frequency dispersion, the FET with the gate width of 18mm exhibited 1.5dB larger output power than that of the FET without undoped InGaP layer.

Moreover, in π/4 shift-QPSK modulation, the FET exhibited 10dB smaller adjacent channel leakage power than conventional FET at output power of 30dBm.

FREQUENCY DISPERSION OF GaAs MESFET

The frequency dispersion of the FET characteristics under a small signal can be explained as follows [2, 3, 5].   It is well known that number of trapping states exist on the surface of GaAs.   The electrons trapped by this surface states form the depletion layer.   As the applied gate voltage is modulated, the charges of the surface states are varied.   So that the thickness of the surface depletion layer in the vicinity of the gate is modulated as shown in Fig. 1 (a).   When the modulation frequency is higher than the time constant of the capture and emission of the electrons, the charges of the surface states cannot follow.   In general, the time constant for the emission of the electrons is larger by several orders of magnitude than the capture of the electrons [4].   For this reason, only the electron capture occurs at high frequency resulting in the surface depletion widening in the vicinity of the gate as shown in Fig. 1 (b).   As a result, the channel current is not controlled by the gate voltage but determined by the depleted channel in the vicinity of the gate.

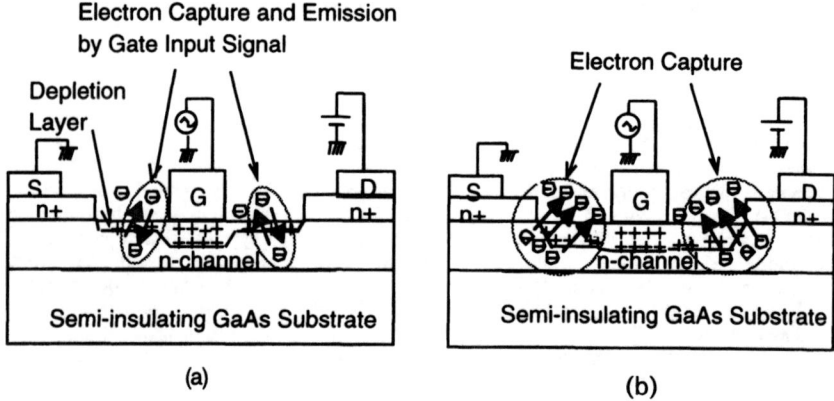

Fig. 1 Schematic of the effect of the surface trap on GaAs MESFET (a) at low frequency and (b) high frequency.

Under large signal operation such as power amplification, the gate voltage of the FET varies over a wide range from +0.5V to the negative value lower than the pinch-off. To analyze the frequency dispersion under large signal operation, the V_{gs} $-I_d$ characteristic at high frequency was measured.   Figure 2 shows the cross section of measured GaAs FET.   The epitaxial structures were grown by MOCVD on (100) semi-insulating GaAs surface.   The channel of the FET was formed by Si-doped GaAs layers over AlGaAs buffer layer.   The channel layer was covered with n^+-doped GaAs cap layer to form ohmic contact.   The measured FET has the gate length of 0.8 μ m and the gate width of 1 mm.   The SiO_2 film was used as the surface passivation. Figure 3 shows the measurement configuration for frequency dispersion characteristics of GaAs FET.   The square wave was used as the input signal to the gate.   The supplied drain voltage was 3V.   The time response of the drain current (Ids) was

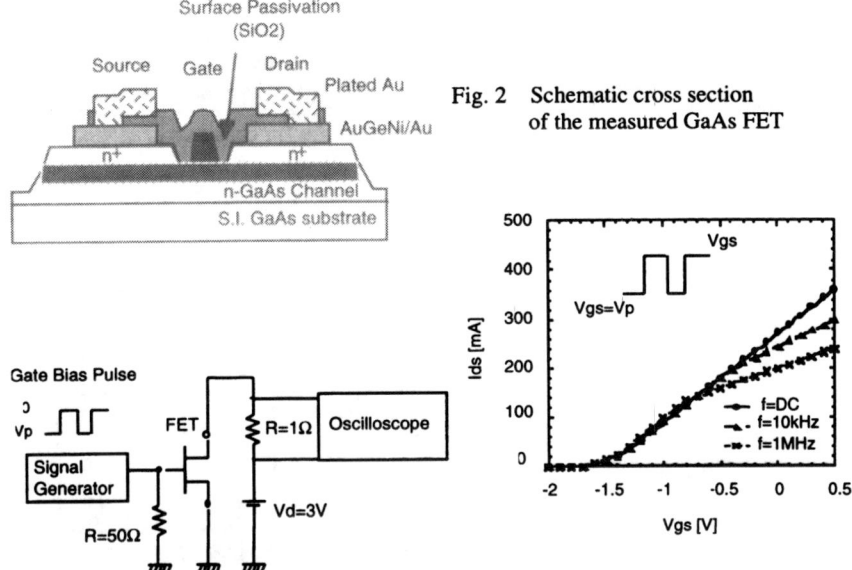

Fig. 2   Schematic cross section of the measured GaAs FET

Fig. 3   Measurement Configuration for Frequency Dispersion of GaAs FET Characteristics

Fig. 4   Measured Frequency Dispersion of Vgs-Ids Characteristics of GaAs FET

measured by monitoring the voltage drop across a 1 Ω resistor in series with the drain. Figure 4 shows the measured frequency dispersion of Vgs–Ids characteristics obtained by changing the input signal frequency.   At the low input signal frequency of 10 Hz, Ids was increased linearly with Vgs from the pinch–off voltage to +0.5 V.   On the other hand, the drain current was decreased at high frequency of 10 kHz and 1 MHz where Vgs was from –1 V to +0.5 V.

Such reduction of the drain current can be explained as follows.   When the gate is biased to a deep negative voltage, the electrons are captured by the surface states in the vicinity of the gate.   Thus the surface depletion layers in these regions become wider.   As the frequency of the input signal becomes higher, the electrons cannot be emitted and the surface depletion layer keeps widening.   As a result, the channel becomes narrower and the drain current is reduced.   The reduction of the drain current is more conspicuous in the region of Vgs from –1 V to 0.5 V can be understood as follows.   When Vgs is increased to 0 V, the depletion layer under the gate becomes narrower than the surface depletion layer due to the electrons captured by the surface states.   In such region, the drain current is determined by the channel thickness narrowed by the surface depletion layer.   For this reason, the reduction of the drain current compared to dc value is more conspicuous in this region.

From the aforementioned, the Vgs –Ids characteristic of the GaAs FET has frequency dispersion under large signal operation.   The nonlinearity of the Vgs-Ids characteristics is increased in high frequency region over several hundred Hz.

Next, the effect of the nonlinearity of the Vgs –Ids characteristic with respect to its contribution to degrade distortion characteristics is investigated.   On the basis of the extracted Vgs-Ids characteristics, the third order intermodulation (IM3) of the FET was

simulated.   In this simulation, the bias point was set at 10% of Idss.   The Vgs –Ids
characteristic extracted at high frequency and at dc was used.   A schematic flow chart
and results of the simulation are shown in Fig. 5.   It is revealed that the most important
IM3, which corresponds to cross talk to the adjacent channel, is significantly degraded
for nonlinear Vgs –Ids characteristic extracted at high frequency.   The IM3 and IM5
are larger by one to two orders of magnitude than that for DC characteristics.

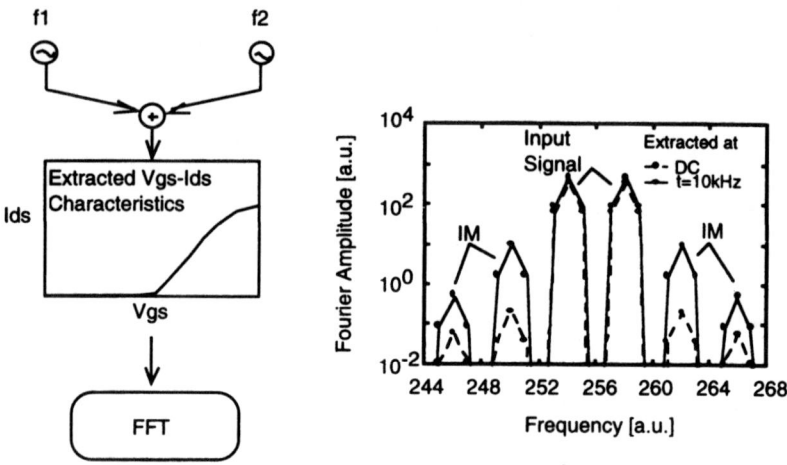

Fig. 5   Schematic flowchart and simulation results of the intermodulation
            distortion using the extracted Vgs-Ids characteristics of GaAs FETs.

From these results, it is explained that the increase of the nonlinearity of the
Vgs-Ids characteristic is due to the frequency dispersion.   The nonlinearity
significantly degrades distortion characteristics of the FET.

SUPRESSION OF THE FREQUENCY DISPERSION

The frequency dispersion of GaAs FET originates from the surface trapping
states.   Thus, the intermodulation distortion can be improved by following two ways.
One is reducing the electron trap itself by surface passivation.   It is well known that
the surface state density can be suppressed by SiN passivation [6, 7].   However, the
FET with SiN passivation still has large frequency dispersion of the drain current at
1MHz compared to dc condition.   The other is making surface insensitive to the
surface trapping effect.   To overcome the surface problem of GaAs FET, new device
design approaches have been studied [7-10].   An undoped GaAs layer or an undoped
AlGaAs layer is utilized as surface passivation on the n-GaAs channel.   However,
these FETs still have frequency dispersion of drain current at high frequency.   They
also have the disadvantage of the process complexity.   We considered the electron
trapping at the surface of the undoped layer to cause the poor frequency dispersion.   It
is well known that InGaP has extremely low surface density compared to GaAs or
AlGaAs.   Thus, the FET with undoped InGaP layer is expected to exhibit better
frequency dispersion than conventional FET.

Device Structure

Figure 6 shows the cross section of fabricated GaAs FET with undoped InGaP layer as the surface passivation. The epitaxial structures were grown by MOCVD on (100) semi-insulating GaAs surface. The channel of the FET was formed by Si-doped GaAs layers over AlGaAs buffer layer. The channel layer was covered with undoped InGaP layer. The FET has the gate length of 0.8 μm and the gate width of 18 mm. The FET with undoped AlGaAs layer as the surface passivation was also fabricated for comparison.

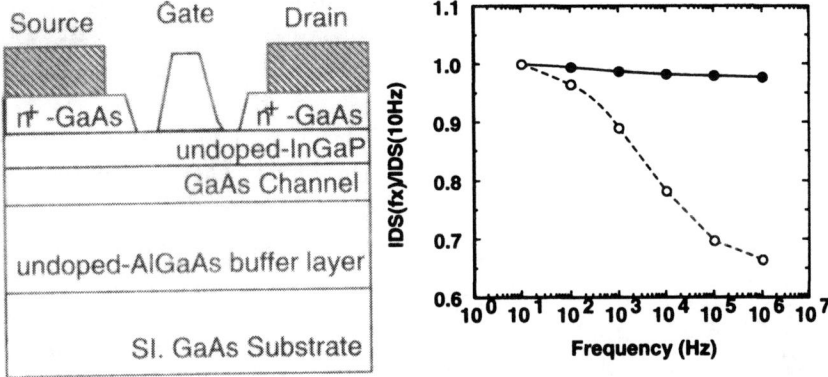

Fig. 6 Cross section of the FET with undoped InGaP layer as a surface passivation.

Fig. 7 Results of the frequency dispersion of Ids for power FET with undoped InGaP (●)and with undoped AlGaAs layer(○).

Frequency Dispersion Characteristics

The frequency dispersion of Ids for fabricated FET with undoped InGaP and undoped AlGaAs passivation layer is shown in Fig. 7. As shown in this figure, the FET with undoped AlGaAs passivation shows approximately 35 % smaller Ids at 1 MHz than at 10Hz. On the other hand, Ids decreases only 2 % at 1 MHz compared to 10Hz for undoped InGaP passivated FET. This result shows that the power FET with undoped InGaP passivation layer becomes insensitive to surface trapping effects. So, they are suitable for low distortion power amplifiers in $\pi/4$ shift-QPSK modulation systems.

The power performance of the device was measured by using automated tuner system at 950 MHz. The measured power FET has a total gate width of 18 mm with a unit gate width of 250 μm. The wafer was thinned to 100 μm and mounted metal was evaporated on the backside to reduce thermal resistance. The device was operated under class AB condition with a quiescent bias current of 0.3 A and the drain bias of 3.5 V. The input-output characteristics of these FETs are shown in Fig. 8. The input circuit was tuned to the maximum gain point. The output matching was tuned to obtain maximum output power. As shown in Fig. 8, the linearity between input and output power is remarkably improved for the FET with undoped InGaP passivation layer. The 1dB gain compression point (P1 dB) of the FET with undoped InGaP

RF Characteristics

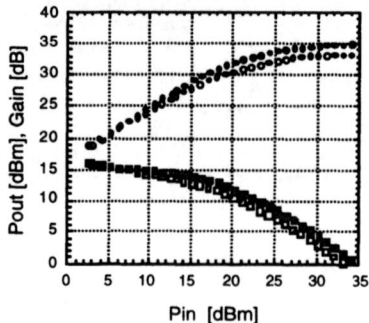

Fig. 8 Input-output characteristics of the power FET with undoped InGaP(●/■)and with undoped AlGaAs layer(○/□) at 950MHz.

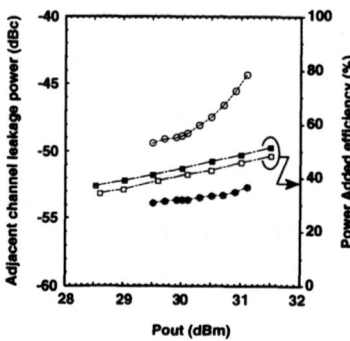

Fig. 9   Adjacent channel leakage power and drain efficiency of the amplifired QPSK signal with the channel separation of 50 kHz from 980MHz comparing the FETs with undoped InGaP(●/■)and with undoped AlGaAs layer(○/□).

layer was 31.5dBm.   On the other hand, the FET with undoped AlGaAs layer has the P1 dB of 29.0dBm.   This improved linearity is due to the decrease of the surface trapping effect achieved by undoped InGaP passivation layer.   The FET with undoped InGaP layer also exhibited high saturation power (Psat) of 34.8dBm.   This Psat was 1.5dB higher than that of the FET with undoped AlGaAs layer.   This so-called no premature power saturation is also due to the decrease of the surface trapping effect.

In the π/4 shift-QPSK digital modulation, the distortion characteristic of power amplifier is measured as adjacent channel leakage power (ACP) of the amplified QPSK signal for a channel separation of 50 kHz.   Figure 9 shows distortion modulation characteristics and power-added efficiency comparing the FET with undoped InGaP layer and undoped AlGaAs layer.   The two types of FETs were measured under same measurement conditions as the idle current of 300mA.   Input and output matching were tuned at the conditions that gave lowest distortion characteristics for each FET. Consequently, the ACP of the FET with undoped InGaP layer is 10 dB smaller than that of the FET with undoped AlGaAs layer at the output power of 30dBm.

CONCLUSION

The surface passivation of GaAs power FET has been investigated. Intermodulation distortion of GaAs power FET was found to be affected by frequency dispersion which originates from electron trap at the surface in the vicinity of the gate. To suppress the frequency dispersion, surface passivation that is less sensitive to the surface trapping effect is studied.   We found that FET with undoped InGaP layers on the n-GaAs channel is free from surface trapping effects.   It showed only 2% frequency dispersion of drain current at 1MHz compared to dc condition.   Because of this small frequency dispersion, the FET with the gate width of 18mm exhibited 1.5dB larger output power than that of the FET without undoped InGaP layer.   Moreover, in $\pi/4$ shift-QPSK modulation, the FET exhibited 11dB smaller adjacent channel leakage power than the conventional one at the output power of 30dBm.   The undoped layer acts as an ideal passivation layer for the channel.

ACKNOWLEDGMENTS

The authors would like to thank M. Kazumura and K. Itoh for their support of this work.   We also would like to thank K. Kanazawa and K. Tateoka for their technical discussion.

REFERENCES

[1] T. S.Tan, K. Kotzebue, D.M. Braun, J.Centanni and D. Mcquate, IEEE Trans. Microwave Theory & Tech., **MTT-36**, 6, p. 1023 (1988).
[2] J. Graffeuil, Z. Hadjoub, J. P. Fortea and M. Pouysegur, Solid-State Electron. , **29**, 10 p. 1087 (1986).
[3] P. H. Ladbrooke and S. R. Blight, IEEE Trans. Electron Devices, **ED-35**, 3, p. 257 (1988).
[4] I. Son and T. W. Tang, IEEE Trans. Electron Devices, **ED-36**, p. 632 (1989)
[5] J. H. Zhao, R. Hwang and S. Chang, Solid-State Electron. , 12 p. 1665 (1993).
[6] K. Miyatsuji, H.Furukawa and D. Ueda, Electronics and Communications in Japan, Part 2, **vol. 77**, No. 7, 1994.
[7] H. Furukawa, K Tateoka, K. Miyatsuji, A. Sugimura and D. Ueda, IEEE Trans. Electron Devices, **ED-43**, 2, p. 193 (1996).
[8] S. Sriram, R. C. Clarke, R. L. Messham, T. J. Smith and M. C. Driver, IEEE Cornell University Conference, p. 218 (1989).
[9] M. Takikawa, K. Kasai, M. Ozeki, Y. Hirachi and A. Shibatomi, Semi-Insulating III-V Materials, p. 603 (1986)
[10] T. Tanaka, H. Furukawa, H. Takenaka, T. Ueda, T. Fukui and D. Ueda, IEEE Trans. Electron Devices, **ED-44**, 3, p. 354 (1997)

MBE GROWTH OF OXIDES FOR III-N MOSFETS

B. GILA, K. N. LEE, J. LAROCHE*, F. REN*, S. M. DONOVAN, C. R. ABERNATHY and
J. HAN**
Department of Materials Science, Rhines Hall, University of Florida, Gainesville, FL 32611
*Department of Chemical Engineering, University of Florida, Gainesville, FL 32611
**Sandia National Laboratory, Albuquerque, NM

ABSTRACT

Reproducible fabrication of high performance metal oxide semiconductor field effect transistors (MOSFETs) from compound semiconductors will require both good interfacial electrical characteristics and good thermal stability. While dielectrics such as SiO_2, AlN, and $GdGaO_x$ have demonstrated low to moderate interface state densities, questions remain about their thermal stability and reliability, particularly for use in high power or high temperature widebandgap devices. In this paper we will compare the utility of two potential gate dielectric materials: GdO_x and GaO_x. GdO_x has been found to produce layers with excellent surface morphologies as evidenced by surface roughness of less than 1 nm. Stoichiometric films can be easily obtained over a range of deposition conditions, though deposition temperatures of 500°C appear to offer the optimum interfacial electrical quality. By contrast GaO_x films are quite rough, polycrystalline and show poor thermal stability. Further they exhibit a range of stoichiometries depending upon deposition temperature, Ga flux and oxygen flux. This paper will describe the relationship between deposition conditions and film characteristics for both materials, and will present electrical characterization of capacitors fabricated from GdO_x on Si.

INTRODUCTION

A number of GaN field effect transistors (FETs) and AlGaN/GaN heterostructure FETs have been reported showing excellent device breakdown characteristics[1-11]. To date however all show evidence of performance degradation due to the presence of high parasitic resistances which arise from the unique processing limitations imposed by the physical characteristics of GaN.[4] These problems can be overcome by using a metal oxide semiconductor FET (MOSFET) approach of the type recently reported for GaAs and InGaAs. In the $Ga_2O_3(Gd_2O_3)$/GaAs MOSFET, a mid-gap surface state density of 2×10^{10} $cm^{-2}eV^{-1}$ was obtained.[12] As a result, both n- and p-type enhancement mode MOSFETs could be demonstrated.[13,14] This method has recently been applied to GaN as well, resulting in demonstration of the first GaN MOSFET.[15]

While preliminary device results are encouraging, little is yet known about the optimum oxide composition, structure and deposition method. A number of investigators have explored oxidation of III-V surfaces in general and more recently of GaN. Invariably this leads to the formation of a poly-crystalline Ga_2O_3 layer which typically exhibits poor surface or interface smoothness and poor capacitor performance.[16-18] A more successful approach, and the one responsible for the MOSFETs referenced above, has been deposition of the oxide films using e-beam evaporation from a single crystal of gadolinium gallium garnet (GGG) in an MBE chamber.[19-22] This has been reported to yield amorphous Ga_2O_3 films with trace amounts of Gd which are not uniform in the growth direction. Deposition of the oxide from individual sources of Ga and oxygen would of course eliminate this contamination. Promising results were obtained by Callegari et. al.[23] using an RF oxygen plasma to produce GaO_x films, though

247

possible damage to the semiconductor surface induced by the high energy plasma remains a potential limitation to this approach. In this paper the feasibility of using individual sources of metal and oxygen to deposit both Ga_2O_3 and Gd_2O_3 films has been investigated using a low energy electron cyclotron resonance (ECR) plasma source to minimize potential surface damage.

EXPERIMENTAL

Samples were grown in a Riber MBE 2300 on n-GaN (0.5 µm)/sapphire samples prepared by MOCVD or on n-Si substrates. Substrate temperature was monitored using the substrate thermocouple. Si substrates were cleaned ex-situ by dipping in buffered oxide etch (BOE) for 30 seconds. The n-GaN was cleaned ex-situ by etching in HCl for 3 min. and then exposing the sample to UV/O_3 for 25 min. Prior to deposition of the oxide layer, the GaN is heated to 650°C under ECR nitrogen plasma and then cooled to the oxide deposition temperature. Solid Ga was heated in a standard effusion oven to temperatures of 883°C to 1100°C while solid Gd was heated to 1200°C. All of the oxide layers were grown under an oxygen plasma obtained from a Wavemat MPDR 610 electron cyclotron resonance (ECR) plasma source operating at 200W forward power. Films were characterized using scanning electron microscopy (SEM), cross-sectional transmission electron microscopy (XTEM), atomic force microscopy (AFM), Auger electron spectroscopy (AES) and electrically via capacitance voltage and I-V analysis. Capacitors for electrical characterization were fabricated by Pt/Au (300Å/1500Å) top contact metallization and In solder for backside ohmic contact.

RESULTS AND DISCUSSION

Using the ECR-MBE system, diodes fabricated from oxides deposited from elemental Ga and the oxygen plasma were fabricated on both Si and GaN. As expected, AES indicates that the Oxygen/Gallium ratio in the oxide layers is a sensitive function of a) Ga cell temperature, b) oxygen flux and c) deposition temperature, as shown in Figure 1. The deposition rate decreases as the cell temperature, and hence Ga flux, is reduced. For a given oxygen flux, this reduction results in an increase in the amount of oxygen in the film. Surprisingly, decreasing the substrate temperature dramatically increases the Ga uptake rate for a given flux, and hence decreases the oxygen concentration in the film. In most cases, the amount of oxygen in the films is higher than that reported for e-beam derived films, 1.5[19-22] and higher than that observed from Ga_2O_3 powder. This suggests that the oxygen flux reaching the surface is more than sufficient to deposit stoichiometric material. While the deposition rates can be varied over a wide range, 0.18 – 3.0 µm/hr, the lower deposition rates give the highest oxygen concentrations and smoothest films as determined by stylus profilometry and SEM. Similarly, lower deposition temperatures tend to produce the best surface morphologies when the deposition rate is corrected for the enhanced Ga uptake, as shown in Figure 2. In fact the smoothest films have been obtained at the lowest temperature tested, 300°C. Even at 300°C however, the films appear quite rough in both the SEM and AFM, Figure 3, with an RMS roughness of 21.6nm on Si and 32.1nm on GaN.

Diodes fabricated on p-GaN using GaO_x deposited at 400°C show significant leakage when analyzed by CV and low breakdown voltage. The low breakdown has been traced to voids in the dielectric which arise from islanding during deposition. When metallized, the metal penetrates through the voids to the semiconductor. Annealing of these structures in hydrogen improved the C-V characteristics significantly. This suggests that much of the leakage is due to

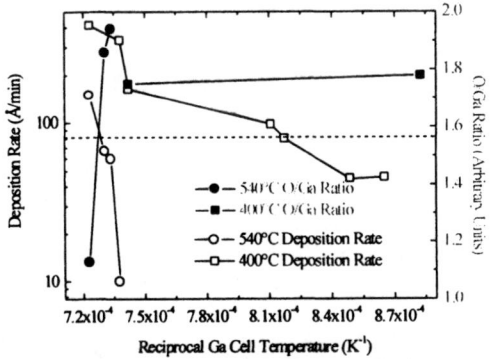

Figure 1. GaO$_x$ deposition rate and O/Ga ratio as a function of Ga cell temperature for two different p-Si substrate temperatures. Dashed line represents O/Ga ratio determined from powder sample of Ga$_2$O$_3$.

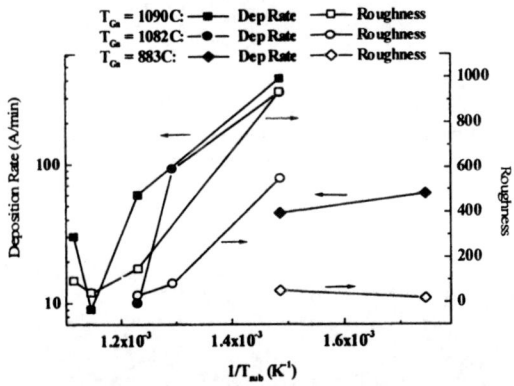

Figure 2. Effect of substrate temperature on GaO$_x$ deposition rate and surface roughness for various Ga cell temperatures. Layers were deposited on n-Si substrates.

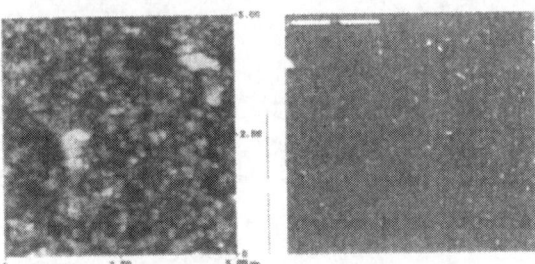

Figure 3. Surface morphology of GaO$_x$ deposited on p-GaN at 300°C as measured by AFM (at left) and SEM (at right). SEM scale marker represents 5 microns.

defects within the dielectric. While powder X-ray diffraction shows no evidence of crystalline diffraction peaks, there may be small regions of crystallinity. The leakage might then be due to dangling bonds at the surfaces of these small regions.

By contrast to the GaO_x, GdO_x layers deposited from elemental Gd and the ECR oxygen plasma exhibit excellent surface morphology, as shown in Figure 4. The surface roughness as measured by AFM is only 0.5-0.6 nm and is not apparent at an SEM magnification of 10,000x. Also unlike GaO_x, the deposition rate did not appear to be a function of substrate temperature, with a rate of 0.1Å/sec for a Gd cell temperature of 1200°C. Similarly, the O/Gd ratio did not appear to change significantly with substrate temperature and like the GaO_x gave an oxygen level higher than that observed for Gd_2O_3 powder. XTEM analysis of the layers shows the material to be poly-crystalline with a columnar grain structure, as in Figure 5.

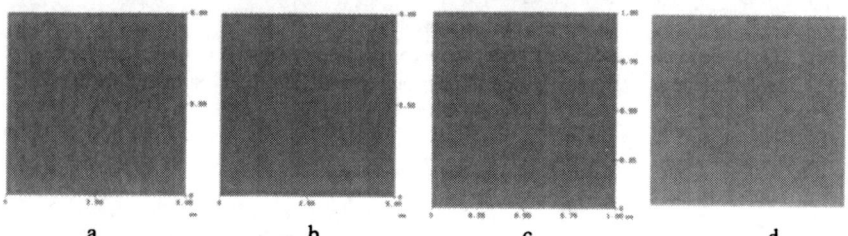

| a | b | c | d |

Figure 4. Surface morphology as measured by AFM (a,b,c) and SEM (d) of GdO_x films deposited at a) 300°C on Si, b and d) 500°C on Si and c) 300°C on GaN. The RMS roughness values were a) 0.6 nm, b) 0.5 nm and c) 0.8nm. Layer thickness is ~31.0 nm.

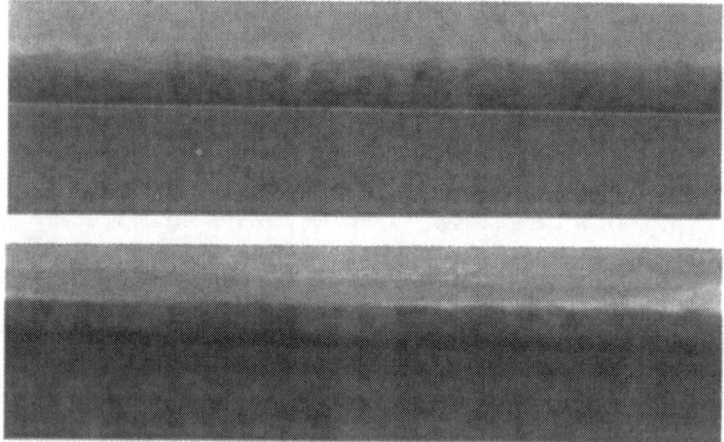

Figure 5. XTEM image of GdO_x layer deposited on n-Si at 300°C (top) and 500°C (bottom).

Diodes were fabricated on n-GaN using GdO_x deposited at 300°C. These diodes show significant leakage when analyzed by CV and low breakdown voltage. Diodes were also fabricated on n-Si using GdO_x deposited at 300°C and 500°C. The 300°C deposited GdO_x on n-

Si showed similar results to those on n-GaN. The structure deposited at 500°C, however, shows a breakdown field of 0.55MV/cm (measured at 0.5mA/cm^2) and a flatband voltage shift of -0.38V when measured at 1kHz.

CONCLUSION

Steps have been taken to study the deposition and properties of GaO$_x$ and GdO$_x$ from elemental Gallium and Gadolinium and an ECR Oxygen plasma. The GdO$_x$ exhibits excellent surface morphologies and maintains stoichiometry over a range of deposition temperatures. The GaO$_x$, however, shows poor surface morphologies and a range of compositions dependent on deposition temperature and source fluxes. Diodes fabricated on n-Si using GdO$_x$ deposited at 500°C show a breakdown field of 0.55MV/cm and a flatband voltage shift of -0.38V.

ACKNOWLEDGMENTS

The authors gratefully acknowledge the U. S. Office of Naval Research under Contract No. N00014-98-1-0204, and DARPA/EPRI under contract MDA972-98-1-0006 for support of this work.

REFERENCES

1. P. M. Asbeck, E. T. Yu, S. S. Lau, G. J. Sullivav, J. Van Hove and J. M. Redwing, Electron. Lett. **33**, p.1230 (1997).

2. S. C. Binari, W. Kruppe, H. B. Dietrich, G. Kelner, A. E. Wickenden and J. A. Freitas, Solid State Electron. **41**, p.1549 (1997).

3. M. S. Shur, Mat. Res. Soc. Symp. Proc. **483**, p.15 (1998).

4. J. Burm, K. Chu, W. J. Schaff, L. F. Eastman, M. A. Khan, Q. Chen, J. W. Yang and M. S. Shur, IEEE Electron. Dev. Lett. **18**, p.141 (1997).

5. R. Gaska, Q. Chen, J. Yang, A. Osinsky, M. A. Khan and M. S. Shur, IEEE Electron. Dev. Lett. **18**, p.492 (1997).

6. M. A. Khan, Q. Chen, C. J. Sun, M. S. Shur and B. Gelmark, Appl. Phys. Lett. **67**, p.1429 (1995).

7. Y.-F.Wu, B. P. Keller, P. Fini, S. Keller, T. J. Jenkins, L. T. Kenias, S. P. DenBaars and U. K. Mishra, IEEE Electron. Dev. Lett. **19**, p.50 (1998).

8. Y. F. Wu, B. P. Keller, S. Keller, D. Kapolneck, P. Kozodoy, S. P. DenBaars and U. K. Mishra, Appl. Phys. Lett. **69**, p.1438 (1996).

9. G. J. Sullivan, M. Y. Chen, J. A. Higgins, J. W. Yang, Q. Chen, R. C. Pierson and B. T. McDermott, IEEE Electron. Dev. Lett. **19**, p.198 (1998).

10. A. T. Ping, Q. Chen, J. W. Yang, M. A. Khan and I. Adesida, IEEE Electron. Dev. Lett. **19,** p.54 (1998).

11. O. Akatas, Z. F. Fan, A. Botcharev, S. N. Mohammad, M. Roth, T. Jenkins, L. Kehias and H. Morkoc, IEEE Electron. Dev. Lett. **18, p.**293 (1997).

12. M. Passlack, M. Hong, J. P. Mannaerts, R. L. Opila, S. N. G. Chu, N. Moriya, F. Ren and J. Kwo, IEEE Trans. Electron. Dev. **44,** p.214 (1997).

13. F. Ren, M. Hong, J. M. Kuo, W. S. Hobson, J. R. Lothian, H. S. Tsai, J. Lin, J. P. Mannaerts, J. Kwo, S. N. G. Chu, Y. K. Chen and A. Y. Cho, 1997 IEEE GaAs IC Symposium, Anaheim, CA Oct. 12-15, 1997.

14. F. Ren, J. M. Kuo, M. Hong, W. S. Hobson, J. R. Lothian, J. Lin, H. S. Tsui, J. P. Mannaerts, J. Kwo, S. N. G. Chu, Y. K. Chen and A. Y. Cho, IEEE Electron. Dev. Lett. **19,** p.309 (1998).

15. F. Ren, M. Hong, S. N. G. Chu, M. A. Marcus, M. J. Schurman, A. Baca, S. J. Pearton, C. R. Abernathy, Appl. Phys. Lett., **73,** p.3893 (1998)

16. S. D. Wolter, B. P. Luther, D. L. Waltenyer, C. Onneby, S. E. Mohney, R. J. Molnar, Appl. Phys. Lett. **70,** p.2156 (1997).

17. S. D. Wolter, S. E. Mohney, H. Venugopalan, A.E. Wickenden, D. D. Koleske, J. Electrochem. Soc. **145,** p.629 (1998)

18. E. D. Readinger, S. D. Wolter, D. L. Waltenyer, J. M. Delucca, S. E. Mohney, B. I. Prenitzer, L. A. Giannuzzi, R. J. Molnar, J. Electronic Mater. **28,** p.257 (1999).

19. M. Passlack, E. F. Schubert, W. S. Hobson, M. Hong, N. Moriya, S. N. G. Chu, K. Konstaninidis, J. P. Mannaerts, M. L. Schnoes, G. J. Zydzik, J. Appl. Phys., **77,** p.686 (1995)

20. M. Hong, M Passlack, J. P. Mannaerts J. Kwo, S. N. G. Chu, N. Moriya, S. Y. Hou, V. J. Fratella, J. Vac. Sci. Tech., **B 14,** p.2297 (1996)

21. M. Hong, M. A. Marcus, J. Kwo, J. P. Mannaerts, A. M. Sergent, L. J. Chou, K. C. Hsieh, K. Y. Cheng, J. Vac. Sci. Tech., **B 16,** p.1395, (1998)

22. M. Hong, F. Ren, J. M. Kuo, W. S. Hobso, J. Kwo, J. P. Mannaerts, J. R. Lothian, Y. K. Chen, J. Vac. Sci. Tech., **B16,** p.1398 (1998)

23. A. Callegari, P. D. Hoh, D. A. Buchanan, D. Lacey, Appl. Phys. Lett. **54,** p.332 (1989).

INTERFACE STATE DENSITIES FOR SiN$_x$:H ON CLEAVED GaAs AND InP(110)

D. LANDHEER, J. E. HULSE, AND K. RAJESH
Institute for Microstructural Sciences, National Research Council of Canada,
Ottawa, Canada K1R 0R6, dolf.landheer@nrc.ca

ABSTRACT

Silicon nitride was deposited *in-situ* by electron-cyclotron resonance plasma chemical-vapour deposition (ECR-CVD) on (110) surfaces formed by cleaving GaAs and InP(100) substrates in an ultra-high vacuum processing system. Capacitors formed by depositing Al gates on the facet surfaces were analyzed by the high-low frequency capacitance-voltage (CV) technique. The minimum interface-state densities obtained for the cleaved GaAs (110) surfaces were 1-2 ×10^{12} eV^{-1}cm^{-2}. For cleaved InP facets the measured minimum interface state densities were a factor of two higher; however, they exhibited a smaller hysteresis in the CV characteristics and a smaller modulation in the surface potential. The interface state densities did not change significantly for the GaAs(110) facets if a Si interface control layer 0.8-2 nm thick was deposited prior to silicon nitride deposition; however, a larger effect was observed for the hysteresis and flatband voltage shift of the CV characteristics. The effect of annealing on the interfaces with Si was investigated and the performance compared with published results for GaAs(100) surfaces prepared by molecular-beam epitaxy.

INTRODUCTION

Surface recombination through interface states is believed to initiate catastrophic optical damage at the facets of III-V laser devices [1], and since the surface recombination velocity can ultimately be calculated from the interface-state distribution [2], it is worth performing measurements of the interface-state densities on cleaved facets of GaAs and InP. The SiN$_x$:H/GaAs interfaces produced by ECR-CVD of silicon nitride using nitrogen and silane have interface state densities in the low 10^{12} eV^{-1} cm^{-2} range [3]. The interface formed during ECR-CVD deposition of the silicon nitride film is formed by plasma nitridation which dominates over CVD during the initial stages of the nitride deposition, and this interface is not abrupt [4]. According to the disorder-induced gap state model of Hasegawa and Ohno [5], interface states are produced in this defect-laden layer between the insulator and semiconductor and a Si interface-control layer (ICL) should create a more abrupt interface with fewer states within the GaAs bandgap. An improvement in the interface state density was observed if a silicon ICL was deposited on S-passivated GaAs(110) surfaces prior to nitride deposition [6]. Lower interface-state densities have been obtained on GaAs(100) surfaces prepared by molecular beam epitaxy (MBE) [7,8], a technique which is not viable for device applications such as laser facet passivation.

This paper describes the electrical characteristics of interfaces formed on cleaved (110) surfaces of GaAs or InP by depositing silicon nitride by ECR-CVD *in-situ* (without breaking vacuum). Al gates have been deposited on the facets after nitride deposition, and the resultant metal-insulator-semiconductor (MIS) structures have been examined by CV measurements to estimate the interface-state density. The effect on the measured CV characteristics of silicon ICLs evaporated between the GaAs(110) surfaces and the deposited silicon nitride layer has also been investigated.

EXPERIMENT

The depositions were done in a chamber which is part of a multi-chamber ultra-high vacuum (UHV) processing system that has been described previously [9]. GaAs or InP(100) substrates, 0.4 mm thick and doped 4-6×10^{15} cm^{-3}, were mounted in a cleaving fixture prior to insertion into

Mat. Res. Soc. Symp. Proc. Vol. 573 © 1999 Materials Research Society

the load-lock of the system. The substrates were then transferred to and cleaved in a chamber which contained a rod-fed (Thermionics) electron-beam deposition system with a 99.9999% pure Si rod.

The principal residual gas in the cleaving chamber prior to Si evaporation, apart from hydrogen, was CO with a partial pressure of 5×10^{-12} Torr which rose to 10^{-9} Torr during Si deposition. Since silicon was evaporated at a rate of ~0.15 nm/s, the total maximum exposure to background CO during the first monolayer of Si deposition was <0.002 Langmuir. Assuming that the sticking coefficient for CO is comparable for the GaAs(100) and (110) surfaces (10^{-5} for GaAs(100)- Ref. [10]), this level of CO exposure should not shift the surface Fermi energy or contribute to the measured interface-state density.

The analysis chamber on the UHV system contains a PHI x-ray photoelectron spectroscopy (XPS) system which was used to perform analysis after film deposition. From the ratio of the XPS signals from the Si 2p peaks to those of the As 3d and Ga 3d peaks and using the electron escape depths in bulk Si and GaAs, the thickness of silicon layers deposited on GaAs was obtained [11]. This was used to calibrate the thickness monitor in the Si deposition/cleaving chamber.

Silicon nitride layers 30 nm thick were deposited using a system described previously [9]. Briefly, In the ECR-CVD system, microwaves at 2.45 GHz pass through the quartz window of the Astex AX-4400 ECR source and plasma is generated using 15 sccm of molecular nitrogen introduced in a ring near the quartz window. The deposition temperature was controlled at 250 °C using quartz-halogen heaters and a thermocouple pressed against the back of the stainless-steel cleaving fixture. Silane (5 sccm of a 1:3 SiH_4:He mixture) was introduced through a 15 cm diameter gas distribution ring at a downstream position 15 cm from the sample stage. The chamber pressure was 0.6 mTorr. The substrates were left electrically floating and 425 W of absorbed microwave power was used for this work. These conditions have previously been shown to produce silicon nitride films with optimal electrical properties, low leakage currents and interface state densities $<10^{11}eV^{-1}cm^{-2}$ on Si substrates [12].

Al gates were deposited through a shadow mask and the samples were subsequently annealed at 450 °C for 5 min in forming gas. CV measurements were made using an HP Model 4140B Picoammeter and HP Model 4275 Multi-frequency LCR meter. Quasistatic CV (QS-CV) measurements were performed by scanning the gate potential from inversion to accumulation at a rate of 0.1 V/s. High-frequency CV (HF-CV) measurements were performed by stepping the gate at the same rate in the same direction, waiting for 5 seconds, and then scanning in the reverse direction. The interface state density was calculated from the QS-CV and the 1MHz HF-CV characteristics using the high-low frequency CV method [13].

RESULTS AND DISCUSSION

The 1MHz HF-CV curves, normalized by their accumulation capacitance, C_{max}, obtained on cleaved GaAs(110) surfaces are shown in Fig. (1) as a function of Si ICL thickness. The reverse CV trace exhibited a clockwise hysteresis caused by a slow charging and discharging of states near the interface. This hysteresis decreased dramatically when the thickness of the Si ICL reached 1.0 nm. When a Si layer is deposited on GaAs some of it will be converted to silicon nitride during the initial stages of silicon nitride deposition. From Ref. [12] it is estimated that the thickness of Si consumed will be in the range of 0.7-1.1 nm, a value that is consistent with the thickness that causes a reduction in the observed hysteresis. The flatband voltage shift also decreased, indicating a decrease in trapped negative charge, and then increased as the Si ICL thickness increased further.

One of the samples with a 1.5 nm ICL was given a 550 °C anneal after silicon nitride deposition prior to Al gate deposition. This did not result in significant changes in the electrical properties. Furthermore, no evidence of any Si crystallization at the interface was observed by high-resolution transmission electron micrographs of the interface, as was reported for MBE-prepared GaAs(100) in Ref. [7].

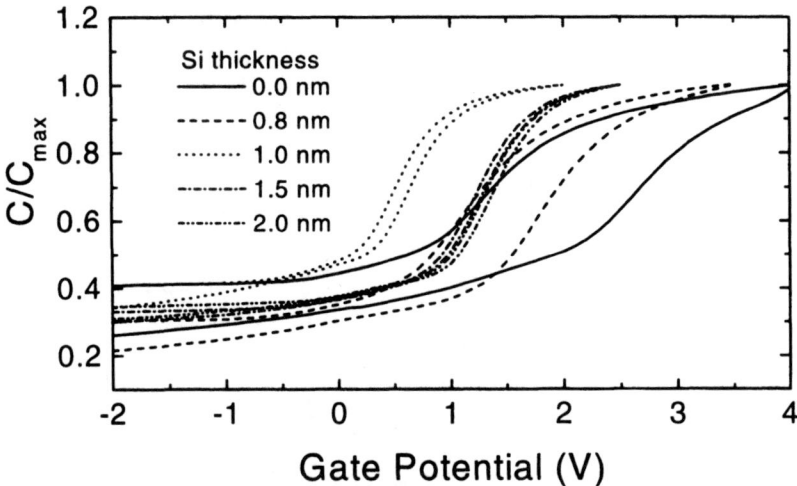

Figure 1. 1 MHz capacitance normalized by the accumulation capacitance, C_{max}, for SiN$_x$:H/Si/GaAs(110) capacitors vs gate potential and Si ICL thickness. The silicon nitride thickness was 30 nm.

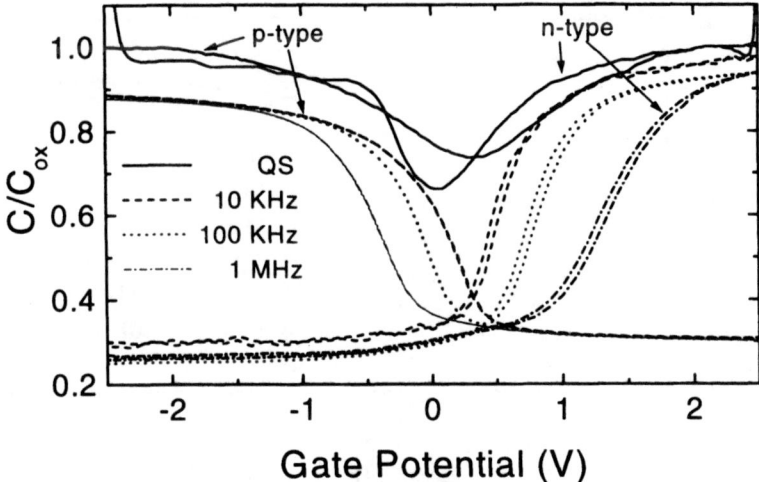

Figure 2. QS- and HF-CV capacitance, normalized by the QS accumulation capacitance, C_{ox}, vs gate potential for (110) facets cleaved from n- and p-GaAs with 1.5 nm Si ICL layers and 30 nm SiN$_x$:H.

Fig. (2) shows the QS- and HF-CV curves obtained on facets cleaved from n- and p-GaAs (both doped ~5×10^{16} cm^{-3}) with Si layers 1.5 nm thick and 30 nm silicon nitride layers. The offset between the low and high frequency curves for both doping types indicates a high density of interface states near both the conduction- and valence-band edges [1] but the p-type samples showed very small hysteresis (~ 5 mV counterclockwise) in the HF-CV curves.

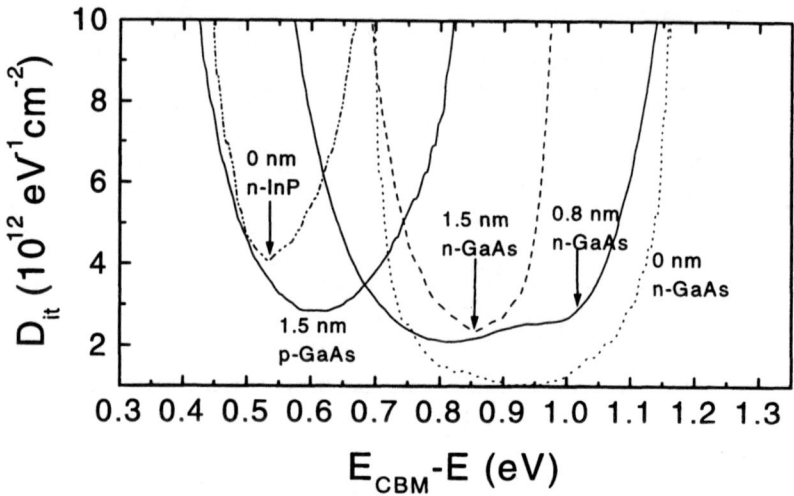

Figure 3. Interface-state density as a function of energy from the conduction band edge for MIS capacitors on GaAs(110) with Si ICL thicknesses of: (——) 1.5 nm , (----) 0.8 nm, (····) 0 nm. The results for a SiN$_x$;H/InP(110) capacitor with no ICL are also shown (·-·-·-).

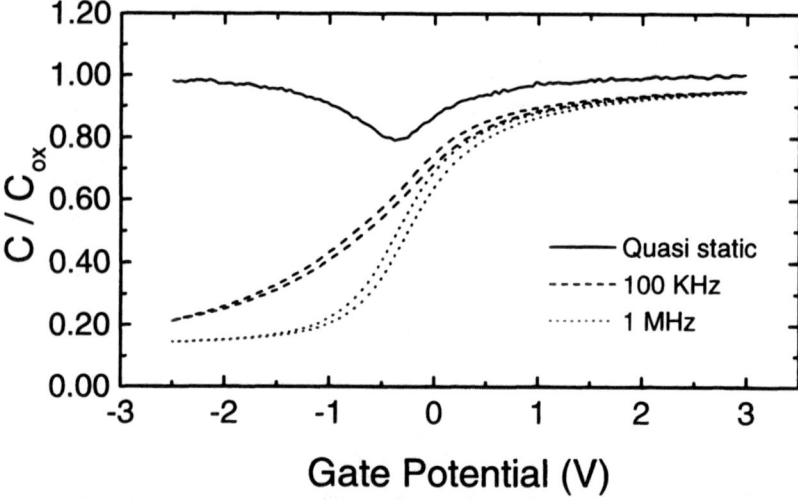

Figure 4. QS- and HF-CV characteristics, normalized by the QS accumulation capacitance, C$_{ox}$, for MIS capacitors formed by the *in-situ* deposition of 30 nm of silicon nitride on cleaved InP(110).

Fig. (3) shows the interface-state density obtained by the high-low frequency CV method as a function of energy from the GaAs conduction band edge for the *n*-type substrates with Si ICLs 0, 0.8 and 1.5 nm thick. Minimum interface-state densities, D_{itm}, with values of $1-3 \times 10^{12}$ eV^{-1} cm^{-2} were obtained for all three samples. The assumption that the 1MHz CV characteristics represent the high-frequency limit, the principal limitation of the method, can introduce errors in D_{itm} with magnitudes $\Delta D_{itm} \sim 0.7\, D_{itm}$ [1], thus a variation of this order may not be significant. The range over which the interface-state density is close to the minimum interface state density decreases for ICL thicknesses > 0.8 nm, a value close to the thickness of Si that will consumed by the initial nitridation reaction [12]. This is consistent with a position of the Si conduction band edge 0.75 eV below that of GaAs [14], which results in an increasing density of Si conduction band states in the upper half of the GaAs bandgap with increasing Si thickness.

The minimum interface-state density for a p-type substrate with a 1.5 nm thick Si ICL, shown in Fig. (3), is 3×10^{12} eV^{-1} cm^{-2}, close to that obtained for the n-type substrate, as expected. However, the position of the minimum in the bandgap is slightly different for the n and p-type samples. This discrepancy may be due to the error introduced by assuming that the 1MHz CV characteristics represent the high-frequency limit.

The effective interface charge densities for the samples with 1.5 nm Si ICLs, calculated from the flatband voltage shift, are -7×10^{11} cm^{-2} and -2×10^{12} cm^{-2} for the p- and n-type samples, respectively. The charge measured in this way contains a contribution from filled acceptor interface-states below the Fermi energy and empty donor interface-states above the Fermi energy. According to Adamowicz and Hasegawa [2] the states between the GaAs neutral level 0.96 eV below the conduction band minimum (E_{cbm}) and E_{cbm} are acceptors. Since the n-type sample has its surface Fermi energy in this range and that of the p-type sample is almost within this range, the difference in negative charge calculated for the n- and p-type samples should be an approximate measure of the density of acceptor states between these Fermi energies.

InP(100) substrates, doped 5×10^{15} cm^{-3} n-type, were also cleaved in vacuum and the resultant (110) facets covered with 30 nm of silicon nitride. The CV characteristics are shown in Fig. (4) and the interface-state densities are shown in Fig. (3). InP exhibited a smaller hysteresis in the CV characteristics and a smaller modulation in the surface potential when compared to GaAs with no ICL. The minimum interface state density is 4×10^{12} eV^{-1} cm^{-2}, within a factor of two of that measured previously on sulfur-passivated or HF-cleaned InP(100) [15,16] . The minimum occurs at 0.5 eV below E_{cbm} for InP, close to the position of the neutral level, 0.37 eV below E_{cbm}, described in Ref. [2].

CONCLUSIONS

The interface state density has been measured on cleaved GaAs(110) surfaces for the first time. The ECR plasma produces an interfacial layer by rapid plasma nitridation during the initial stages of silicon nitride deposition which has fair electrical properties. Deposition of a Si ICL prior to silicon nitride deposition with an ECR source reduces the flatband voltage and the hysteresis in the CV characteristics but does not dramatically effect the minimum interface state density. In fact, the interface state densities are comparable to those reported (low 10^{12} eV^{-1} cm^{-2} range) for GaAs(110) wafers given a gas-phase sulfur passivation prior to Si ICL and silicon nitride deposition [3].

Park *et al.* [17] have deposited 1 nm thick Si layers on GaAs(100) surfaces prepared by MBE. *In-situ* STM analysis showed that the morphology of the Si surface was irregular unless the surfaces were annealed. It thus appears likely that annealing is crucial to the formation of low interface state densities. Although the samples were given a post-metallization anneal at 450 °C for 5 min, the fact that the cleaving fixture used for this work was not heated during Si deposition and could not be heated above 250 °C may have limited the effectiveness of the Si ICL. Different stress profiles in our silicon nitride layers and less effective hydrogen termination of midgap states in our e-beam evaporated Si layer as compared with the ECR-CVD deposited Si layers in Ref. [17] may also be factors. Heating of the films to 550 °C for 1 min after SiN$_x$:H deposition did not result in significant changes in the electrical properties and did not result in crystallization at the Si/GaAs interface. The failure to induce crystallization at the interface may be due to different stress profiles or it may be due to the differences between the stoichiometric cleaved (110) facet of GaAs and the As-terminated (100) surface prepared by MBE.

For InP(110) the interface state density measured without a Si ICL is only slightly higher than for GaAs and the minimum occurs close to the neutral level [2] as it did for GaAs. The effect of depositing Si ICLs on cleaved InP(110) facets prior to silicon nitride deposition will be investigated in the future.

ACKNOWLEDGMENTS

We thank S. Ingrey and W.M. Lau for inspiration and many useful discussions and E. Estwick and M. Tomlinson for technical assistance.

REFERENCES

1. W.C. Tang, H.J. Rosen, P. Vettiger and D.J. Webb, Appl. Phys. Lett. **59**, 1005 (1991).
2. B. Adamowicz and H. Hasegawa, Jpn. J. Appl. Phys. **37**, 1631 (1998).
3. D. Landheer, Z.-H. Lu, J.-M. Baribeau, L. J. Huang, and W. M. Lau, J. Electron. Mat. **23**, 943 (1994).
4. D. Landheer, K. Rajesh, J.E. Hulse, G.I. Sproule, J. McCaffrey, T. Quance, and M.J. Graham, to be published.
5. H. Hasegawa and H. Ohno, J. Vac. Sci. Technol. B **4**, 1130 (1986).
6. L.J. Huang, K. Rajesh, W.M. Lau, X.Z. Wu, D. Landheer, J.-M. Baribeau, and S. Ingrey, Appl. Phys. Lett. **71**, 237 (1997).
7. D.G. Park, Z. Chen, A.E. Botchkarev, S.N. Mohammad and H. Morkoç, Phil. Mag. B **74**, 219 (1996).
8. M. Hong, M. Passlack, J. P. Mannaerts, J. Kwo, S. N. G. Chu, N. Moriya, S. Y. Hou, and V. J. Fratello, J. Vac. Sci. Technol. B **14**, 2297 (1996).
9. D. Landheer, J.E. Hulse, and T. Quance, Thin Solid Films **293**, 52 (1997).
10. J.P. Chamberlain, *The adsorption of carbon monoxide on silicon and gallium arsenide surfaces*, p. 29, PhD Dissertation, Georgia Institute of Technology. (1993).
11. Z.-H. Lu and J.-M. Baribeau, Appl. Phys. Lett. **70**, 1989 (1997).
12. D. Landheer, K. Rajesh, D.P. Masson, J.E. Hulse, G.I. Sproule, and T. Quance, J. Vac. Sci. Technol A **16**, 2931 (1998).
13. R. Castagné and A. Vapaille, Surface Sci. **28**, 157 (1971).
14. T. Hashizume, K. Ikeya, M. Mutoh, and H. Hasegawa, Applied Surface Science **123/124**, 599 (1998).
15. D. Landheer, G.H. Yousefi, and J.B. Webb, J. Appl. Phys. **75**, 3516 (1994).
16. S. Garcia, I. Martil, G. Gonzalez Diaz, E, Castan, S. Duenas, and M. Fernandez, J. Appl. Phys. **83**, 600 (1998).
17. D.G. Park, M. Tao, J. Reed, K. Suzue, A.E. Botchkarev, Z. Fan, G.B. Gao, S.J. Chey, J. Van Nostrand, D.G. Cahill, and H. Morkoç, J. Cryst. Growth **150**, 1275 (1995).

WET AND DRY ETCHING CHARACTERISTICS OF ELECTRON BEAM DEPOSITED SiO AND SiO$_2$

J.R. LaRoche[1], F. Ren[1], J.R. Lothian[2], J. Hong[3], S.J. Pearton[3], and E. Lambers[3]

1 Department of Chemical Engineering, University of Florida, Gainesville, FL 32611
2 Multiplex Inc., South Plainfield, NJ 07080
3 Department of Materials Science and Engineering, University of Florida, Gainesville, FL 32611

We have studied the thermal stability and etching characteristics of E-beam deposited SiO and SiO$_2$. Dry etch rates were studied using SF$_6$ and NF$_3$ discharges in a Plasma Therm inductively coupled plasma system. Wet etch rates were assessed with buffered HF and HF/H$_2$O solutions. SiO$_2$ etched faster than SiO under all etch conditions. Dry etch rate of SiO$_2$ was comparable with PECVD SiO$_2$. Auger analysis indicated that SiO$_2$ maintained excellent thermal stability after annealing to 700°C. The Si/O ratio of SiO in the film increased when annealed to 700°C. Ellipsometry also revealed greater refractive index variance across the sample for SiO, as compared to SiO$_2$. However, thickness variation of both films was $\leq$ 2% across the wafer. Ellipsometry data also showed great thermal stability of SiO and SiO$_2$. There was <4% change after 700°C annealing.

Introduction

It is well known that hydrogen can unintentionally contaminate Si or III-V crystalline semiconductors at almost any step in the growth or fabrication process (such as plasma chemical vapor deposition or wet chemical etching). For example, hydrogen can be present in the form of water vapor, OH species in oxides, in silicon nitride, and in photoresists. These hydrogen sources may be an intentional process component (as in C$_2$H$_6$-H$_2$-Ar plasmas), or may result from leaks or reactor out-gassing [1]. Hydrogen exposure leads to shallow acceptor and donor impurities passivation in Si and GaAs [2]. Naturally, this results in increased resistivity and lower device switching speeds. Furthermore, in silicon, hydrogen (in the absence of plasma or radiation damage) has also been shown to induce microcracks (platelets) in the near surface region, and to form deep levels. In compound semiconductors, platelet formation has been observed in GaAs after proton implantation and high temperature annealing [3]. Therefore, as thermal budget and device dimensions continue to decrease, the control of unintentional hydrogen incorporation in device processing will become increasingly more important to device reliability and operation.

Previously, we studied the effectiveness of E-beam deposited SiO and SiO$_2$ in blocking the trench etch effect by passivating the metal contacts on TLM structures on GaAs [4]. Essentially, this work showed that the etch current generated by the edge of the TLM contact metal was negligible as compared to the exposed surface of the top of the TLM contact. Therefore, by passivating the upper surface of the metal contact, the etch trench effect could be greatly reduced. TLM measurements indicated that the trench etch effect was suppressed in SiO and SiO$_2$ passivated contacts for periods up to 80 minutes during wet etching in H$_2$O and 20/1; H$_2$O/HCl. During the same period unpassivated samples changed substantially. Auger analysis confirmed that a sharp interface was maintained between the AuGe metal contacts and SiO and SiO$_2$ interfaces. Wet etch results showed that SiO$_2$ etched faster in both BOE and 20/1; H$_2$O/HCl

Mat. Res. Soc. Symp. Proc. Vol. 573 © 1999 Materials Research Society

solutions. It was suggested that the enhanced etch rate of SiO_2 over SiO was due to the presence more highly polar Si-O bonds in SiO_2 acting as insertion sites for acid/base attack. We concluded that SiO and SiO_2 were effective in blocking the trench etch effect on GaAs substrates.

In this work, an E-beam deposited hydrogen free amorphous SiO or SiO_2 was used to passivate GaAs substrates to assess the film characteristics of the SiO and SiO_2 and to study process compatibility. Specifically, the GaAs passivated substrates were used to study the dry etch rates of SiO and SiO_2, and the refractive index of the deposited layers. The passivation of the metal contacts was done to assess the film morphology and pattern transmission.

Experimental

To study the etching characteristics of SiO and SiO_2 dry etching, 3000 Å of SiO and SiO_2 was deposited on Si substrates and AZ 1818 was used as the etch mask. For dry etching, 96% CF_4 balanced with O_2, $5SF_6/10Ar$, and $5NF_3/10Ar$ gas mixtures were utilized and Micro-RIE as well as Plasma Therm 770 inductively coupled plasma systems were used to etch the dielectrics.

Thermal stability of the films was studied with Auger, AFM, and ellipsometry. Auger was used to assess the Si/O ratio of the as-deposited films, and the films after they were annealed to 700 °C. AFM found the root mean square (rms) roughness of the as-deposited films and the 700 °C annealed films. Ellipsometry was then performed with an AutoEl IV on the SiO and SiO_2 samples to compare the refractive index of the as-deposited films and films annealed at 300, 500, and 700 °C for 20 seconds in an N_2 ambient.

Results and Discussion

Scanning electron microscope (SEM) photos reveal good morphology and edge definition before and after annealing for both SiO and SiO_2 (Fig. 1).

    As-Deposited   **SiO**   **SiO₂**

Figure 1. SEMS of SiO/SiO_2 film deposition on TLM contacts

The dry etching of SiO and SiO_2 was also studied, as illustrated in Figures 2 and 3. The SiO layer has slower etch rates in both SF_6 and NF_3 discharges under all conditions. Similar to our wet etch results (Table I), the slower etch rate of SiO_x has been attributed to more Si-Si bonds than that of SiO_2.

Table I. Etch rates of SiO_2/SiO in dilute HF and BOE solutions.

Evaporated Oxide	Solution	Etch Rate (Å/min)
SiO	DI Water/HF (20/1)	109.8
SiO_2	DI Water/HF (20/1)	162
SiO	BOE	50.36
SiO_2	BOE	108

It is believed that the polar Si-O bonds make the passivation layer more vulnerable to positive ion attack in the ICP plasma [4,5]. In figure 2 increasing the D.C. bias causes a linear increase in etch rate due to a linear increase in energy of the ion bombardment. In figure 3 the etch rate of both SiO_2 and SiO saturate quickly with increasing ICP power. This is because the increasing ICP power quickly saturates the amount of ions generated in the plasma. SF_6 etched SiO_2 and SiO faster than NF_3 because it is able to generate more F atoms.

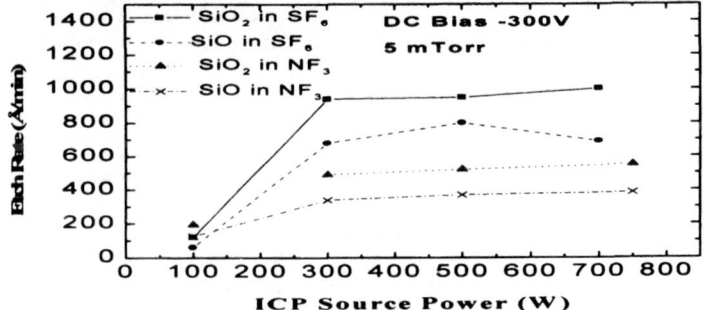

Figure 2. SiO/SiO_2 Film Etch Rate vs. DC Bias in SF_6 and NF_3

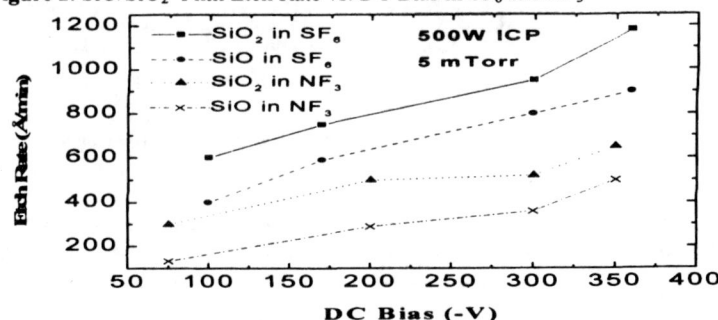

Figure 3. SiO/SiO_2 Film Etch Rate vs. ICP Source Power in SF_6 and NF_3

Auger analysis of the layers indicates that upon annealing to 700°C the Si/O ratio of SiO increases in the film (fig. 4), while the surface Si/O ratio decreases from .62 to .54. This is consistent with oxygen flux to the surface during annealing. Figure 5 shows that the film Si/O ratio of SiO_2 remains constant both before and after annealing to 700°C. The surface SiO/O ratio remained constant as well.

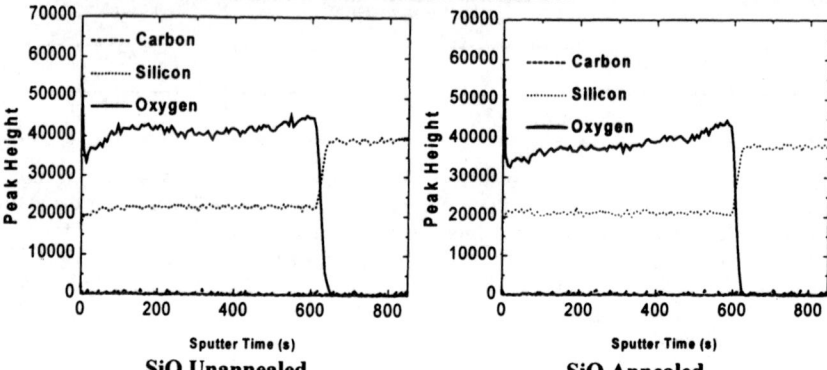

SiO Unannealed   SiO Annealed

Figure 4. Auger Depth Profiles of 3000 Å SiO on Si as-deposit and after annealing 700°C

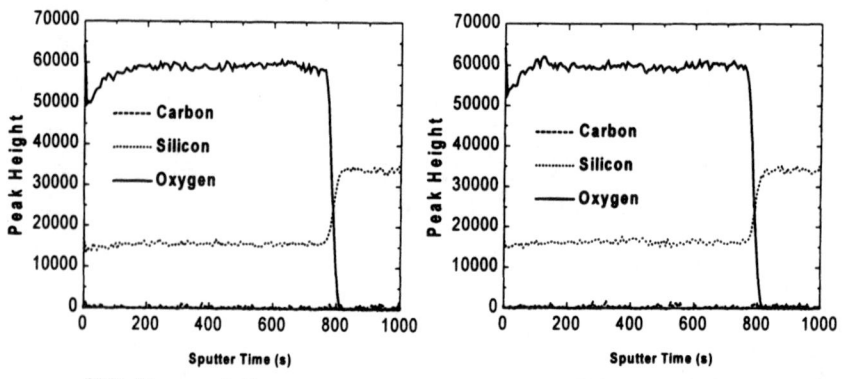

SiO₂ Unannealed   SiO₂ Annealed

Figure 5. Auger Depth Profiles of 3000 Å SiO_2 on Si as-deposit and after annealing 700°C

AFM of SiO (fig. 6) reveals that the root mean square (rms) surface roughness decreases upon annealing to 700 °C from .364 nm to .263nm. This decrease in surface roughness may be related to the oxygen flux to the surface indicated in the Auger analysis. However, for SiO_2 film (fig. 7), the rms roughness increases from .876 nm to 1.01 nm

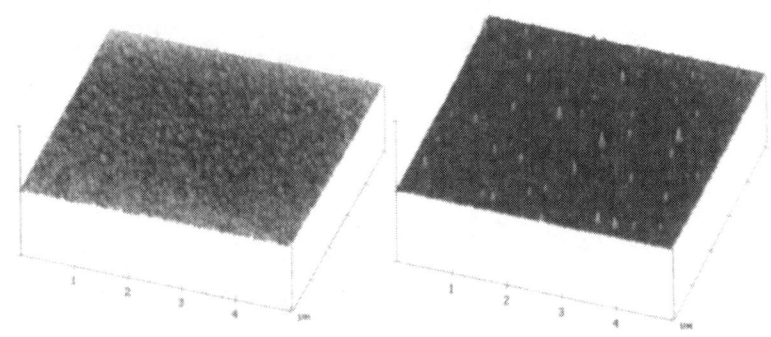

SiO Unannealed   **SiO Annealed 700 °C**

Figure 6. 5 μm AFM of 3000 Å SiO on Si as-deposit and after annealing 700°C

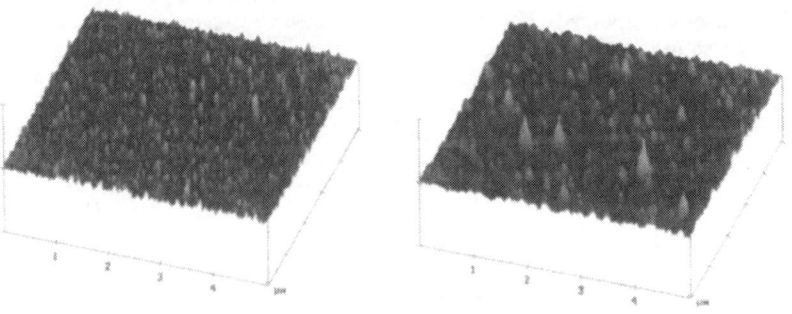

SiO₂ Unannealed   **SiO₂ Annealed**

Figure 7. 5 μm AFM of 3000 Å SiO_2 on Si as-deposit and after annealing 700°C

Ellipsometry was performed to assess the change in refractive index and thickness at different positions on the wafer, and the refractive index upon annealing. The refractive index of the as-deposited SiO varies considerably across the wafer (+/- .025), while the SiO_2 remains more constant (+/- .001). It has been shown previously, that e-beam deposition of SiO always results in Si/O≈1 regardless of deposition pressure and rate [6]. Therefore, spatial variance in refractive index may be more structural (such as surface roughness) than chemical. Both samples showed excellent thickness uniformity across the wafer varying at ≤ 2%. Figure 8 shows the refractive indexes of the annealed samples. The magnitude of the changes is relatively the same, and minor (approximately +/-3% at 700 °C). However, the refractive index of SiO increases, while that of SiO_2 decreases over the temperature range. The increase in SiO refractive index has been linked to the increase in Si/O ratio upon annealing that is evident in Auger analysis of the layers on silicon. The decrease in SiO_2 refractive index was attributed to the increase in surface roughness of the film upon annealing.

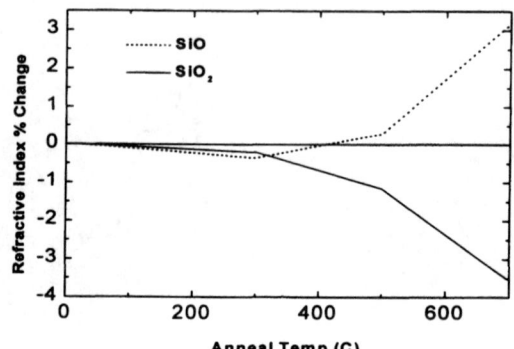

Figure 8. Refractive index change vs. anneal temperature for SiO/SiO₂ films.

Conclusions

The characteristics of hydrogen free SiO and SiO₂ passivation layers used to passivate metallization and prevent the Galvanic etching effect were studied. Both passivation layers showed good morphology and edge definition on metal contacts upon annealing. E-beam deposited SiO₂ and SiO had dry etch rates comparable to PECVD SiO₂. They also had similar magnitudes of refractive index change upon annealing. SiO showed more spatial variation in refractive index between samples and less thermal stability than SiO₂. However, SiO was smoother before and after annealing. Therefore, we conclude that the use of E-beam SiO₂ and SiO is compatible with current processing technologies, but that the superior dielectric would be application dependent in view of the mixed thermal stability and roughness results.

References

1. S.J. Pearton, C.R. Abernathy, W.S. Hobson, F. Ren, T.R. Fullowan, J. Lopata and U. K. Chakrabarti, Materials Science and Engineering, **B13,** p. 171 (1992)
2. S.J. Pearton, Appl. Phys. **A43,** p. 153 (1987)
3. N.M. Johnson, C. Doland, F. Ponce, J. Walker and G. Anderson, Physica B, **170,** p.3 (1991)
4. J. R. LaRoche, F. Ren, J.R. Lothian, J. Hong, S.J. Pearton, and E. Lambers, *The Use of Amorphous SiO and SiO₂ to Passivate AuGe Based Contacts for GaAs Ics,* Journal of the Electrochemical Society, in press
5. G. Higashi, The Physics and Chemistry of SiO2 and the Si-SiO₂ Interface **2,** p.187 (1993).
6. Pivot, Thin Solid Films, **89,** p. 175 (1982)
7. P. C. Chao, P. M. Smith, S. Wanuga, W. H. Perkins, and E. D. Wolf, IEEE Electron Dev. Lett., **EDL-4,** p. 326 (1983).
8. C. S. Wu, F. Ren, S. J. Pearton, M. Hu, C. K. Pao, and R. F. Wang, IEEE Trans. Electron Dev. **42,** p. 1419 (1995).
9. F. Ren, A. B. Emerson, V. J. Scarpelli, S. J., Pearton, and J. Brown, Appl. Phys. Lett., **58,** p. 1030 (1991).
10. M. Hagio, J. Electrochem. Soc., **140,** p. 2402 (1993).
11. W. Strifler and W. Yeung, 1994 US Conf. On GaAs Man. Tech., Las Vegas, NV, p.155.

Anodic sulfidation and model characterisation of GaAs (100) in $(NH_4)_2S_x$ solution

R. F. Elbahnasawy*, J. G. McInerney*, P. Ryan**, G. Hughes** and M. Murtagh***

*Department of Physics, National university of Ireland, University College, Cork, Ireland.
**Department of Physics, Dublin City University, Ireland.
***National Microelectronics Research Center, National University of Ireland, University College, Cork, Ireland.

Abstract

Electrochemical sulfidation of n-type GaAs (100) has been investigated under anodic conditions with a view to surface passivation for improved electronic and optical properties. This treatment has successfully removed the native oxide and formed a thick layer of gallium and arsenic sulfides displaying high durability against oxidation and optical degradation compared to conventional dipping treatment using $(NH_4)_2S$ solution. X-ray photoelectron spectroscopy (XPS), Auger electron spectroscopy (AES), secondary ion mass spectroscopy (SIMS) and atomic force microscopy (AFM) have been used to characterize the treated surfaces. These studies have been used to devise a structural model of the near-surface region. The results of Raman backscattering spectroscopy measurements indicate that there is a 35% reduction of the surface barrier height compared to the untreated surface. This passivation technique has been shown to be effective in reducing surface band bending on GaAs (100) and enhancing the chemical stability of the surface, making it more suitable for electronic and optoelectronic device applications.

1: Introduction

For numerous semiconductor devices, it is important to fabricate surfaces/interfaces with low densities of states.[1] For example defects in heterojunction bipolar transistors (HBT) cause high rates of recombination along mesa surfaces resulting in lower current gain. In metal-semiconductor-metal (MSM) photodiodes, they cause large reverse leakage currents and lower breakdown voltages. Always the performance of GaAs devices is limited by poor surface passivation, high surface recombination velocity and large interface state density. In semiconductor laser sources, surface defects are particularly harmful for reducing reliability and reducing power output dramatically. A poorly passivated laser facet creates nonradiative recombination centers, which lowers laser light output and heats up laser facets.[2] Moreover cubic GaS formation exhibits properties that make it suitable for waveguide fabrication.[2] It is generally believed that treatment with sulfide solutions etch surface oxides and terminate the GaAs surface with S-Ga/As bonds, which temporarily protect the surface from ambient air oxidation.[3,4]

2: Experimental arrangement

The samples used were n-type $0.6 \sim 2.5 \times 10^{18}$ cm^{-3} GaAs (100) single crystal wafer (Si-doped). The wafer was cleaved into 4x4 and 9x9 mm^2 for dipping and anodic treatments, respectively. The sample was held using a vacuum clamp and a silicon O-ring. Electrical contact was made by connecting the isolated rear of the wafer to the anode of a potentiostat using silver paint and wiring through a glass tube as shown in Figure 1. The exposed surface area of the sample was 1.2 cm^2. The potentiostat was a German made (Bank Electronik) Potentio-Galvano-Scan 25V/2A Wenking PGS95 with PC-control and SPK-RP software. The sample functioned as an anode, with an AgCl-reference electrode (saturated calomel electrode) and Platinum standard gauze basket as a counter electrode, all aligned in a rectangular-shaped glass container in 3M $(NH_4)_2S_x$ (x=5g S/100ml) solution. The samples were degreased and stripped of oxides by dipping into $(NH_4)_2S_x$ solution for 3 minutes followed by anodic polarization in the same solution.

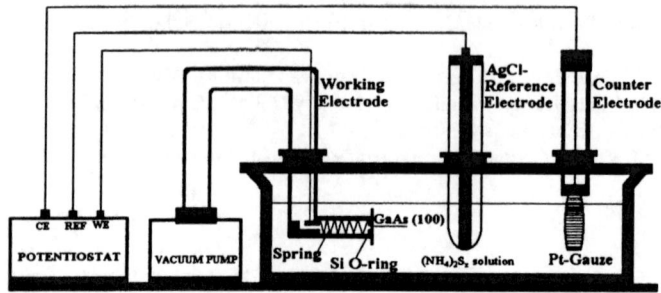

Figure 1: Experimental arrangement for anodic passivation.

The XPS scans were performed using a VG-Microtech x-ray source (Al K_α). The depth profiling was performed using Ar^+ bombardment at a milling rate of 2 nm/min. SIMS depth profiling was also performed on a SIMS analyzer (cameca IMS-3f) with primary ion beam 50 nA (14.5 keV) and a heavy bombardment of cesium ions. Raman backscattering was recorded in the $z(x, x + y)\bar{z}$ geometry using the 512 nm line of an 30 mW Ar^+ laser with a 2 μm^2 probe area. AFM was recorded in contact mode, yielding the average deviation of the average height, R_a.

3: Experimental results and discussion

In this work molarity and sulfur-ion concentration of $(NH_4)_2S_x$ solution including the anodic scan direction have profound effects upon the anodic passivation of n-type GaAs (100). Careful control was necessary to achieve both a chemically and electronically stable anodic passivation. The following cases have been considered for anodic characterization: (i) At high anodic potential as shown in the reverse anodic scan (Figure 2, Region II), XPS characterization has shown no evidence of sulfur deposition after DI water rinse and blow-dry in nitrogen. In this case the anodic potential and sulfur-ion concentration were high, with the result that the anodic treatment was not better than the conventional dipping of GaAs surfaces in $(NH_4)_2S$ solutions, leaving behind only an atomic layer of sulfur which is removed by DI water rinsing.

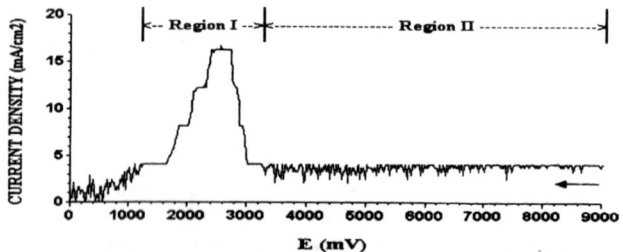

Figure 2: Cyclic voltammogram of n-type 1×10^{18} cm^{-3} GaAs (100) in 3M $(NH_4)_2S_x$ solution (x=5g S/100 ml) at anodic scan rate 20 mV/s.

(ii) However, at the sulfur characteristic peak (Figure 2, Region I), anodic passivation takes place in characteristic steps at which the thickness of the sulfide overlayer can be controlled by time and/or current density. This characteristic peak is mainly dependent upon the sulfur-ion concentration and positioned according to the electrochemical cell parameters for a given substrate doping concentration. It has been found that the reverse anodic scanning i.e. from high potential to low potential, is the best way to achieve good surface quality with low surface roughness compared to the forward scan (from low to high anodic scan potential) as shown in AFM later. The structure of the deposited overlayer has been analyzed using XPS and SIMS depth profiling and AES. The depth profiling revealed the atomic concentration of Ga, As, O, S and C. Rinsing the sample in DI water after anodic passivation has dissolved the arsenic compounds at the top of the deposited layer. As shown in Figure 3, Ga3d and As3d core level spectra show evidence of strong Ga-S and As-S bonding at approximate depth of 120 nm. The chemical shifts at 1.36 eV and 2.8 eV were attributed to Ga-S bond and 1.61 eV and 3.0 eV to As-S bond. The gallium and arsenic sulfide chemical shifts were consistent with previous work as summarized in Table I. however the presence of Ga/As-oxides can not be ruled out. Both oxygen and sulfur are present throughout the depth profiling of the anodic layer with constant gallium concentration and the arsenic signal increases as the interface with the GaAs substrate is approached. The SIMS depth profile was in broad agreement with the XPS data given that the elemental sensitivities are different. The estimated overlayer thickness were 200 ~ 300 nm.

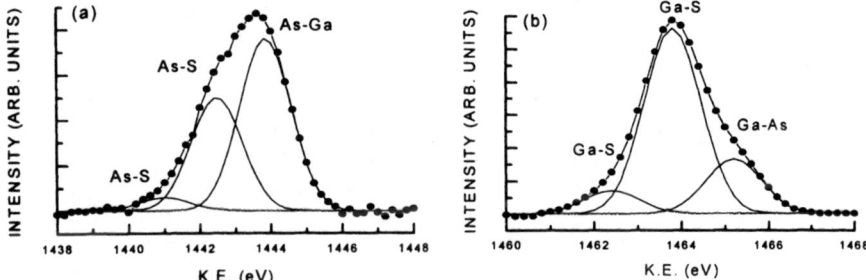

Figure 3: Core level spectra (60 minutes Ar$^+$-milling) for anodically passivated n-type (Si) 1×10^{18} cm^{-3} GaAs (100) in 3 M (NH$_4$)$_2$S$_x$ (x=5g/100 ml) solution. (a) As 3d and (b) Ga3d

Table I: XPS chemical shifts for anodically treated GaAs (100) in (NH$_4$)$_2$S solution. The present work is for n-type (Si) GaAs sample degreased and anodically treated in 3M (NH$_4$)$_2$S$_x$ (x=5g/100 ml) solution at room temperature.

Ga-S bond	As-S bond	
1.36 ± 0.15 & 2.8 ± 0.2 eV (Ga3d)	1.61 ± 0.2 & 3.0 ± 0.3 eV (As3d)	Present work
1.65 & 3.25 eV (Ga3d)	1.8 & 3.5 eV (As3d)	Hou et al.[13]
1.35 & 2.9 eV (Ga2p)	1.6 & 3.3 eV (As2p)	Ershov et al.[12]
1.3 & 2.8 eV		Cai et al.[11]

The surface quality and the effects of sulfide overlayer thickness on the surface roughness have been assessed by atomic force microscopy (AFM). Figure 4 illustrates the effect and sensitivity of current density on surface roughness at the I-V characteristic peak (Figure 2, Region I), where surface roughness (R$_a$) increased from 21 nm at (5 mA/cm^2, 3 V) to 100 nm at (14 mA/cm^2, 2.8 V). The surface was flat and smooth at low current density as shown in Figures 4a and 4b. At higher current density (12 mA/cm^2), the deposited layer cracked as shown in Figure 4c, probably because of the high mechanical stresses created by the thick sulfide layer. By increasing current densities the thickness and surface roughness increases causing more mechanical stresses and severe cracks as shown in Figure 4d.

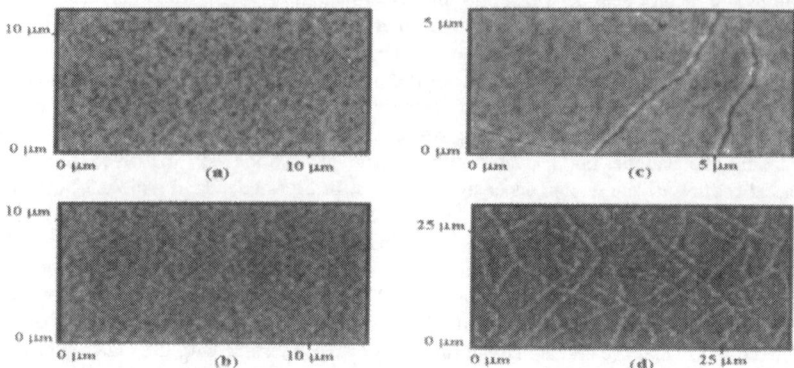

Figure 4: AFM of anodically passivated n-type (Si) 1×10^{18} cm^{-3} GaAs (100) as shown in Figure 2 (Region I). (a) Surface roughness (R_a) 21 nm (5 mA/cm^2, 3 V), (b) R_a 44 nm (8 mA/cm^2, 3 V), (c) R_a 63 nm (12 mA/cm^2, 2.8 V) and (d) R_a 100 nm (14 mA/cm^2, 2.8 V).

4: Raman analysis

Raman spectroscopy is well suited to the study of surface electronic properties of semiconductors provided that the light beam is absorbed within the space-charge layer.[5-8] In heavily doped GaAs (100), the Raman band at 270 cm^{-1} (Figure 5) is due to scattering by the coupled plasmon-phonon mode in the bulk (ω.) where the free carriers exist, while the line at 290 cm^{-1} is due to scattering by longitudinal optical phonons (ω_{LO}) arising from the surface depletion region. The relative reduction of ω_{LO}/ω. intensity indicates a reduction of the surface depletion width and hence a lowering of the surface barrier[9], here implying a shift in the surface Fermi level towards the conduction band after sulfur passivation. Raman spectra of both treated and untreated n-type (Si) 1×10^{18} GaAs (100) are shown in Figure 5. The shift in the surface Fermi level is evident after anodic treatment. Note that the relative invariance of the ω. mode is consistent with the recorded smooth surfaces.

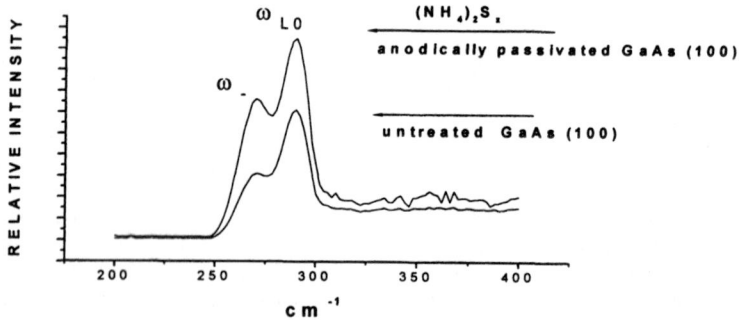

Figure 5: Relative Raman spectral intensity of anodically passivated n-type (Si) 1×10^{18} cm^{-3} GaAs (100) in $(NH_4)_2S_x$ solution. Passivated (top), untreated (bottom) reveals a 35% reduction in the surface barrier height which exhibits surface roughness of 20 ~ 40 nm.

The calculation were carried out at doping concentration $N_D = 1\times10^{18}$ cm^{-3}, assuming a surface barrier height of untreated GaAs $V_b = 0.78$ eV.[18] After anodic sulfidation (Figure 2, Region I) the Raman backscattering measurements recorded a reduction in the surface barrier height, which was attributed to a lowering of the surface state density. The surface barrier height is reduced to $V_b = 0.50$ eV (35% reduction in the surface barrier height of the untreated GaAs). This reduction remained unchanged for an hour after exposure to laser illumination at a power density 5mW/μm^2. This shows good potential for optoelectronic device applications. Detailed Raman investigations have also been performed to examine the surface roughness upon the backscattered data. An apparent doubling of the actual reduction in surface barrier height has been measured for surface roughness between 80 and 1200 nm. The investigation showed that surface roughness should not exceed 60 nm to extract barrier height dependent Raman measurements.

5: Structural models

Based on XPS, AES and SIMS experimental data, a structural model has been proposed: Under the action of a strong electric field (Figure 2, Region I) and in a high molarity sulfur-saturated $(NH_4)_2S_x$ aqueous solution, the oxides and the defective GaAs (100) layer are removed and replaced by thick sulfide overlayer as shown schematically in Figure 6. At the electrolyte/GaAs interface, the concentration of sulfur ions becomes quite large and the high anodic potential effectively renders the n-type GaAs (100) surface locally p-type like, this accelerates sulfur species reactions with the GaAs (100) substrate and activates sulfur deposition as shown in equation (1). The reaction of the hydroxyl group with the substrate as in equation (2) is made insignificant as a result of its lower concentration compared to the sulfur ions.

$$GaAs + 5\,S^{2-} + 6\,h^+ \longrightarrow GaS_3^{3-} + AsS_3^- \qquad (1)$$
$$GaAs + 10\,OH^- + 6\,h^+ \longrightarrow GaO_3^{3-} + AsO_2^- + 5\,H_2O \quad (2)$$

The XPS results suggest that the anodically grown layer consists of a mixed chemical phase region with the presence of Ga-As, Ga-S, As-S bonds and possibly As/Ga-O bonds as shown in Figure 6c. Due to ambient air contamination and DI water rinsing after anodic passivation, a thin layer of mixed compounds could be formed at the top of the deposited layer.

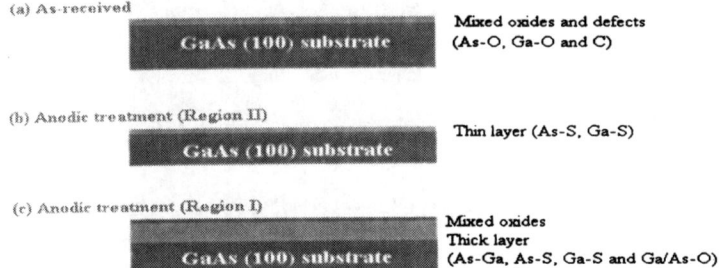

Figure 6: Illustrated model of GaAs (100) surface. (a) Native oxides and defective layer cover the surface, (b) anodic passivation in 3M $(NH_4)_2S_x$ involves oxide removal and a sulfide atomic layer is formed (Figure 2, Region II), and (c) thick layer of mixed phases with thickness depending upon the I-V position on the step-peak (Figure 2, Region I). A small layer of oxide is assumed to cover the deposited layer after DI water rinse.

The model suggests that the anodic sulfide growth proceed in the ammonium sulfide solution via the following steps: At the sulfur characteristic step-peak (Figure 2, Region I) the negatively charged sulfur species in aqueous $(NH_4)_2S_x$ solution extract both Ga and As atoms from an initial sulfide overlayer, producing negatively charged vacancies. These vacancies near the anodic sulfide surface are driven by the anodizing field toward the sulfide/GaAs interface where they extract both Ga and As atoms from the GaAs (100) substrate. This extraction creates a new sulfide overlayer and carries away the negative charges through the GaAs (100) substrate to complete the anodization circuit. The migration of the vacancies toward the sulfide/GaAs interface produces a flow of Ga and As atoms in the opposite direction by the vacancy exchange mechanism.[10]

6: Conclusions

The sulfur ion concentration in aqueous $(NH_4)_2S_x$ solution plays a very important role in passivating n-type GaAs (100). Turning the n-type material effectively to p-type like by biasing the circuit at high anodic potential is another stage that appears vital, with high sulfur ion concentration producing chemically stable anodic passivation. Effective passivation removes surface states, prevents oxygen desorption and terminates the dangling bonds. In this study SIMS and XPS have provided evidence of Ga, As, C, S and O with significant As-S and Ga-S bonding. Compared with the unstable dipping treatment in ambient air the anodic passivation has provided chemically stable surfaces against oxidation for at least four months in ambient air with Raman barrier height data revealing enhanced electronic properties of the GaAs surface, indicating considerable promise for optoelectronic device applications.

References

[1] V. Malhotra, Electrochem. Soc. Proc. **2**, 233 (1996).

[2] V. N. Bessolov, M. V. Lebedev and D. R. T. Zahn, J. Appl. Phys. **82**(5), 2640 (1997).

[3] A. Kapila and V. Malhotra, IEEE, 0-7803-3374-8, 275 (1997).

[4] V. N. Bessolov, E. V. Konenkova and M. V. Lebedev, Tech. Phys. Lett. **22**(9), 749 (1996).

[5] M. S. Carpenter, M. R. Melloch and T. E. Dungan, Appl. Phys. Lett. **53**(1), 66 (1988).

[6] Z. S. Li, X. Y. Hou, W. Z. Cai, W. Wang, X. M. Ding, X. Wang, J. App. Phys. **78**(4), 2764 (1995).

[7] Z. S. Li, X. Y. Hou, W. Z. Cai, W. Wang, M. Zhang, G. S. Dong, X. Jin and X. Wang, Mat. Res. Soc. Symp. Proc. **284**, 607 (1993).

[8] X. Y. Hou, W. Z. Cai, Z. Q. HE, P. H. Hao, Z. S. Li, X. M. Ding and X. Wang, App. Phys. Lett. **60**(18), 2252 (1992).

[9] D. J. Olego, R. Schachter, J. A. Baumann, Appl. Phys. Lett. **45**(10), 1127 (1984).

[10] C. C. Chang, P. H. Citrin, B. Schwartz, J. Vac. Sci. Technol. **14**(4), 943 (1977).

[11] W. Z. Cai, Z. S. Li, R.Z. Su, G. S. Dong, D. M. Huang, X. M. Ding, X. Y. Hou and X. Wang, Appl. Phys. Lett. **64**(25), 3425 (1994).

[12] S. G. Ershov, A. F. Ivankov, V. V. Korablev and V. Yu. Tyukin, Tech. Phys. Lett. **22**(7), 561 (1996).

[13] X. Hou, X. Chen, Z. Li, X. Ding and X. Wang, Appl. Phys. Lett. **69**(10), 1429 (1996).

HIGH-DENSITY PLASMA-INDUCED ETCH DAMAGE OF GaN

R. J. Shul,[*] L. Zhang,[*] A. G. Baca,[*] C. G. Willison,[*] J. Han,[*] S. J. Pearton,[**] F. Ren, [**] J. C. Zolper,[***] L. F. Lester[****]

[*]Sandia National Laboratories, Albuquerque, NM 87185-0603, rjshul@sandia.gov
[**]University of Florida, Department of Materials Science and Engineering, Gainesville, FL 32611
[***]Office of Naval Research, Arlington, VA 22217
[****]University of New Mexico, Center for High Technology Materials, Albuquerque, NM 87106

ABSTRACT

Anisotropic, smooth etching of the group-III nitrides has been reported at relatively high rates in high-density plasma etch systems. However, such etch results are often obtained under high dc-bias and/or high plasma flux conditions where plasma induced damage can be significant. Despite the fact that the group-III nitrides have higher bonding energies than more conventional III-V compounds, plasma-induced etch damage is still a concern. Attempts to minimize such damage by reducing the ion energy or increasing the chemical activity in the plasma often result in a loss of etch rate or anisotropy which significantly limits critical dimensions and reduces the utility of the process for device applications requiring vertical etch profiles. It is therefore necessary to develop plasma etch processes which couple anisotropy for critical dimension and sidewall profile control and high etch rates with low-damage for optimum device performance. In this study we report changes in sheet resistance and contact resistance for n- and p-type GaN samples exposed to an Ar inductively coupled plasma (ICP). In general, plasma-induced damage was more sensitive to ion bombardment energies as compared to plasma flux. In addition, p-GaN was typically more sensitive to plasma-induced damage as compared to n-GaN.

INTRODUCTION

Plasma-induced damage has become more relevant to the group-III nitride materials as interest in electronic devices has increased. For example, with the recent reports of GaN/AlGaN heterojunction bipolar transistors (HBTs)[1-3] low damage etch processes are required to form the collector and base contacts. For HBTs, plasma-induced damage can increase surface recombination currents in the base-emitter junction and surface generation currents in the base-collector junction. Furthermore, the active regions of many electronic devices of interest are often shallow thus requiring low damage plasma processes to ensure optimum device performance.

To date, the majority of plasma etch development for the group-III nitrides has been for optoelectronic devices where mesa structures are etched to depths often greater than 1 μm. The etch requirements typically include high rate (~ 1 μm/min), smooth sidewall morphologies, and anisotropic profiles. With the increased interest in electronic devices, etch requirements must also include slow, controlled etch rates, selectivity of one material over another, and low-damage. High-density plasma (HDP) etch systems, and in particular inductively coupled plasma (ICP) etch systems, have shown encouraging results for the development of versatile, well-controlled etch processes. For example, GaN etch rates ranging from ~100 Å/min to >1 μm/min have been reported in ICP etch systems with anisotropic profiles and smooth etch morphologies.[4-10] Unfortunately, the best etch results are often obtained under energetic ion bombardment and/or high plasma flux conditions which can increase plasma-induced damage.

Plasma-induced damage can take many forms causing degradation of electrical and optical properties of the device. Several damage mechanisms are summarized below.[11-13]

a) Energetic ion bombardment can create lattice defects or dislocations, formation of dangling bonds on the surface, or implanted etch ions. These defects often act as deep level states and produce compensation, trapping, or recombination in the material. Damage as deep as 1000 Å has been reported in GaAs.[13]

b) The presence of hydrogen during the etch process due to either its use in the plasma chemistry, residual water vapor in the chamber, or other sources including erosion of the photoresist mask, can unintentionally passivate the dopants present in the material up to depths of several thousand angstroms.

c) Polymer deposition may occur due to the use of plasma chemistries containing CH_x radicals or reactions of photoresist masks with the plasma.

d) Non-stoichiometric surfaces may be formed due to preferential loss of one of the lattice constituents. This often occurs due to large differences in the volatility of the respective etch products, leading to enrichment of the less volatile species, or preferential sputtering of the lighter element. Non-stoichiometric depths are typically < 100 Å.

Since GaN is more chemically inert than GaAs and has higher bonding energies, it may be reasonable to use higher ion energies during the etch process with potentially less damage to the material. However, reports of plasma-induced damage of the group-III nitrides have been limited. Pearton and co-workers have reported plasma-induced damage results for InN, InGaN, and InAlN in an ECR-generated plasma where the damage increased as a function of ion flux and ion energy.[11] Ren and coworkers also studied the effect of plasma-induced damage on the electrical characteristics of InAlN and GaN FET structures for ECR BCl_3, BCl_3/N_2, and CH_4/H_2 plasmas.[14] Several trends were observed: 1) in the presence of hydrogen, passivation of the doping in the channel layer can occur; 2) ion bombardment energies can create deep acceptor states that compensate the material; and 3) preferential loss of N can produce rectifying gate characteristics on etched surfaces. Ping and coworkers studied GaN Schottky diodes for Ar and $SiCl_4$ RIE plasmas and observed more damage in pure Ar plasmas and a strong dependence on dc-bias.[15]

In this paper we report the effect of plasma-induced-damage on n- and p-type GaN by monitoring changes in the sheet resistance (R_{sh}) and specific contact resistance (r_c) of thin conducting layers under a variety of ICP plasma conditions simulating those used during device etching. Sheet resistance and specific contact resistance determined from the circular transmission line model (TLM) are used to evaluate plasma-induced damage.

EXPERIMENT

The GaN films etched in this study were grown by either metal organic chemical vapor deposition (MOCVD)[16] or RF-molecular beam epitaxy (MBE).[17] Both n- and p-GaN were evaluated for plasma-induced damage using an ICP reactor. The n-GaN was ~ 1.2 μm thick while the p-GaN was 0.3 μm thick and grown over a 1 μm undoped GaN buffer layer (see Figure 1). Hall measurements were performed on these samples to determine the carrier concentration and mobility. For the MOCVD and MBE n-GaN, the electron concentrations were 8×10^{17} cm^{-3} and 3×10^{17} cm^{-3} and the mobility was 315 cm^2/Vs and 263 cm^2/Vs, respectively. For the MBE p-GaN the hole concentration was 1×10^{17} cm^{-3} and the mobility was 5 cm^2/Vs.

Ohmic contacts for circular TLM patterns were formed by metal evaporation and liftoff using standard photolithography techniques. In order to evaluate changes in sheet resistance (ΔR_{sh}), GaN samples were first metallized and then exposed to the Ar plasma. Sheet resistance was measured at the same position before and after the plasma exposure. Therefore the effect of the doping nonuniformity across the wafer was minimized. To evaluate the specific contact resistance (r_c) as a function of the plasma conditions, the samples were initially patterned with AZ 5214 photoresist, exposed to the Ar plasma and then metallized over areas exposed to the plasma. Measurements were taken at minimum of 4 positions on each sample. The unannealed contacts were 300 Å Ti and 2500 Å Al for n-GaN and 300 Å Ni and 2500 Å Au for p-GaN.

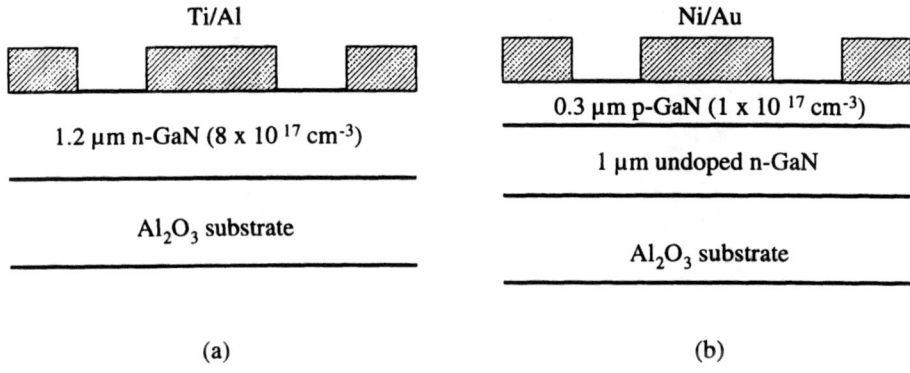

Figure 1. Schematic diagram of n- and p-GaN with metal contacts in place.

The ICP reactor was a load-locked Plasma-Therm SLR 770, which used a 2 MHz, 3 turn coil ICP source. The ion energy or dc-bias was defined by superimposing a rf-bias (13.56 MHz) on the sample. The standard ICP etch parameters used in this study were: 45 sccm of Ar, 25°C electrode temperature, 2 mTorr total pressure, 500 W of ICP source power, and –100 V dc-bias. All samples were mounted using vacuum grease on an anodized Al carrier that was clamped to the cathode and cooled with He gas. Samples were exposed to ICP Ar plasmas for 30 seconds. Samples used to calculate etch depth were patterned using AZ 4330 photoresist. Etch depths were measured with a Dektak stylus profilometer after the photoresist was removed with an acetone spray and were consistently < 150 Å independent of plasma conditions.

Plasma-etch-induced damage effects were evaluated by calculating R_{sh} and r_c with the circular transmission line-method analysis.[18] Ohmic metals were deposited with gap spacings (d) of 2.5, 5, 10, 15, 20, and 25 μm (see Figure 2). The radius of the outer metal contact (r_2) was 100 μm while the inner radius (r_1) was defined by the radius of the outer contact minus the gap spacing for the individual structures. The resistance between contacts as a function of gap spacing was measured using a HP-4145B Semiconductor Parameter Analyzer. For n-GaN samples, the contacts exhibited ideal ohmic I-V characteristics and the resistance was taken at an injection current of 20 mA. However, a slightly rectified I-V curve was observed for Ni/Au contacts to p-GaN, therefore it was necessary to define the current at which the resistance was measured. Due to the relatively large Mg acceptor ionization energy (~170 meV), only small portions of Mg acceptors were ionized at room temperature and the hole concentration of p-GaN was sensitive to the temperature. In order to minimize the effect of temperature variation on the measurement due to local resistive heating of each device, the resistance was measured at a fixed power of 100 mW. The gap spacing, which was used in the analysis, was measured by SEM imaging. Linear regression of measured resistance versus spacing yielded excellent fits with correlation factors > 0.995. The data reported is the average of the measured devices.

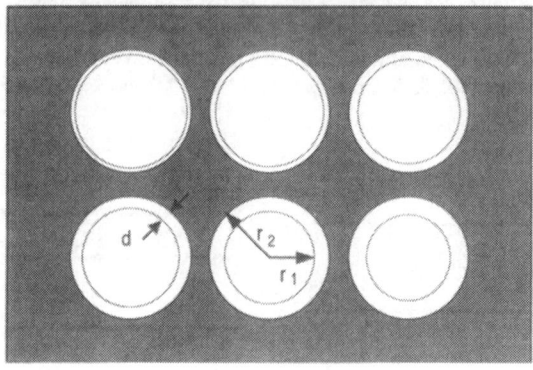

Figure 2. Schematic diagram of the circular metal contact mask used to calculate R_{sh} and r_c. The outer radius r_2 was held constant at 100 μm.

RESULTS AND DISCUSSIONS

It is important to realize that the use of a pure Ar plasma creates a worse case scenario for plasma-induced damage due to the lack of chemical interactions. With the introduction of reactive gases to the plasma for a given ion energy and plasma density, the damage will be reduced when compared to a sputter mechanism since damaged material is typically being removed at a higher rate, leaving a shallower damage depth.

Sheet resistance prior to plasma exposure (R_{sh}), following plasma exposure, and the net change (ΔR_{sh}) are shown in Figures 3 and 4 as a function of dc-bias. The percent variation in sheet resistance ($\Delta R_{sh}/R_{sh}$) is also displayed as a function of dc-bias. This data was obtained for MOCVD (Figure 3) and MBE (Figure 4) grown n-GaN exposed to a 30 sec Ar ICP plasma. ICP conditions were 500 W ICP-source power, 2 mTorr chamber pressure, 40 sccm Ar flow, and 25°C cathode temperature. As dc-bias increased, the net change in sheet resistance for both MOCVD and MBE n-GaN increased monotonically due to higher ion bombardment energies. However, the percent variation in sheet resistance was relatively low, less than 20% at –350 V dc-bias. In contrast, for MBE p-GaN samples plasma-induced damage was much more sensitive to dc-bias. In Figure 5, the percent variation in sheet resistance was ~95% at –350 V dc-bias. The much stronger effect of plasma-induced damage on p-GaN was attributed to the preferential loss of lighter nitrogen atoms under energetic ion bombardment conditions, which can create n-type N-vacancies and compensate the p-GaN. In addition, thinner p-GaN layers (~0.3 μm as compared to ~1.2 μm for n-GaN) and lower hole concentrations (~1 x 10^{17} cm^{-3} as compared to 3-8 x 10^{17} cm^{-3} electron concentration in n-GaN) made the conductivity of the p-GaN film more sensitive to the plasma-induced damages at the surface. It is worth noting that the plasma-induced damage did not produce a lower sheet resistance for n-GaN despite higher electron concentrations at the surface due to n-type N-vacancies generated by the plasma. Higher sheet resistances observed in n-GaN at higher dc-biases may be attributed to either a thinner n-GaN layer caused by Ar ion sputter removal (typically <150 Å), lower electron mobility in the damaged n-GaN layer, or trapping of electrons by the plasma-induced defects.

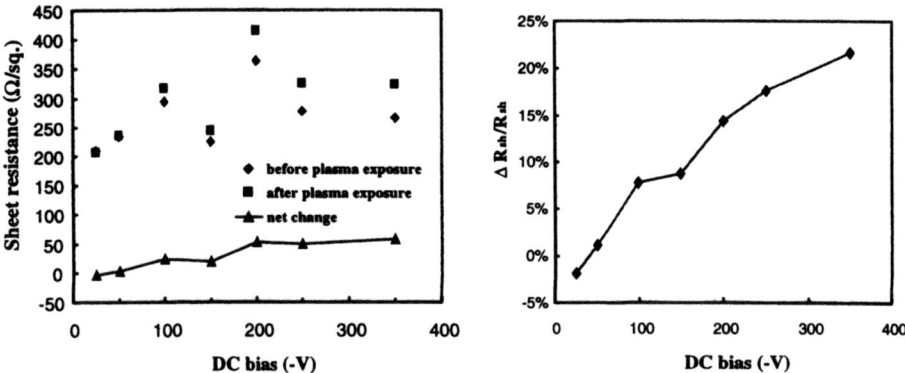

Figure 3. MOCVD n-GaN sheet resistance before plasma exposure, after exposure, the net change, and the % variation ($\Delta R_{sh}/R_{sh}$) is plotted as a function of dc-bias. ICP conditions were 500 W ICP–source power, 2 mTorr chamber pressure, 40 sccm Ar flow, and 25°C substrate temperature.

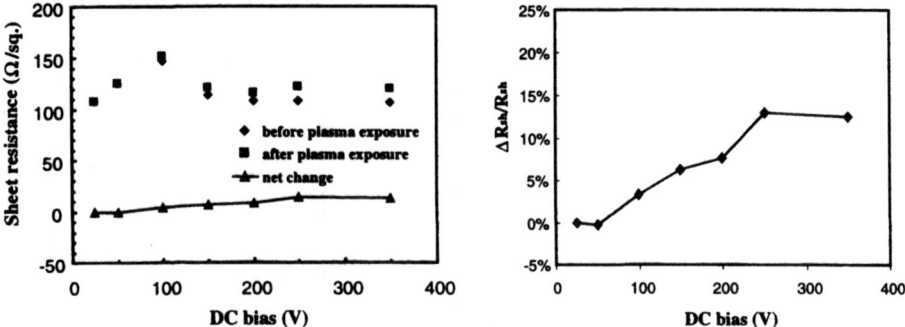

Figure 4. MBE n-GaN sheet resistance before plasma exposure, after exposure, the net change, and the % variation ($\Delta R_{sh}/R_{sh}$) is plotted as a function of dc-bias. ICP conditions were 500 W ICP–source power, 2 mTorr chamber pressure, 40 sccm Ar flow, and 25°C substrate temperature.

Sheet resistance was also studied as a function of plasma flux for MOCVD and MBE n- and MBE p-GaN. As a function of source power, the concentration of reactive species typically increases which increases the chemical component of the etch mechanism. In addition, the ion flux increases which increases the bond breaking efficiency and the sputter desorption component of the etch mechanism. Thus by carefully controlling the ICP-source power and the ion bombardment energy in the plasma, the chemical and physical components of the etch mechanism can be balanced. In Figures 6 and 7, MOCVD and MBE n-GaN sheet resistance and percent variation in sheet resistance is shown as a function of ICP-source power. Sheet resistance for n-GaN samples showed a very weak dependence on ICP-source power. For these experiments, the dc-bias was held constant at –100V. The percent variation in sheet resistance showed less than a 10% change independent of plasma flux. The small amount of change in sheet resistance was attributed to the low ion bombardment energies generated at –100 V dc-bias. In comparison, Figure 8 shows sheet resistance and percent variation in sheet resistance plotted as a function of ICP-source power for MBE grown p-GaN. Once again, the p-GaN material was

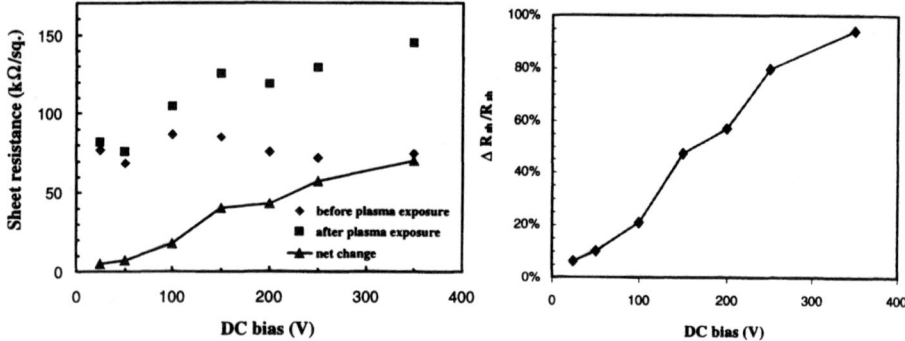

Figure 5. MBE p-GaN sheet resistance before plasma exposure, after exposure, the net change, and the % variation ($\Delta R_{sh}/R_{sh}$) is plotted as a function of dc-bias. ICP conditions were 500 W ICP–source power, 2 mTorr chamber pressure, 40 sccm Ar flow, and 25°C substrate temperature.

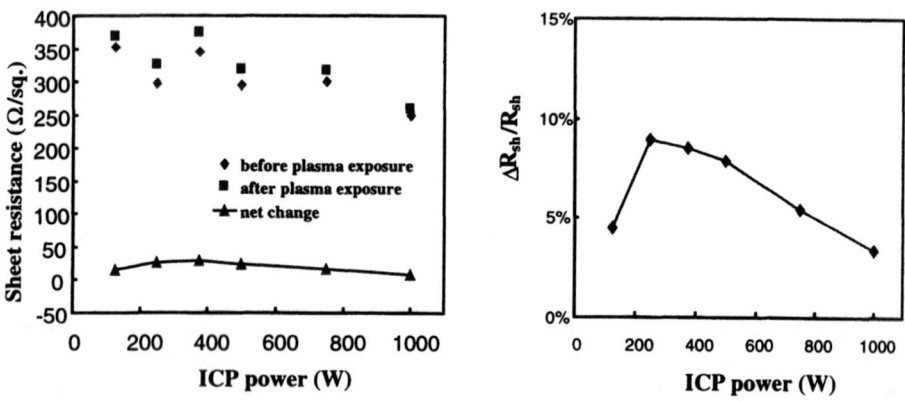

Figure 6. MOCVD n-GaN sheet resistance before plasma exposure, after exposure, the net change, and the % variation ($\Delta R_{sh}/R_{sh}$) is plotted as a function of ICP-source power. ICP conditions were –100 V dc-bias, 2 mTorr chamber pressure, 40 sccm Ar flow, and 25°C substrate temperature.

much more sensitive to plasma-induced damage due to preferential loss of lighter nitrogen atoms which created n-type N-vacancies and compensated the p-GaN, thinner p-GaN layers, and lower hole concentrations. The percent variation in sheet resistance increased to ~25% at 375 W and then decreased as the ICP-source was increased further. This trend may be attributed to the balance between the formation and removal of the damaged layers under high plasma flux conditions. Typically, <150 Å of GaN was removed under these plasma conditions.

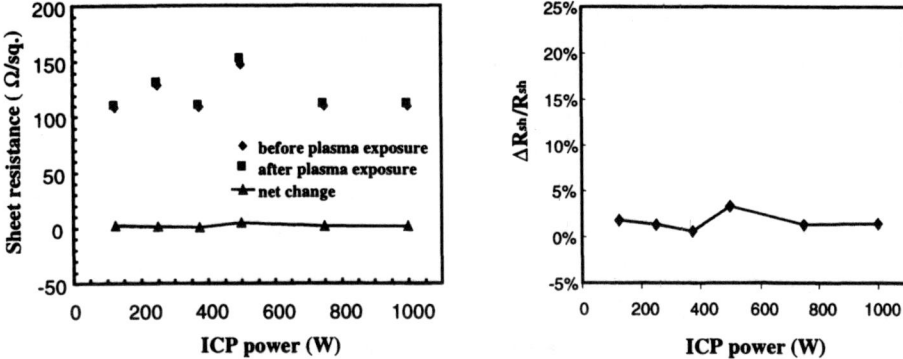

Figure 7. MBE n-GaN sheet resistance before plasma exposure, after exposure, the net change, and the % change is plotted as a function of ICP-source power. ICP conditions were -100 V dc-bias, 2 mTorr chamber pressure, 40 sccm Ar flow, and 25°C substrate temperature.

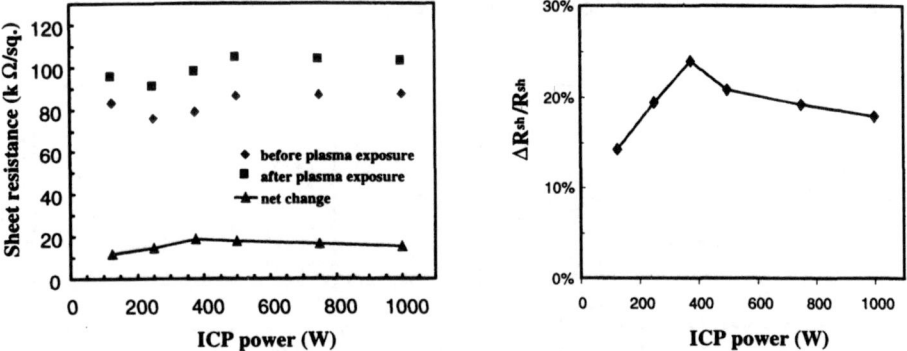

Figure 8. MBE p-GaN sheet resistance before plasma exposure, after exposure, the net change, and the % variation ($\Delta R_{sh}/R_{sh}$) is plotted as a function of ICP-source power. ICP conditions were -100 V dc-bias, 2 mTorr chamber pressure, 40 sccm Ar flow, and 25°C substrate temperature.

Contact resistance was also used to evaluate ICP damage for MOCVD n-GaN and MBE p-GaN. In Figure 9, the contact resistance is plotted as a function of dc-bias and ICP-source power for MOCVD grown n-GaN. The contact resistance increased by almost 3 orders of magnitude as the dc-bias increased from –25 V to –350 V. The increase in damage was attributed to the more physical nature of the etch process under high ion bombardment energy conditions. As the ICP-source power increased from 125 W to 375 W, the contact resistance increased by less than an order of magnitude. The dc-bias was held constant at −100 V. Therefore, as the plasma flux increased, the damage was less severe as compared to the more physical ion bombardment component of the etch mechanism. Above 375 W ICP-source power the contact resistance remained relatively constant possibly due to more effective removal of the damaged layer. Thus, the balance between the formation and removal of the damaged layers are critical to the plasma-induced damage mechanism. The contact resistance is also plotted as a function of dc-bias and ICP-source power for MBE grown p-GaN in Figure 10. Despite the fact that the contacts to p-GaN were not Ohmic, the contact resistance for p-GaN was again more sensitive to the increase in ion bombardment energies as compared to plasma flux.

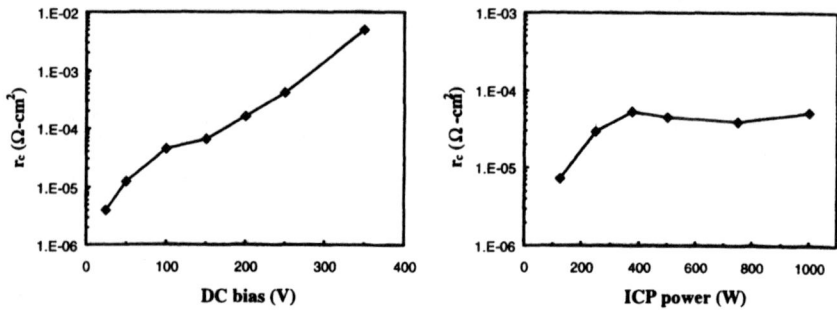

Figure 9. MOCVD n-GaN contact resistance as a function of dc-bias and ICP-source power. ICP conditions were either 500 W ICP-source power, or -100 V dc-bias, 2 mTorr chamber pressure, 40 sccm Ar flow, and 25°C substrate temperature.

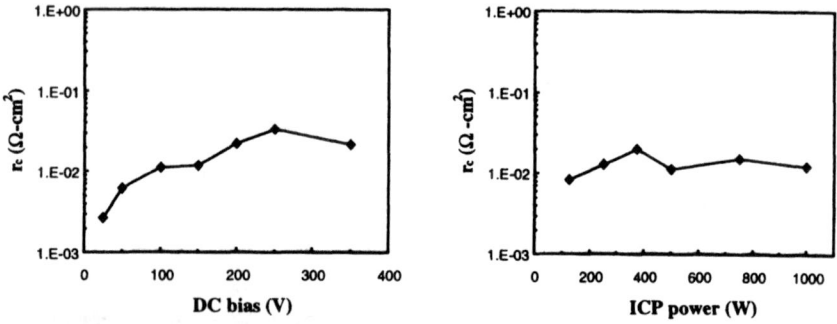

Figure 10. MBE p-GaN contact resistance as a function of dc-bias and ICP-source power. ICP conditions were either 500 W ICP-source power, or -100 V dc-bias, 2 mTorr chamber pressure, 40 sccm Ar flow, and 25°C substrate temperature.

It is interesting to note that damage generated as a function of ICP-source power or plasma flux, showed recovery or at least stabilization under high power conditions (see Figures 6-10). This trend may demonstrate the balance between depth of the induced damage and the ability to remove the damaged material. Under low to moderate ICP-source powers, the damaged area was not removed efficiently. However, under high power conditions there is a competition between generation of damage and sputter removal of the damaged layers. Less than 150 Å of GaN was removed under these conditions, which implies that the damage is quite shallow.

CONCLUSIONS

In summary, damage to GaN devices during plasma processing can be significant. Under all conditions, plasma induced damage was more sensitive to dc-bias as compared to ICP-source power implying that ion bombardment energy plays a significant role in the damage mechanism. For sheet resistance, the p-GaN samples showed much more damage than the n-GaN possibly due to the preferential loss of lighter nitrogen atoms that created n-type N-vacancies and compensated the p-GaN. In addition, thinner p-GaN layers and lower hole concentrations made the conductivity of p-GaN film more sensitive to the plasma-induced damages at the surface. Higher sheet resistances observed in n-GaN at higher dc-biases may be attributed to either a thinner n-GaN layer caused by Ar ion sputtering, lower electron mobility in the damaged n-GaN layer, or trapping of electrons by the plasma-induced defects. It is important to note that this study presents a worst case scenario since material is not being removed at rates typical in a standard etch process. Additionally, much of this damage may be removed by post-etch processes including high-temperature anneals or wet chemical etch processes.

ACKNOWLEDGMENTS

Sandia is a multiprogram laboratory operated by Sandia Corporation, a Lockheed Martin Company, for the United States Department of Energy under contract DE-ACO4-94AL85000. Two of the authors (L. F. Lester and L. Zhang) are also supported by a National Science Foundation CAREER Grant EC5-9501785. The work at the University of Florida is partially supported by a DARPA/EPRI grant (MDA 972-98-1-0006) and by NSF (DMR-9732865).

REFERENCES

1. L. S. McCarthy, P. Kozodoy, S. P. DenBarrs, M. Rodwell, and U. K. Mishra, 25th Int. Symp. Compound Semicond., Oct. 1998, Nara, Japan.
2. F. Ren, C. R. Abernathy, J. M. Van Hove, P. P. Chow, R. Hickman, J. J. Klassen, R. F. Kopf, H. Cho, K. B. Jung, J. R. LaRoche, R. G. Wilson, J. Han, R. J. Shul, A. G. Baca, and S. J. Pearton, MRS Internet J. Nitride Semicond. Res. 3, 41 (1998).
3. J. Han, A. G. Baca, R. J. Shul, C. G. Willison, L. Zhang, F. Ren, A. P. Zhang, G. T. Dang, S. M. Donovan, X. A. Cao, H. Cho, K. B. Jung, C. R. Abernathy, S. J. Pearton, and R. G. Wilson, Appl. Phys. Lett, submitted 1/98.
4. S. A. Smith, C. A. Wolden, M. D. Bremser, A. D. Hanser, and R. F. Davis, Appl. Phys. Lett. 71, 3631, 1997.
5. Y. H. Lee, H. S. Kim, W. S. Kwon, G. Y. Yeom, J. W. Lee, M. C. Yoo, and T. I. Kim, J. Vac. Sci. Technol. A16, 1478 (1998).

6. Hyun Cho, J. Hong, T. Maeda, S. M. Donovan, C. R. Abernathy, S. J. Pearton, R. J. Shul, and J. Han, J. Electron. Mats. 27, 915 (1998).
7. R. J. Shul, C. G. Willison, M. M. Bridges, J. Han, J. W. Lee, S. J. Pearton, C. R. Abernathy, J. D. MacKenzie, S. M. Donovan, L. Zhang, and L. F. Lester, J. Vac. Sci. Technol A16, 1621 (1998).
8. R. J. Shul, C. G. Willison, M. M. Bridges, J. Han, J. W. Lee, S. J. Pearton, C. R. Abernathy, J. D. MacKenzie, S. M. Donovan, Mat. Res. Soc. Symp. Proc. Vol. 483, 155 (1998).
9. R. J. Shul, C. I. H. Ashby, C. G. Willison, L. Zhang, J. Han, M. M. Bridges, S. J. Pearton, J. W. Lee, and L. F. Lester, Mats. Res. Soc. Symp. Proc. 512, 487 (1998).
10. R. J. Shul, L. Zhang, C. G. Willison, J. Han, S. J. Pearton, J. Hong, C. R. Abernathy, and L. F. Lester, MRS Internet J. Nitride Semicond., submitted 11/98.
11. S. J. Pearton, Appl. Surf. Sci. 117/118, 597 (1997).
12. S. J. Pearton and R. J. Shul, in *Defects in Optoelectronic Materials*. ed. S. W. Pang and O. Wada (Gordon and Breach, The Netherlands, in press).
13. S. W. Pang, J. Electrochem. Soc. 133, 784 (1986).
14. F. Ren, J. R. Lothian, S. J. Pearton, C. R. Abernathy, C. B. Vartuli, J. D. MacKenzie, R. G. Wilson, and R. F. Karlicek, J. Electron. Mater. 26, 1287 (1997).
15. A. T. Ping, A. C. Schmitz, I. Adesida, M. A. Khan, O. Chen, and Y. W. Yang, J. Electron. Mater. 26, 266 (1997).
16. T.-B. Ng, J. Han, R. M. Biefeld, and M. V. Weckwerth, J. Electron. Mat. 27, 190 (1998)
17. J. M. Van Hove, P. P. Chow, J. J. Klaassen, R. Hickman, A. M. Wowchak, D. R. Croswell, C. Polley, Mater. Res. Soc. Symp. Proc. 468, 51 (1997).
18. L. F. Lester, J. M. Brown, J. C. Ramer, L. Zhang, S. D. Hersee, and J. C. Zolper, Appl. Phys. Lerr 69, 2737 (1996).

Selective Dry Etching Of The GaN/InN/AlN, GaAs/AlGaAs And GaAs/InGaP Systems

D. C. Hays*, C. R. Abernathy*, W. S. Hobson**, S. J. Pearton*, J. Han***, R. J. Shul***, H. Cho*, K. B. Jung*, F. Ren****, and Y. B. Hahn*

*Department of Materials Science and Engineering, University of Florida, Gainesville, FL 32611
**Bell Laboratories, Lucent Technologies, Murray Hill, NJ 07974
***Sandia National Laboratories, Albuquerque, NM 87185
****Department of Chemical Engineering, University of Florida, Gainesville, FL 32611

ABSTRACT

Selective etching of InN over GaN and AlN, and of GaAs over both AlGaAs and InGaP was examined with a number of different plasma chemistries under inductively coupled plasma conditions. Selectivities up to 55 for InN/GaN and 20 for InN/AlN were achieved in ICl/Ar discharges. For GaAs/AlGaAs, maximum selectivities of 75(with BCl_3/SF_6) were obtained while for GaAs/InGaP values of 80(with BCl_3/SF_6) and 25(with BCl_3/NF_3) were achieved. Selective etching of InGaP over GaAs is possible with either CH_4/H_2 or BI_3. The selectivity is a strong function of ion flux and ion energy, and can result from two factors – either formation of a nonvolatile etch product, or a difference in bond strength between the two materials.

INTRODUCTION

There are a number of device processing steps where high selectivity etch removal for GaAs over AlGaAs or InGaP is required[1]. The first is the gate recess for high electron mobility transistors, where the ohmic contact layer must be removed to expose the donor layer for subsequent gate metal deposition. The second is in the emitter mesa process for heterojunction bipolar transistors, where it is desirable to remove the GaAs contact layer to the wide bandgap emitter first, prior to etching the emitter and exposing the base for contacting. There is an extensive literature on selective wet and dry etching of GaAs over AlGaAs and InGaP[2-11]. For wet etching, most attention for the GaAs/AlGaAs system has focused on the NH_4OH/H_2O_2, the peroxide ammonia, or PA etch[2,5,6], and citric acid/H_2O_2 formulations[1,2,8]. The advantage of the latter is that it does not attack the copolymer resist often used to produce the T-gate pattern for submicron HEMT devices. For the GaAs/InGaP system, $H_3PO_4/HCl/H_2O$ solutions provide selective removal of InGaP[10], while virtually any other set of acids that do not contain HCl will selectively etch GaAs from an underlying InGaP layer[10].

For dry etching, GaAs can be removed from AlGaAs using plasma chemistries involving a combination of Cl_2, which produces the etching reaction and F_2, which provides the etch stop reaction with Al[2-4,11-8]. Typical gas chemistries include BCl_3/SF_6, $SiCl_4/SiF_4$ and, before its restriction due to its ozone depleting properties, CCl_2F_2. There is currently no selective etch for AlGaAs over GaAs. For the GaAs/InGaP system, GaAs can be selectively etched in any Cl_2-based plasma[10], while CH_4/H_2 can remove InGaP from GaAs. Virtually all of the selective dry etching has been performed under reactive ion etching(RIE) conditions, and there is little information available for the newer high density tools, particularly Inductively Coupled Plasma(ICP) systems.

The III-nitrides are of great current interest for applications in full-color displays, undersea communications, data storage, UV solar-blind detectors and high power/high temperature electronics[12-15]. Little attention has been paid to development of selective dry etching processes for nitride materials, since most of the work to date has focused on fabrication of mesas for photonic devices[12]. In this case etching is performed in a non-selective fashion through GaN/AlGaN/InGaN/AlGaN/GaN structures. In the case of electronic devices however, there is a need to selectively remove In-based ohmic contact layers from underlying GaN or AlGaN layers[13-15].

In this paper we report on a detailed study of selective etching of GaAs of AlGaAs and InGaP in different plasma chemistries under ICP conditions. The chemistries investigated include BCl_3/SF_6, BCl_3/NF_3, IBr, ICl, BI_3, and BBr_3. In all cases we find it is necessary to restrict both the ion flux and ion

281

Mat. Res. Soc. Symp. Proc. Vol. 573 ° 1999 Materials Research Society

energy to maximize etch selectivity, and in general the selectivities under ICP or high density plasma conditions are lower than those under RIE conditions. We also report on etch selectivity for InN over GaN and AlN in ICP discharges based on interhalogen compounds. ICl and IBr are attractive candidates for use in high density reactors because of their weak bonding and absence of both polymer deposition or hydrogen passivation of dopants.

EXPERIMENT

Device quality epitaxial layers of $Al_{0.22}Ga_{0.78}As$ and $In_{0.5}Ga_{0.5}P$ 1μm thick were grown on undoped GaAs substrates by either Metal Organic Molecular Beam Epitaxy[16] of Metal Organic Chemical Vapor Deposition[17]. The AlGaAs and InGaP layers were nominally undoped ($n\sim10^{16}$ cm^{-3}). The GaAs was semi-insulating, (100) oriented substrates. GaN, AlN and InN layers were grown on c-plane Al_2O_3 substrates by Metal Organic Chemical Vapor Deposition (GaN)[18] or Metal Organic Molecular Beam Epitaxy (MOMBE)[16]. In etch rate experiments they were masked with Apiezon wax, which was removed in acetone and the step height measured by stylus profilometry. The vapor from ICl and IBr, contained in heated stainless-steel vessels, was introduced into the reactor through electronic mass flow controllers at a total gas load up to 15 standard cubic centimeters per minute (sccm).

The dry etching was performed as a function of ICP source power, percentage in input gas mixture, chamber pressure, and rf chuck power. The reactor used was a Plasma Therm 790 ICP system, in which a 2MHz, 3 turn coil generates the plasma, and the sample is mounted on a rf-powered (13.56MHz), He backside cooled chuck.

RESULTS

(a)BCl$_3$/SF$_6$

Under RIE conditions selectivities for GaAs over both AlGaAs and InGaP of >80 were achieved for low dc chuck biases (< -80V) for discharge compositions comprising ~20% SF$_6$. This is the composition where the Cl^0 concentration and hence GaAs etch rate is highest. With the addition of 200W ICP source power(Figure 1) the etch rates for all three materials increase by at least an order of magnitude for similar discharge composition. However, peak selectivities of ~45 for GaAs/AlGaAs and ~8 for GaAs/InGaP are obtained. Note the essentially non-selective etching for pure BCl$_3$, as expected.

ICP source power was varied to more fully investigate its role on etch rates and selectivities for a fixed discharge composition(12BCl$_3$/3SF$_6$). When source power is first added, the etch rates of all three materials decreases, probably due to a strong increase in atomic fluorine in the discharge which blocks the surface to chlorine neutral adsorption. As source power is increased above 100W the etch rates go through a maximum. This is commonly observed in ICP etching and is suggested to result from a competition between increased ion and radical flux, and reduced ion energy at high source power. Note that the etch selectivities rapidly decrease above 200-300W, since sputter-assisted desorption of the AlF$_3$ and InF$_3$ will be efficient at high ion fluxes.

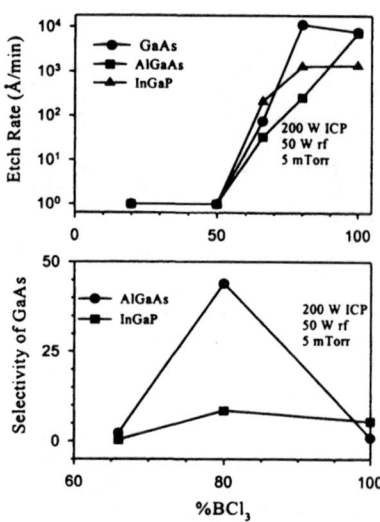

(Total flow of BCl$_3$/SF$_6$ is 15 sccm)

Figure 1. Etch rates of GaAs, AlGaAs, and InGaP(top) and resultant selectivities for GaAs over AlGaAs and InGaP(bottom) as a function of BCl$_3$ percentage in ICP BCl$_3$/SF$_6$ discharges.

The effect of rf chuck power, which controls incident ion energy, was also studied for a fixed discharge composition and source power (12BCl$_3$/3SF$_6$, 200W source power). The etch rates tend to saturate above ~50W for the chosen parameter space, since the etching probably becomes reactant-limited, i.e. the adsorbed chlorine coverage is less than necessary to achieve the highest rate of etch product formation and ion-assisted desorption. The selectivities are maximized at low rf powers and decrease at higher values due to ion-assisted desorption of the aluminum and indium fluoride and the indium chloride. Thus, both ion energy and ion flux must be controlled to achieve good etch selectivities in both heterostructure systems.

(b) BCl$_3$/NF$_3$

NF$_3$ is a less desirable choice than SF$_6$ from the point of view of cost and corrosion issues, but it is less strongly bonded and might be expected to produce more atomic fluorine at low ICP source powers. We performed the same runs with BCl$_3$/NF$_3$ in order to compare with the more common BCl$_3$/SF$_6$ chemistry.

Figure 2 shows the influence of discharge composition for low ICP source power (200W) conditions. The resultant etch selectivities for GaAs/AlGaAs and GaAs/InGaP are significantly lower than achieved with BCl$_3$/SF$_6$ under the same conditions. This was true even for RIE conditions, where maximum selectivities of typically <10 were obtained.

The effect of ICP source power on etch rates and selectivity for a fixed plasma composition of 12BCl$_3$/3NF$_3$ was examined. The maximum selectivities are ~50 for GaAs/AlGaAs and ~25 for GaAs/InGaP, which again are lower than achieved with BCl$_3$/SF$_6$. The selectivities decreased as rf chuck power was increased, and were also lower than those obtained with BCl$_3$/SF$_6$.

(c) IBr and ICl

The interhalogen compounds are alternative etch chemistries for high density reactors because of their weak bonding and absence of polymer formation. Moreover, InI$_3$ is much more volatile the InCl$_3$, leading to the possibility of selective etching of InGaP over GaAs.

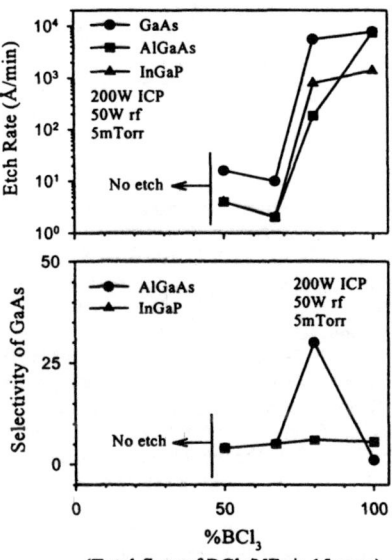

(Total flow of BCl$_3$/NF$_3$ is 15sccm)

Figure 2. Etch rates of GaAs, AlGaAs, and InGaP(top) and resultant selectivities for GaAs over AlGaAs and InGaP(bottom) as a function of BCl$_3$ percentage in ICP BCl$_3$/NF$_3$ discharges.

The effects of ICl and IBr percentage on etch rates and selectivity was investigated under high source and chuck power conditions . The etch rates for AlGaAs are low over the whole range of ICl percentages investigated, producing selectivities up to ~30(ICl) and ~40(IBr) for GaAs over AlGaAs. This chemistry produces non-selective etching of the GaAs/InGaP heterostructure under these conditions, suggesting that at these biases and fluxes InCl$_x$ desorption is not the limiting step. Lower biases and fluxes might produce better selectivity.

We also explored the effect of increasing source power at fixed plasma composition (i.e. 2ICl/13Ar ; 2IBr/13Ar), and found maximum selectivities <10 for GaAs/AlGaAs and typically 0.7-2 for GaAs/InGaP. The etched surface morphology was very good for these chemistries for a broad range of conditions, with root-mean-square(RMS) roughness in the range of 0.9-1.6nm (i.e. similar to unetched controls).

The rates are a strong function of ion energy (rf chuck power), but the etching for both GaAs/AlGaAs and GaAs/InGaP is basically non-selective at low chuck powers. These do not appear to be attractive candidates for selective ICP etching of GaAs/AlGaAs and GaAs/InGaP.

Etch rates for the binary nitrides and resultant selectivities for InN over GaN and AlN are shown in Figure 3 as a function of discharge composition at fixed source power(750W) and rf chuck power(250W). The etch rates for AlN and GaN increase with ICl percentage, indicating some degree of chemical enhancement. The etch rate for InN is basically dominated singly by sputtering and is independent of discharge composition. Selectivities of 13 and 14, respectively, for InN/AlN and InN/GaN are achieved at low ICl percentages.

Source power was investigated at the optimum discharge composition. The etch rates saturate beyond ~300W source power, producing similar behavior in the selectivity. In device fabrication a selectivity of >10 is desirable for process repeatability, and this is achieved for both InN/GaN and InN/AlN. These results also show the value of using high density plasma conditions for etching of the nitrides, since the rates are 1-2 orders of magnitude higher than for RIE conditions (i.e. 0W source power)[20].

The relatively high bond strength for the nitrides (7.72eV/atom for InN, 8.92eV/atom for GaN, and 11.52eV/atom for AlN)[21] means that bond-breaking prior to formation of the etch products is often the rate-limiting step. The incident ion energy is basically controlled by the rf chuck power. The rates saturate above ~150W(corresponding to dc self-bias of ~ -125V), where the etching becomes reactant-limited. The selectivity for InN over both materials is quite good over a broad range of rf chuck power.

The same experiments were performed for the IBr chemistry. Figure 4 shows the effect of plasma composition on the material etch rates and selectivity. The results are fairly similar to those obtained with the ICl mixtures (Figure 3) and maximum selectivities >10 were achieved for both InN/GaN and InN/AlN.

The effect of ICP source power, and rf chuck power, on the etching with IBr/Ar are similar to those obtained with ICl/Ar, except the selectivities are slightly lower than obtained with the ICl/Ar chemistry

(d) BI$_3$ and BBr$_3$

These compounds are the iodine and bromine analogs of BCl$_3$, and have only recently begun to be explored as etching chemistries for III-V semiconductors. Figure 5 shows the effect of source power in 4BI$_3$/6Ar discharges on GaAs and InGaP etch rates and the resultant selectivity for InGaP over GaAs. Note that values ~60 can be

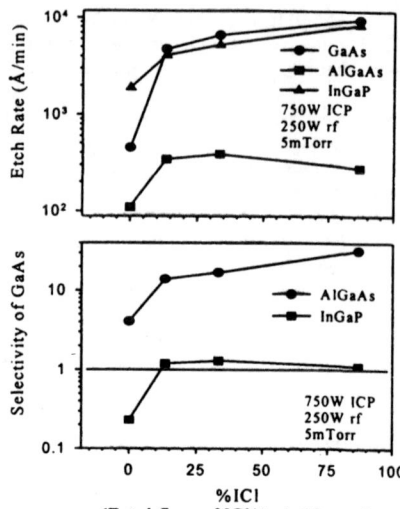

Figure 3. Etch rates of binary nitrides(top) and selectivity of InN over GaN and AlN (bottom) as a function of ICl percentage in ICP discharges.

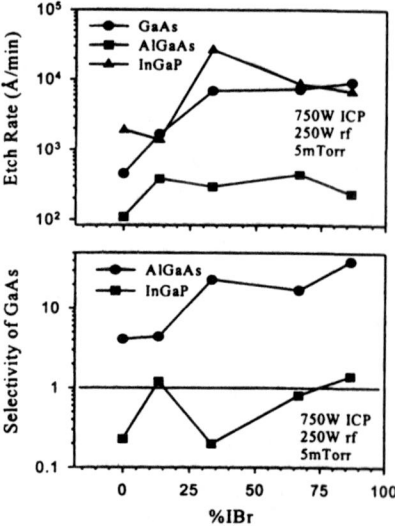

Figure 4. Etch rates of binary nitrides(top) and selectivity of InN over GaN and AlN (bottom) as a function of IBr percentage in ICP discharges.

achieved. This is an important result, since previously only CH_4/H_2 has been found to produce selectivity for InGaP over GaAs. However, the use of CH_4/H_2 has many attendant problems, including hydrogen passivation of near-surface dopants and polymer deposition on the mask material and within the chamber. The GaAs etch rate at low BI_3 content, flux and ion energy is a sensitive function of ion/neutral ratio.

For the BBr3/Ar chemistry we had to employ much higher rf chuck power with this chemistry to achieve measurable etch rates. The data show that BBr_3 is essentially non-selective for GaAs/InGaP.

We summarize the selectivities for GaAs/AlGaAs and GaAs/InGaP achieved with the different chemistries under both RIE and ICP conditions in Table I.

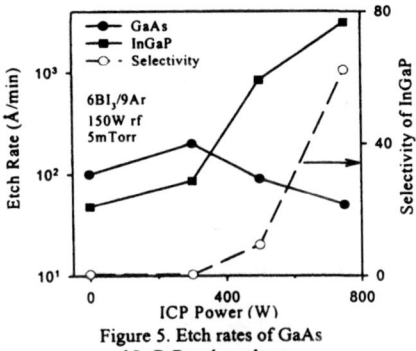

Figure 5. Etch rates of GaAs and InGaP and resultant selectivity of InGaP over GaAs as a function of BI_3 percentage.

Table I. Dry etch selectivities for GaAs/AlGaAs and GaAs/InGaP under conventional RIE conditions and under high density plasma conditions in different plasma chemistries.

Chemistry	GaAs/AlGaAs		GaAs/InGaP	
	RIE	ICP	RIE	ICP
BBr_3/Ar	-	-	-	0.5-1
BI_3/Ar	-	-	-	0.02-2
ICl/Ar	2	4-30	2	0.2-1
IBr/Ar	1	1-40	0.5	0.2-1
BCl_3/SF_6	Infinite	2-75	400	50-80
BCl_3/NF_3	1-50	1-30	1-75	1-25

CONCLUSIONS

While ICP reactors produce much higher etch rates than capacitively-coupled tools, as expected, etch selectivity for the GaAs/AlGaAs and GaAs/InGaP systems is generally lower under ICP conditions relative to conventional RIE because of the more efficient sputter-assisted desorption of the etch products that produce the etch-stop reactions, i.e. AlF_3, InF_3, and $InCl_3$. An obvious strategy to improve selectivity is simply to shut off the source power toward the end of the etch process and remove the last few hundred angstroms of material using RIE conditions.

Most of the past work on nitride etching has reported selectivities for InN/GaN and InN/AlN in the range 0.5-5, using Cl_2 or BCl_3 plasma chemistries[22-27]. The highest values for InN over both GaN and AlN have been achieved using BI_3 discharges under ICP conditions, where values of 40-100 were obtained[28-29]. The results reported here for ICl and IBr do not reach this level, but are nonetheless superior to those achieved with conventional Cl_2 or BCl_3 plasma chemistries. In contrast to the GaAs/AlGaAs and GaAs/InGaP, the maximum selectivities for InN/GaN and InN/AlN are obtained in moderate source and rf chuck powers and moderate pressures, rather than low powers and high pressures. These differences reflect the need to have fairly high ion fluxes and energies in order to achieve practical etch rates for the nitrides.

ACKNOWLEDGMENTS

The work at UF is partially supported by a DOD MURI monitored by AFOSR (H. C. Delong), contract no. F49620-1-96-0026. Sandia is a multiprogram laboratory operated by Sandia Corporation, a Lockheed-Martin Company for the Department of Energy under contract DEAC04-94AL85000. YBH acknowledges the support of the Korean Research Foundation for Faculty Research Abroad.

REFERENCES

1. see for example HEMT's and HBT's, ed F. Ali, Artech House, Dedham, MA.(1991); also, Topics in Compound Semiconductors, S. J. Pearton, C. R. Abernathy, and F. Ren, (World Scientific, Singapore 1996)
2. M. Tong, D. G. Ballegeer, A. Ketterson, E. J. Roan, K. Y. Cheng and I. Adesida, J. Electron. Mater. **21**, 9(1992).
3. K. Hikosuka, T. Miamura and K. Joshin, Jap. J. Appl. Phys. **20**, L847(1981).
4. S. Salimian, C. B. Cooper III, R. Norton and J. Beam, Appl. Phys. Lett. **51**, 1083 (1987).
5. J. –H. Kim, D. H. Lim and G. M. Yang, J. Vac. Sci. Technol. **B16**, 558 (1998).
6. G. C. DeSalvo, W. F. Tseng and J. Comas, J. Electrochem. Soc. **139**, 831 (1992).
7. R. Zhao, W. S. Lao, T. C. Chong and M. F. Li, Jap. J. Appl. Phys. **35**, 22 (1996).
8. C. Juang, K. J. Kuhn and R. B. Darling, J. Vac. Sci. Technol. **B8**, 1122 (1990).
9. D. H. Lim, G. M. Yang, J. –H. Kim and H. J. Lee, Appl. Phys. Lett. **71**, 1915(1997).
10. S. J. Pearton, Int. J. Mod. Phys. **B8**, 1781 (1994).
11. S. J. Pearton, W. S. Hobson, C. R. Abernathy, F. Ren, T. Fullowan, A. Katz and A. Perley, Plasma Chem. Plasma Proc. **13**, 311 (1993).
12. S. Nahamura, M. Senoh, S. Nagahama, N. Iwasa, T. Yamada, T. Matsushito, H. Kiyoku and V. Sugimotu, Jap. J. Appl. Phys. **35**, L74 (1996).
13. Y. F. Wu, S. Keller, P. Kozodoy, B. P. Keller, P. Parikh, D. Kapolneck, S. P. De Baars and U. K. Mishra, IEEE Electron. Dev. Lett. **18**, 290 (1997).
14. M. A. Khan, Q. Chen, M. S. Shur, B. T. McDermott, J. A. Higgins, J. Burn, W. J. Scharff and L. F. Eastman, Electron. Lett. **32**, 357 (1996).
15. O. Aktas, Z. Fan, S. N. Mohammad, A. Botcharev and H. Morkoc, Appl. Phys. Lett. **69**, 25 (1996).
16. C. R. Abernathy, Mat. Sci. Eng. Rep., **R14** (1995).
17. W. S. Hobson, Mat. Res. Soc. Symp. Proc., **300**, 75 (1993).
18. J. Han, J. G. Fleming and D. M. Follstaedt, Mat. Res. Soc. Symp. Proc. **512**, 53 (1998).
19. D.C. Hays, H. Cho, K. B. Jung, Y. B. Hahn, C. R. Abernathy and S. J. Pearton, Appl. Surf. Sci. (submitted).
20. R. J. Shul, S. P. Kilcoyne, M. Hagerott-Crawford, J. E. Parmeba, C. B. Vartuli, C. R. Abernathy and S. J. Pearton, Appl. Phys. Lett. **66**, 1761 (1995).
21. W. A. Harrison, Electronic Structure and Properties of Solids (Freeman, San Francisco, (1980).
22. J. W. Lee, H. Cho, D. C. Hays, C. R. Abernathy, S. J. Pearton, R. J. Shul, G. A. Vawter and J. Han, IEEE J. Selected Top. In Quantum Electronics **4**, 557 (1998).
23. R. J. Shul, G. B. McClellan, R. D. Briggs, D. J. Rieger, S. J. Pearton, C. R. Abernathy, J. W. Lee, C. Constantine and C. Burratt, J. Vac. Sci. Technol. **A15**, 633 (1997).

24. C. B. Vartuli, S. J. Pearton, J. D. MacKenzie, C. R. Abernathy and R. J. Shul, J. Electrochem. Soc. **143**, L246 (1996).
25. S. A. Saith, C. A. Wolden, M. D. Bremser, A. D. Hansen and R. F. Davis, Appl. Phys. Lett. **71**, 3631 (1997).
26. R. J. Shul, C. G. Willison, M. M. Bridges, J. Han, J. W. Lee, S. J. Pearton, C. R. Abernathy, J. D. MacKenzie, S. M. Donovan, L. Zhang and L. F. Lester, J. Vac. Sci. Technol. A. **16**, 1621 (1998).
27. R. J. Shul, C. I. H. Ashby, C. G. Willison, L. Zhang, J. Han, M. M. Bridges, S. J. Pearton, J. W. Lee and L. F. Lester, Mat. Res. Soc. Symp. Proc. **512**, 495 (1998).
28. H. Cho, J. Hong, T. Maeda, S. M. Donovan, J. D. MacKenzie, C. R. Abernathy, S. J. Pearton, R. J. Shul and J. Han, MRS Internet J. Nitride Res. **3**, 5 (1998).
29. H. Cho, J. Hong, T. Maeda, S. M. Donovan, C. R. Abernathy, S. J. Pearton, R. J. Shul and J. Han, J. Electron. Mater. **27**, 915 (1998).

KINETICS OF THE INTERACTION OF ATOMIC SPECIES WITH (100) GALLIUM ARSENIDE SURFACES

LIGIA GHEORGHITA, ELMER OGRYZLO*
Advanced Materials and Process Engineering Laboratory (AMPEL), University of British Columbia, 2355 East Mall, Vancouver, BC, Canada V6T 1Z4, *ogryzlo@chem.ubc.ca

ABSTRACT

The interactions of atomic hydrogen, deuterium and sulfur with (100) GaAs surfaces have been studied. The atoms were produced in a remote microwave plasma and their effect on carrier recombination velocities was continuously monitored *in situ* by the change in photoluminescence intensity (PLI). It was observed that the PLI increased by about 1-2 orders of magnitude following a few seconds exposure to hydrogen and deuterium atoms. A subsequent treatment with sulfur atoms further increased the PLI. A kinetic analysis of the room temperature hydrogen atom interactions with the (100) GaAs surface was attempted. A similar behavior was observed at higher temperatures when hydrogen and deuterium atoms were allowed to interact with a Si/SiO_2 interface. A comparison of the two systems leads us to conclude that the hydrogen and deuterium atoms can be trapped at interstitial sites near these interfaces. The kinetics of the hydrogen atom loss from these semiconductors is presented and analyzed in terms of a distribution of trapping sites.

INTRODUCTION

Unlike silicon, which forms a hard oxide layer that acts as an effective dielectric material and eliminates a majority of the recombination centers on its surface, the oxides of the III-V materials are not effective in reducing the undesirable carrier-trapping surface-states.

There has been some attempt to identify these carrier traps (i.e. recombination centers) on GaAs surfaces. It has been suggested that "dangling bonds" on the surface atoms may be responsible for the shortened carrier lifetimes observed.[1] It is also possible that the reaction between GaAs and As_2O_3 produces elemental As which could act as a carrier trap on oxidized III-V surfaces.[2]

Gottscho et al. have studied the passivation of GaAs surfaces by hydrogen. They explained their observations by involving both kinds of carrier traps.[3,4,5,6]

We have undertaken a kinetic study of the reactions of H and D and S with GaAs surfaces. In this paper we present some of our results with hydrogen atoms.

EXPERIMENTAL

The GaAs samples were cut from single crystal, semi-insulating (100) wafers grown by molecular beam epitaxy in AMPEL at the University of British Columbia. Other GaAs wafers were supplied by EG&G Optoelectronics Canada (Vaudreuil, Quebec). A configuration, with a layer of wider band gap AlGaAs below the GaAs, was chosen in order to keep the carriers confined to the GaAs layer and thus to maximize the sensitivity of the PLI to the changes on the exposed surface. The samples were undoped or only modestly doped (n-type), in order to reduce the carrier loss by radiative recombination in the bulk.

Mat. Res. Soc. Symp. Proc. Vol. 573 © 1999 Materials Research Society

The Institute for Microstructural Sciences (NRC, Ottawa) provided the silicon samples used in this work. Intrinsic Si was grown on top of either n or p-type (100) Czochralski Si substrates by chemical vapor deposition.

The 5x5 mm GaAs samples and the 5x10 mm exposed surface Si samples, respectively, cut from the wafers were dipped in a 10% HF solution for about 60 seconds, rinsed in deionized water, and quickly inserted into the reactor and dried under a flow of argon.

The photoluminescence intensity (PLI) was used to measure the steady state carrier concentration that is inversely proportional to the concentration of surface states. The GaAs band gap fluorescence at 865 nm was excited by a 10 mW red (632.8 nm) HeNe laser, focused onto the entrance slit of a scanning monochromator, dispersed and detected by an infrared sensitive photomultiplier tube, connected to a photon counter and a PC. Thus the PLI could be measured as a function of real time.

A 200 W 2.45 GHz microwave generator attached to a quarter wave cavity, was used to create the hydrogen or deuterium plasma inside a quartz discharge tube located about 20 cm upstream from the reaction chamber. Molecular hydrogen or deuterium (0.1 Torr) diluted with up to 0.3 Torr of argon flowed through the quartz discharge tube. The GaAs wafer was exposed to the atomic hydrogen for various periods (from a few seconds up to 5 minutes). All the GaAs exposures to atomic hydrogen were performed at room temperature.

Sulphur atoms were produced *in situ* with the two step rapid reaction of hydrogen sulphide with atomic hydrogen, yielding one S atom for each H atom consumed. The technique[7] involved adding a very small amount of H_2S (1-2%) to a stream of H atoms.

A 20 Å layer of SiO_2 was grown at 200°C on Si by exposing a clean surface to atomic oxygen from an upstream microwave discharge in 0.5 Torr of O_2.

Because silicon luminescence is extremely weak, its PLI could not be used. We therefore monitored the changing surface carrier recombination sites by following the laser induced (633 nm HeNe laser) conductivity of an SiO_2 covered Si wafer during the exposure to H atoms. This "contactless" technique, described in detail elsewhere[8], employed a radio frequency signal to measure the change in the conductivity of the Si chip.

The Si/SiO_2 interfaces were exposed to hydrogen or deuterium atoms in a manner similar to that used with GaAs. The only difference was that the working temperature of the Si was typically about 200°C.

RESULTS

Changes in the "surface recombination site" density (i.e. the inverse of the 865 nm PLI signal) on GaAs exposed to hydrogen atoms are shown in Figure 1(a). An irreversible process occurred initially, i.e. the trap density decreased by about a factor of 3 when the discharge was on first turned on and decreased even further (up to 10-100 times the initial trap density, depending on the sample) when the discharge was shut off. Subsequently the trap density reversibly increased when the plasma was switched on, and decreased when it was switched off. The reversible process could be repeated almost indefinitely as long as the exposure to H atoms was kept short (typical exposure times ranged between 30-60 s). Similar results were obtained by exposure to deuterium atoms.

The treatment of GaAs surfaces with sulfur atoms produced an increase in the PLI by only a factor of 2, but with a much greater stability in air.

During these studies of GaAs surface-states passivation, we observed a similar phenomenon on silicon. The Si was covered with a thin oxide layer through which the H (or D) atoms readily diffuse, but which excluded other atomic species such as O and N that also come from the discharge.

Figure 1(b) shows the variation of recombination site densities at the Si/SiO$_2$ interface as they are affected by exposure to H atoms. The behavior is very similar to that displayed for GaAs. There is both an irreversible decrease in trap sites during the first exposure to H atoms and a subsequent reversible creation of recombination sites while the substrates are being exposed to atomic hydrogen.

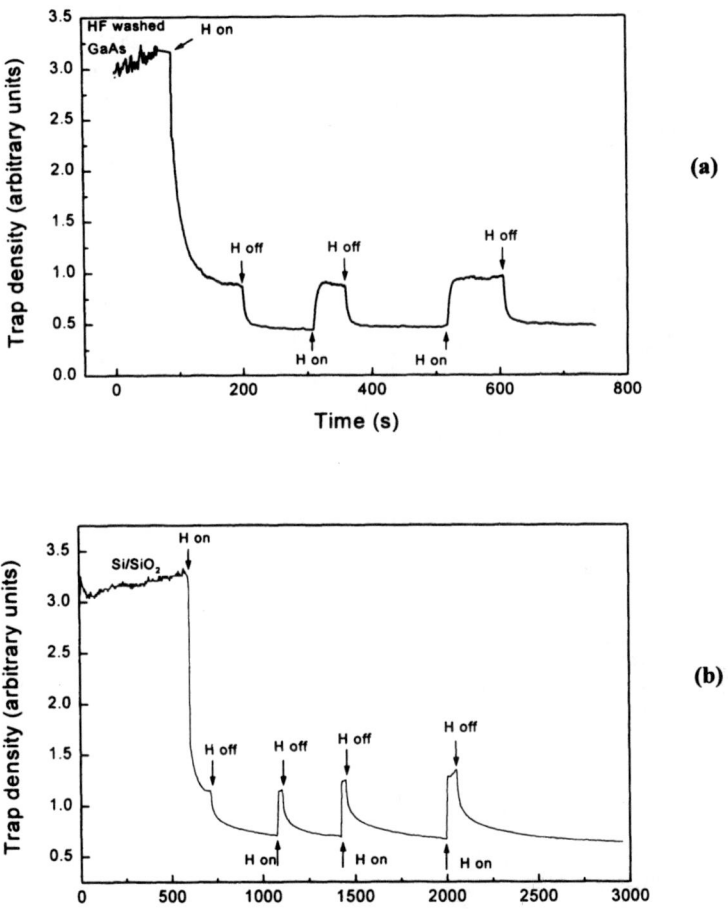

Figure 1 Effect of H atom exposure on trap density of **(a)** (100) GaAs surface at room temperature and **(b)** Si/SiO$_2$ interface at 200°C.

CONCLUSIONS

A simple exponential decay illustrated in Figure 2(a) fits the trap density versus time curve for the irreversible process in GaAs. Therefore this process follows a first order kinetic rate law. This process could be due to (i) increased termination of dangling bonds by H atoms, (ii) surface reconstruction, (iii) the etching away of surface oxides, and/or removal of As[9].

The similar irreversible process observed in the case of the Si/SiO_2 interface must however be due to the removal of dangling bonds at the interface[10].

The reversible processes seen when both GaAs and Si are exposed to H atoms appear to be very similar. Neither can be described by a single exponential. Figure 2(b) illustrates the non-exponential decay of the process. However, this process can be fitted quite well with a triple exponential as shown in Figure 2(b) suggesting that there are at least three kinds of traps with lifetimes ranging from a few seconds to a few minutes at room temperature. The same appears to be true for the loss of traps created at the Si/SiO_2 interface on exposure to H atoms as illustrated in Figure 3.

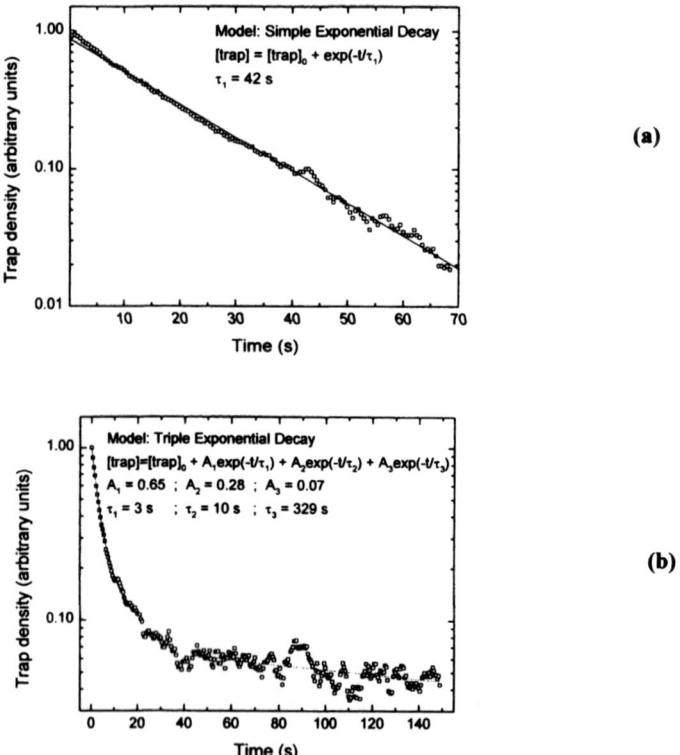

Figure 2   Effect of atomic H on a (100) GaAs surface at room temperature **(a)** irreversible process after exposure of a freshly washed surface and **(b)** reversible process and triple exponential decay best fit after repeated H atom exposures.

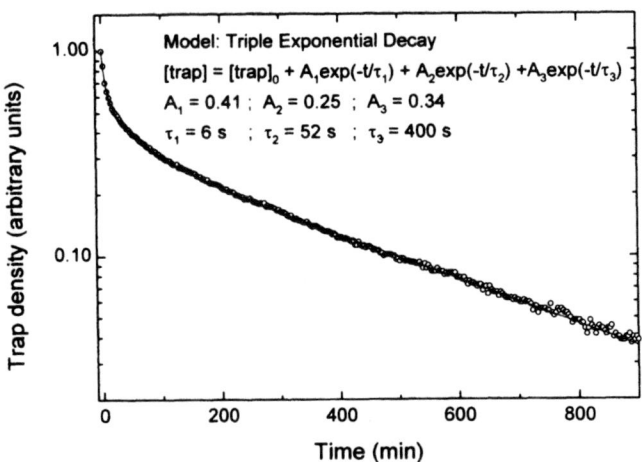

Figure 3   Triple exponential best fit to trap decay for a (100) Si/SiO$_2$ interface exposed to atomic H at 200°C.

An alternative to the triple exponential decay model is the multiple-trapping model of dispersive transport well known in work on amorphous Si [11]. In this model electrons in localized sites must first be thermally excited above the mobility edge before they can move to other sites. Experimentally this means that a power law governs the behavior of the photocurrent as a function of time. In our case H atoms are assumed to be trapped in sites with dissociation energies that are distributed exponentially. Thus, the power law has to apply to the trapped H atoms that are recombination centers at (or near) the surface or interface. This treatment yields the expression [11]

$$[H] = b\,t^{-\alpha} \tag{1}$$

where [H] represents the density of trapped H atoms, b is a constant, and α ($0<\alpha<1$) is a parameter which is determined by the disorder in the material.

However, as illustrated by the non-linearity of the plot in Figure 4 such a power law provides a somewhat unsatisfying fit of our data.

In summary it would appear that H atoms could be physically bound to sites with a distribution of activation energies at both GaAs and Si surfaces. Having observed very similar behavior when a GaAs or a Si/SiO$_2$ interface are exposed to atomic hydrogen we are led to believe that the results can be explained by the occurrence of very similar phenomena in both systems, namely the initial removal of some unsaturated valencies (dangling bonds) at the semiconductor interface and then the temporary trapping of H atoms, which retain their unsaturated valencies, at sites with a distribution of binding energies. A study of these systems at a variety of temperatures should establish the validity of these conclusions.

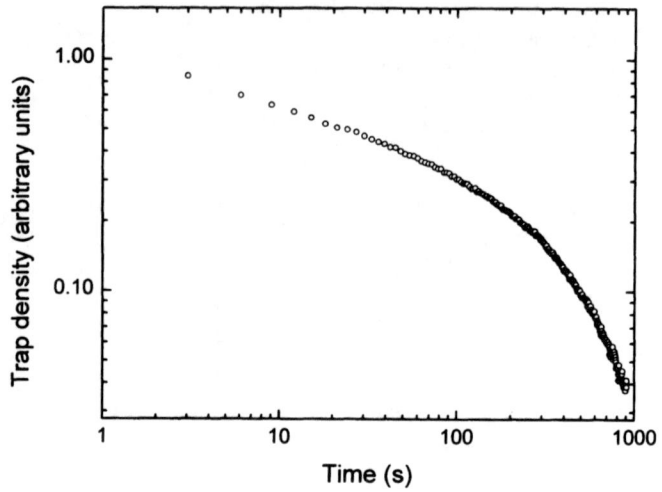

Figure 4   Power law plot of trap density versus time for a Si/SiO$_2$ interface exposed to atomic H at 200°C.

ACKNOWLEDGEMENTS

The research for this paper was funded by NSERC (Canada).

REFERENCES

[1] P. Viktorovitch, F. Benyahia, C. Santinelli, R. Blanchet, P. Leyral, and M. Garrigues, Appl. Surf. Sci. **31**, 317 (1988).

[2] C. Wilmsen, J. Vac. Sci. Technol. **19**, 279 (1982).

[3] R.A. Gottscho, B.L. Preppernau, S.J. Pearton, A.B. Emerson, and K.P. Giapis, J. Appl. Phys. **68**, 440 (1990).

[4] A. Mitchell, R.A. Gottscho, S.J. Pearton, and G.R. Scheller, Appl. Phys. Lett. **56**, 821 (1990).

[5] E. Yoon, R.A. Gottscho, V.M. Donnelly, and H.S. Luftman, Appl. Phys. Lett. **60**, 2681 (1992).

[6] E. Yoon, R.A. Gottscho, V.M. Donnelly, and W.S. Hobson, J. Vac. Sci. Technol. **10**, 2197 (1992).

[7] R.W. Fair and B.A. Thrush, Trans. Faraday Soc. **65**, 1208 (1969).

[8] H. Li and E.A. Ogryzlo, Can. J. Phys. **74**, S233 (1996).

[9] E.S. Aydil, Z. Zhou, K.P. Giapis, Y.Chabal, J.A. Gregus, and R.A. Gottscho, Appl. Phys. Lett. **62**, 3156 (1993).

[10] K.L. Brower, Phys. Rev. B **38**, 9657 (1988).

[11] T. Tiedje, in *The Physics of Hydrogenated Amorphous Silicon*, edited by J.D. Joannopoulos and G. Lucovsky (Springer-Verlag, Heidelberg, 1984), p. 267-281.

AUTHOR INDEX

Abernathy, C.R., 247, 281
Ashby, C.I.H., 203

Baca, A.G., 271
Bertness, K.A., 95
Brodersen, P., 31
Busani, T., 157

Campidelli, Y., 157
Chang, G.S., 213
Charbonneau, S., 227
Chen, Y.K., 219
Cho, A.Y., 219
Cho, H., 281
Cohen, E., 119
Courant, J.L., 189
Csutak, S., 81

Danilov, I., 137
de Barros, Jr., L.E., 137
Deppe, D.G., 81
Devine, R.A.B., 157
Dharma-Wardana, M.W.C., 3, 31
Diniz, J.A., 137
Donovan, S.M., 247
Driad, R., 227

Elbahnasawy, R.F., 145, 265
Ellis, A.B., 15
Etrillard, J., 189

Friedman, D.J., 95
Furukawa, H., 239

Geisz, J.F., 95
Gheorghita, L., 287
Gila, B., 247
Graham, L.A., 81
Gruzdev, V.E., 163
Gruzdeva, A.S., 163
Guliamov, R., 119

Hahn, Y.B., 281
Han, J., 247, 271, 281
Hasegawa, H., 45
Hashizume, T., 45
Hays, D.C., 281
Hernandez, C., 157
Himpsel, F.J., 15
Hitchcock, A.P., 31
Hobson, W.S., 281
Hong, J., 259
Hong, M., 21, 57, 131, 219
Hsieh, K.C., 57
Huang, D.J., 131
Huffaker, D.L., 81
Hughes, G., 145, 265
Hulse, J.E., 253

Hung, W.H., 131
Hwang, J.S., 213

Johnson, D., 69, 183
Jung, K.B., 281

Kibbler, A.F., 95
Kopylov, N., 21
Kortan, A.R., 21, 57
Krajewski, J.J., 219
Kuech, T.F., 15
Kuo, J.M., 219
Kurtz, S.R., 95j
Kwo, J., 21, 57, 131, 219

Laframboise, S., 227
Lambers, E., 259
Landheer, D., 253
LaRoche, J.R., 183, 247, 259
Lay, T.S., 131
Lee, J.W., 69, 183
Lee, K.N., 247
Lester, L.F., 125, 271
Lewis, L.J., 3
Lifshitz, E., 119
Lothian, J.R., 183, 259
Lu, Z.H., 31, 227
Lujan, G.S., 137

MacInnes, A.N., 125
Mackenzie, K.D., 69
Maher, H., 189
Mannaerts, J.P., 21, 57, 131, 219
Matsumoto, F., 175
McAlister, S.P., 227
McInerney, J.G., 145, 265
McKinnon, W.R., 227
Medjdoub, M., 189
Miyatsuji, K., 239
Mochizuki, Y., 107
Murphy, D.W., 57
Murtagh, M., 265

Nair, M.T.S., 169
Nair, P.K., 169
Newell, T.C., 125
Nishizawa, T., 175
Nissim, Y.I., 189

Ogryzlo, E., 287
Olson, J.M., 95
Oshida, Y., 175
Oyama, Y., 175

Pearton, S.J., 69, 183, 259, 271, 281
Plantier, H., 157
Plotka, P., 175
Poole, P.J., 227

293

SUBJECT INDEX

CPSIA information can be obtained at www.ICGtesting.com
Printed in the USA
LVOW06s1353260514

386972LV00010BA/140/P